THE OXFORD ENGINEERING SCIENCE SERIES

GENERAL EDITORS
F.W. CRAWFORD, A.L. CULLEN, J.W. HUTCHINSON
W.H. WITTRICK, L.C. WOODS

Introduction to the theory of thin-walled structures

N.W. MURRAY

Professor of Structural Engineering,
Monash University

CLARENDON PRESS · OXFORD
1984

Oxford University Press, Walton Street, Oxford OX2 6DP

London Glasgow New York Toronto
Delhi Bombay Calcutta Madras Karachi
Kuala Lumpur Singapore Hong Kong Tokyo
Nairobi Dar es Salaam Cape Town
Melbourne Auckland

and associates in
Beirut Berlin Ibadan Mexico City Nicosia

Oxford is a trade mark of Oxford University Press

Published in the United States
by Oxford University Press, New York

British Library Cataloguing in Publication Data

Murray, N.W.
   Introduction to the theory of thin-walled
   structures.—(Oxford engineering science series)
   1. Elastic plates and shells
   I. Title
   624.1′7762   QA935
   ISBN 0-19-856151-2

Library of Congress Cataloging in Publication Data

Murray, Noel W.
   Introduction to the theory of thin-walled structures.

   (The Oxford engineering science series)
   Includes bibliographical references and indexes.
   1. Thin-walled structures. 2. Structures, Theory of. I. Title. II. Series.
TA660.T5M87 1984   624.1′77   83-13279
ISBN 0-19-856151-2

Filmset and printed in Northern Ireland at The Universities Press (Belfast) Ltd.

PREFACE

A thin, flat steel plate is flexible when loads are applied at right angles to the plane of the plate but it is very stiff when loads are applied in the plane. It is this high stiffness which is utilized in the thin-walled steel structures treated in this book. Structures such as box girders, plate girders, box columns, cold-formed sections, and stiffened plates all rely upon the in-plane stiffness of the individual plates from which they are fabricated. Unfortunately the relatively high out-of-plane flexibility of the plate elements means that they may be prone to certain types of buckling with the consequent stress redistribution, loss of stiffness and so on. Furthermore, it is not possible to manufacture ideally flat plates and, generally, to fabricate thin-walled structures without introducing distortions and welding stresses. Both of these factors may have a strong influence on the response of a thin-walled structure to the applied loads.

There has been much research into these and other factors, and over the years some of this work has become incorporated into codes of practice. There appears to be a need for a book which traces the theoretical and experimental background of the code formulae and rules, and then explains how this background has been modified into a more readily usable form for designers. This book is an attempt to fill this gap. A second aim is to provide a basic text for graduate students who wish to commence work in this fascinating field. It assumes a knowledge of structural mechanics presented in an average undergraduate course. The book is not exhaustive, and in writing it the author has had to be very selective. Many workers in this fruitful research field will disagree with the choice of material. However, it is hoped that the conscientious reader will gain some kind of basis for further study.

In attempting to deal with the problems discussed in the previous paragraphs the book starts with a very brief discussion of some basic concepts of structural analysis, such as: buckling; plastic behaviour; first and higher order analysis; the engineers' theory for the determination of bending and shear stresses in beams with open, single-cell, and multi-cell cross-sections; and the Saint Venant theory of torsion with constant torsion moment of such beams. Much of this material is covered in undergraduate texts so the aim of its treatment here is to revise and review rather than to present rigorous analyses.

Chapter 2 develops the elastic non-uniform torsion theory of beams with any shape of cross-section. Because of the generality of this theory it takes many pages of text to derive the governing equations. Chapter 3 shows how this theory can be used to solve torsion problems of beams with single or multiple spans and the torsional buckling of columns and beams.

Chapter 4 concentrates on the linear elastic behaviour and buckling of box girders and stiffened plates. These problems can conveniently be solved by using the finite strip method. It is shown how the behaviour of box girders with and without cross-frames and diaphragms may be studied and how the buckling stresses of stiffened plates are determined. The finite strip method easily accommodates the possible interaction of buckling modes which can cause significant reductions in the buckling stresses. Examples of this phenomenon are given in this chapter.

The structures studied to this point have been assumed to be ideal but because of the important influence of initial imperfections, Chapter 5 deals with the elastic behaviour of thin-walled structures which are not perfect. Furthermore, second-order plate theory is developed so that it can be used to study elastic post-buckling behaviour. In contrast to earlier chapters where buckling was treated only as a linear eigenvalue problem, Chapter 5 concentrates on non-linear problems associated with elastic large-deflection theory.

Post-buckling elasto-plastic behaviour and the determination of the collapse loads of thin-walled structures are important problems for de-signers and Chapter 6 describes ways of studying them. Because of the different approaches to the problem of collapse adopted by various research groups a selection of some groups has been made and a brief description of their philosophy and results is presented. The structures covered by this review are box girders, box columns, stiffened plates, and plate girders.

It is seen that Chapters 1–6 cover material which enables a research worker to derive accurately the response of a given structure to a set of applied loads. The methods the worker would use are too cumbersome for use in design offices and the purpose of Chapter 7 is to provide a link between the theory presented in the earlier chapters of this book and some of the rules in a few codes of practice.

The choice of notation for the large number of parameters considered here has been difficult. In the first three chapters the notation used in Germany has been adopted for many of the variables because it seems to the author to be the most logical notation to use in the study of warping torsion problems. For the remainder of the book, where elastic and plastic buckling problems are studied, that used in Britain has generally been used. Although it is repetitious to do so the symbols are usually defined where they are used. Also in many instances the overuse of symbols has been avoided by writing out expressions in full instead of grouping symbols together and then replacing each group by a new symbol. In Chapter 7, where some codes of practice are reviewed, the notation used in the remainder of this book has been followed rather than that used in the codes. This is because the codes each use their own notation and it would be more difficult to relate the code rules and formulae to the theories developed in earlier chapters.

The author has developed this book from various courses of graduate and undergraduate lectures he has delivered in Australia, Thailand and Germany. The discussions he has had with his colleagues, with students, and other research workers have been most important. He would especially like to thank the many students who have checked and corrected the worked examples and the remainder of the manuscript. Two trips to Germany were financed partly by Deutscher Akademischer Austauschdienst and the Essen, Bochum, Munich, and Monash Universities: these visits were invaluable because they enabled much material to be gathered from sources in that country.

The first typescript was prepared by Mrs Pam Smith and Ms Linda Peach, who are thanked for their patience with a difficult manuscript. The enormous task of translating the typescript to a word processor was carried out by Mrs Joy Helm with great efficiency. Mr Robert Alexander prepared the diagrams. The help of all these persons and of Monash University is gratefully acknowledged.

N.W.M.

Monash University
June 1981

To Gracie

CONTENTS

NOTATION

The symbols used in this book are explained where they are first introduced. The following is a list of the main symbols used.

a	dimension
a_1	initial deflection at centre of strut
A	area enclosed by profile of hollow tube
b	width of plate or element
b_e	effective width of plate panel
B	width of flange panel
c	carryover factor
C	constant of integration
d	depth of stiffener or depth of web plate
D	$= \dfrac{Et^3}{12(1-\nu^2)}$
e	eccentricity of load or distance between moment pair
E	Young's modulus
F	area of cross-section in warping theory
$F_{xx}, F_{yy}, F_{\omega\omega}$, etc.	section properties in warping theory (see pp. 84–107)
G	shear modulus
h	dimension of finite difference net
I	second moment of area
J_p	polar moment of inertia for circular shaft or Saint Venant torsion constant for non-circular shaft
k	plate critical stress factor (see eqn (5.3.24) and Fig. 5.1.4)
K_{bs}	secant effective width factor
K_{bt}	tangent effective width factor
L	length of plate; also indicates Laplace transform
L^{-1}	inverse Laplace transform
L_{euler}	length of stiffened plate at which local and global σ_{cr} are equal
m	applied torque per unit length
M	bending or twisting moment
M_{DS}	warping torsion moment
M_p	full plastic moment of beam
M_p'	reduced plastic moment due to axial load
M_p''	reduced plastic moment due to axial load and plastic hinge inclination
M_{st}	Saint Venant torsion moment

M_ω	bimoment
N	normal force
p	applied distributed load per unit area for struts
P	applied concentrated load
P_{cr}	critical load
P_E	Euler load $\left(=\dfrac{\pi^2 EI}{L^2}\right)$
P_y	squash load $(= \sigma_y \times$ area of cross-section$)$
q_z	applied load in z-direction per unit length
Q	transverse shear force per unit length in plate
R^*	factored strength of structure
s	distance measured around profile from starting point
s_E	distance measured around profile from edge of profile
S	stiffness matrix or spacing of stiffeners
S_{bm}	stiffness matrix for mth Fourier component of bending
S_{pm}	stiffness matrix for mth Fourier component of in-plane loading
S^*	factored design loading
t	thickness of plate
u	displacement
U	total potential energy
U_e	external potential energy
U_i	internal potential energy
v	displacement
V	transverse shear force
w	displacement
x	coordinate
y	coordinate or initial deflection of plate
y_c	deflection at centre of plate or strut
Y	load per unit area in y-direction
z	coordinate
α	angle between systems of axes or $\left(\dfrac{P}{EI}\right)^{\frac{1}{2}}$
β	angle of inclination of a plastic hinge
$\bar{\beta}$	$\left(\dfrac{\sigma_y}{\sigma_{cr}}\right)^{\frac{1}{2}}$
γ	shear strain or partial load factor
γ_s	shear strain due to Saint Venant torsion
γ_w	shear strain due to warping torsion
δ	deflection
Δ	deflection

ε	direct strain
ε_y	direct strain at yield point
ζ	displacement in z-direction
η	displacement in y-direction
λ	$= \left(\dfrac{GJ_p}{EF_{\omega\omega}} \right)^{\frac{1}{2}}$ (see eqn (3.2.5)) or load factor
$\bar{\lambda}$	$= \left(\dfrac{\sigma_y}{\sigma_E} \right)^{\frac{1}{2}}$
ν	Poisson's ratio
ξ	displacement in x-direction
σ	direct stress
σ_{cr}	critical stress
σ_e	equivalent direct stress
σ_m	average stress at maximum load in plate or strut
σ_w	direct stress due to warping
σ_y	yield stress
τ	shear stress
τ_s	Saint Venant torsion shear stress
τ_w	shear stress due to warping
ϕ	angle of twist
Φ	stress function
χ	warping stiffness (see Fig. 3.4.2)
ψ	angle between axes or a factor
Ψ	factor for torsion of multi-cell profiles or effective width factor for shear lag effects
ω, ω_M	warping function with pole at shear centre M
ω_B	warping function with pole at B

1

SOME BASIC IDEAS OF STRUCTURAL THEORY

1.1. INTRODUCTION

It is not easy to define in a precise, quantitative way what a thin-walled structure is. It seems sufficient to say that a thin-walled structure is one which is made from thin plates joined along their edges. The plate thickness is small compared to other cross-sectional dimensions which are in turn often small compared with the overall length of the member or substructure. Structures of this kind are used extensively in steel and concrete bridges, ships, aircraft, mining head frames, and gantry cranes. They are seen in the form of box girders, plate girders, box columns, purlins (Z and channel sections), pallet stacks, and stud walls (Fig. 1.1.1). Thin-walled structures can be designed to exhibit great torsional rigidity, for example, as box girders, or they may have very little torsional rigidity as, for example, in the case of a plate girder. However, one property they all have in common is that they are very light compared with alternative structures and, therefore, they are used extensively in long-span bridges and other structures where weight and cost are prime considerations. They are not always steel structures; a box girder or box column can be constructed in concrete and could be classified as a thin-walled structure. This is especially true when high-strength concrete is specified. Although this book is mainly concerned with the analysis of the behaviour of steel structures, some of the theory may be used for the analysis of thin-walled concrete structures.

There are several reasons why thin-walled structures must be given special consideration in their analysis and design. In a thin-walled beam the shear stresses and strains are relatively much larger than those in a solid rectangular beam. This immediately calls into question the usual assumption of bending theory, that is, that plane sections remain plane over the entire cross-section of, say, a box girder. This assumption is known as the Bernoulli hypothesis, after Jakob Bernoulli (1654–1705). However, as seen in Fig. 1.1.2, it is easily demonstrated that when certain thin-walled structures are twisted there is a so-called *warping* of the cross-section and the Bernoulli hypothesis is violated. The term warping is defined as the out-of-plane distortion of the cross-section of a beam in the direction of the axis. Warping of the cross-section can be greatly inhibited by introducing direct stresses in the axial direction and shear

Fig. 1.1.1. Uses of thin-walled profiles as: (a) box girder bridges and towers; (b) cranes and ships; (c) storage racks.

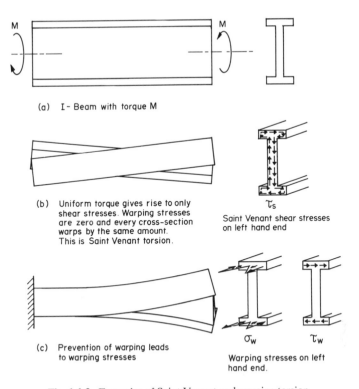

(a)   I- Beam with torque M

(b)   Uniform torque gives rise to only
      shear stresses. Warping stresses
      are zero and every cross-section
      warps by the same amount.
      This is Saint Venant torsion.

τ_s

Saint Venant shear stresses
on left hand end

(c)   Prevention of warping leads
      to warping stresses

σ_w τ_w

Warping stresses on left
hand end.

Fig. 1.1.2. Examples of Saint Venant and warping torsion.

stresses in the cross-section. These stresses are called the *warping stresses*
and are given the symbols σ_w and τ_w. Figure 1.1.2(c) shows an example of
how these stresses act when the beam shown in Fig. 1.1.2(b) is prevented
from warping by a thick plate welded to one end. The latter structure is
found to be very much stiffer than the first.

Because of the importance of shear stresses in the plane of the plates
making up the cross-section of a beam, it is necessary to study how they
are distributed through the cross-section. It is found that the shear
stresses appear to flow through the cross-section as if they were a fluid
(Fig. 1.1.3), so the idea of *shear flow* is introduced, and it is described in
Section 1.4.1. Shear stresses and their associated strains result in another
phenomenon known as *shear lag* (Section 1.4.1). This results in a distribution of direct stresses in the member which is different from that
predicted by the Bernoulli hypothesis. Both of these phenomena occur in
beams which are not twisted but carry only bending loads. When twisting
also occurs warping effects such as warping stresses are found to add to
those arising from bending loads.

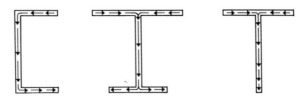

(a)   Shear flow in open profiles due to transverse shear force V.

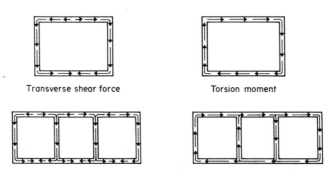

Transverse shear force   Torsion moment

(b)   Shear flow in closed profiles due to transverse shear force V
      and torsion moment M

Fig. 1.1.3. Some examples of shear flow.

The theory of thin-walled beams with open cross-sections (Fig. 1.1.3) was developed by Vlasov (1906–1958). His theory is more general than that based on the Bernoulli hypothesis. In the foreword of his book (Vlasov 1961), he stated:

> A thin-walled beam which has in its natural (unloaded) condition the shape of a cylindrical shell or a prismatic hipped section is considered in this theory as a continuous spatial system composed of plates capable of bearing, in each point of the middle surface, not only axial (normal and shear) stresses but moments as well. The deformation of the beam is not analyzed on the basis of the usual hypothesis of plane sections. In its stead the author uses the more general and natural hypotheses of an inflexible section contour and absence of shear stresses in the middle surface, which constitute the basis for a new law of distribution of longitudinal stresses in the cross section. This law, which the author calls the law of sectorial areas and which includes the law of plane section as a particular case, permits the computation of stresses in the most general cases of flexural–torsional equilibrium of a beam.

Thin-walled structures are also susceptible to local buckling if the in-plane stresses (i.e. stresses in the plane of the plate elements) reach their critical values. If this happens the geometry of the cross-section of

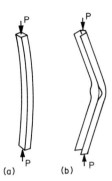

Fig. 1.1.4. (a) Overall or global buckling and (b) local buckling.

the structure changes, in contrast to overall buckling, as in the case of a pin-ended column whose cross-section retains its form (Fig. 1.1.4). However, if a thin-walled column is made sufficiently long it may suffer overall buckling before it buckles locally. This means that thin-walled structures must be designed against both local and overall buckling. Theory and experiments show that these two phenomena can interact and when this happens the buckling load can be depressed below the value of the individual buckling loads. This phenomenon is discussed in Sections 1.3.2 and 3.5 and in Chapters 4 and 6.

A third way in which a thin-walled strut or beam differs from, say, a strut or beam of solid, square cross-section is in the way that stresses attenuate along its length. Let us first consider the strut shown in Fig. 1.1.5 which has a solid, square cross-section and which is loaded at one

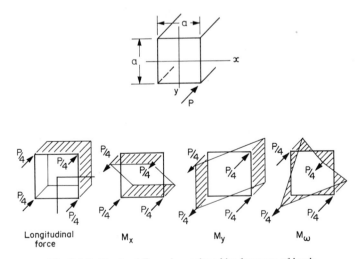

Fig. 1.1.5. The load P can be replaced by four sets of loads.

corner by a load P acting parallel to the axis. This load can be constituted from the four sets of loads as indicated. In the case of the first three sets, which represent axial loading and bending moments about the x- and y-axes, the Bernoulli hypothesis is valid. In the last loading case it is obvious that the cross-section does not remain plane. However, we note that this set of loads is self-equilibrating and make recourse to the Principle of Saint Venant (1797–1886) which is usually stated in the following manner:

> If one applies to a small part of the surface of the body a set of forces which are statically equivalent to zero, then this system of forces will not noticeably affect parts of the body lying away from the above region.

In this case it simply means that the effect of this fourth set of loads can be ignored, i.e. the load P can be replaced by an equivalent axial load equal to P and two bending moments each equal to $Pa/2$. This fourth set of loads gives rise to a set of stresses which attenuate very rapidly towards zero (i.e. in a distance a along the length of the strut they are for all practical purposes equal to zero). When we consider the thin-walled I-beam in the same way (Fig. 1.1.6), it is found that the first three sets of loads can be treated as before and the Bernoulli hypothesis is still valid. However, while the fourth set of loads is also self-equilibrating as before, in this case the stresses attenuate very slowly along the length of the strut. This is because the web acts as if it were some kind of insulator separating the loads into two sub-sets, one in each flange. Each sub-set is not self-equilibrating and causes longitudinal bending stresses in each flange. The cross-section does not remain plane but *warps*, and this set of stresses are called *warping stresses*. The longitudinal stress at a point whose coordinates are (x, y) is shown later to be

$$\sigma = \frac{P}{F} + \frac{M_x y}{F_{yy}} + \frac{M_y x}{F_{xx}} + \frac{M_\omega \omega}{F_{\omega\omega}} \tag{1.1.1}$$

where the first three terms are the axial stress and the bending stresses due to simultaneous bending about two principal axes, and the last term gives the longitudinal warping stress.

M_ω is the so-called bimoment (N mm^2). As the name implies a bimoment consists of two equal and opposite moments acting about the same axis and separated from one another. Its value is the product of the moment and the separation.

$F_{\omega\omega}$ is called the warping constant of the section (mm^6).

ω is the so-called warping function (mm^2).

This warping stress can be as large as or even larger than the bending stresses, and it cannot be ignored. It usually has its maximum value at a free edge of the cross-section, e.g. at the outer edge of a flange of a

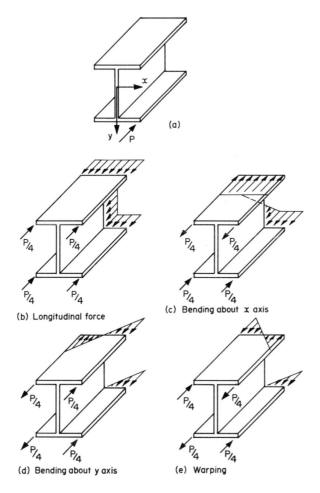

Fig. 1.1.6. Components of single load P result in warping of the profile. Saint Venant's Principle is not valid and the Bernoulli Hypothesis is not applicable for part (e).

channel section, and it therefore plays a significant role in initiating local buckling because free edges are the weak parts of a cross-section.

It is seen from the above discussion that the Principle of Saint Venant should include some statement about limiting the distortions as well as the stresses. Alternatively, in cases such as the I-beam, one must be careful about what is meant by the term 'a small part of the surface of the body'.

To understand the practical significance of a bimoment, some simple examples are now considered. The distortion of an I-column (Fig. 1.1.7(a)) which carries an eccentric load P acting parallel to its axis, is first considered. As shown in Fig. 1.1.6, and indeed in eqn (1.1.1), the single load can be treated as a combination of four sets of loads, one of which represents axial loading, two of which represent bending about the axes of symmetry of the cross-section, and the last is a bimoment. The first three sets of loads will result in the deformation pattern familiar to engineers. The bimoment will result in distortions of the cross-section parallel to the longitudinal axis of the column and in a twisting of the column about its longitudinal axis (Fig. 1.1.7(b)). To see why this is so the web and flanges are separated from one another as shown in Fig. 1.1.7(c), and the pairs of moments $Pa/4$ are applied at the top of each flange. The flanges each bend as shown, while the web remains straight. The integrity of the cross-section can be restored by twisting the web and each of the flanges through an angle $2\Delta/b$ (Fig. 1.1.7(d)) and since these individual plates are very flexible the forces required to do this are likely to be very

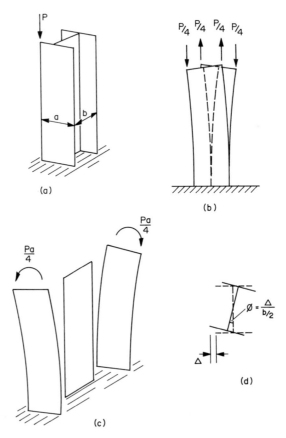

Fig. 1.1.7. A bimoment applied to a column results in twisting and longitudinal stresses.

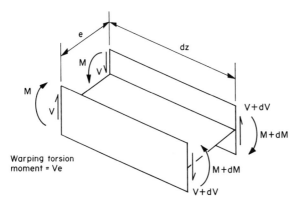

Fig. 1.1.8. A bimoment which varies along a beam results in a torque called a 'warping torsion moment'.

small. This pure twisting of each plate results in another set of stresses which are shear stresses. They are called the *Saint Venant stresses* τ_{st}, and the total torque required to twist all of the plates in this way is the *Saint Venant torque* M_{st}. Thus it is seen that a bimoment can result in the twisting of a column and longitudinal stresses in the flanges, and for the design engineer these can be very important considerations.

As a second example the relationship between bimoment and twisting moment is considered in a simplistic manner. A small element dz of a beam or column is illustrated in Fig. 1.1.8. On its left-hand end a bimoment M_ω is acting. It is represented by a pair of moments M separated by a distance e so

$$M_\omega = Me. \tag{1.1.2}$$

At the other end of the element the bimoment, and hence the moment also, is assumed to have changed, so the bimoment on that end is

$$M_\omega + dM_\omega = Me + e\,dM. \tag{1.1.3}$$

Since the moments have changed, a transverse shear force V should exist in each flange to maintain its rotational equilibrium. This is because the web, being flexible out of its plane, cannot do this. Thus

$$V\,dz = dM \quad \text{or} \quad V = \frac{dM}{dz}. \tag{1.1.4}$$

From Fig. 1.1.8 it is seen that at each end the shear forces form a couple or so-called *warping torsion moment* M_{DS} where

$$M_{DS} = Ve = e\,\frac{dM}{dz} = \frac{d(Me)}{dz} = \frac{dM_\omega}{dz}. \tag{1.1.5}$$

Thus a twisting moment occurs when the bimoment varies along the length of the beam or column. As in simple beam theory there are shear stresses associated with V. They lie in the plane of the cross-section, they are uniformly distributed through the thickness, and they are the warping shear stresses τ_w referred to above. The direct warping stresses σ_w are those stresses arising from the moments M. All of these quantities will be treated in a more formal way in later chapters, and relationships such as those given here in eqns (1.1.1) and (1.1.5) will also be developed in a rigorous manner. The present discussion is intended to serve as an introduction only.

Finally, the behaviour of thin-walled structures is usually sensitive to the nature and magnitudes of initial imperfections which arise inevitably during fabrication. Two so-called identical structures will not respond to a given set of loads in the same manner if their imperfections are of different values. This is one of the topics studied in this book.

The first chapter of this book is concerned with some basic ideas about structural analysis. Many of these topics are treated in undergraduate courses but here they are presented in a manner such that they can be applied in later chapters to thin-walled structures. In the second chapter the analysis of thin-walled beams of both open and closed cross-section is developed, and in Chapter 3 this theory is used to study the bending and twisting of thin-walled beams, the lateral buckling of deep beams and trusses, and the torsional buckling of thin-walled columns. It is here that one sees clearly the effect of the interaction of buckling modes in depressing the buckling load.

There are various methods available for the stress analysis of box girders. Apart from the elementary theory based on the Bernoulli hypothesis they are nearly all applications of the finite element method. Three such applications are described in Chapter 4. As stated above, a thin-walled structure will buckle locally if the in-plane stresses, i.e. the direct and shear stresses, in the plate elements reach their critical value. This aspect of thin-walled structural behaviour is studied in Chapter 5 by first considering isolated plates and seeing how they buckle under various combinations of in-plane direct and shear stresses. It is also possible to study the effect of initial imperfections upon their behaviour. The governing equations are two simultaneous partial differential equations which are also non-linear (the Marguerre equations). While they can be solved for isolated plates by, for example, Galerkin's method or the perturbation technique, it is not yet possible to obtain closed-form solutions for any but the most elementary cases. However, it is possible to obtain the critical load and buckling mode for isolated plates with various edge conditions and for combinations of plates, e.g. box columns, stiffened plates, etc., by using numerical techniques. Two of these methods have recently been applied (finite difference and Runge–Kutta) to obtain buckling loads and modes of a large variety of cross-section, and some of these cases will be presented.

The theory discussed so far is elastic and it is valid only up to the load at which one or more of the fibres in the structure reaches the yield stress of the material. Beyond this point the theory becomes very complicated although there have been attempts to trace the elasto-plastic response of isolated plates. However, once the plastic zones become well-developed a so-called plastic mechanism forms and the load-carrying capacity of the structure diminishes. This part of the structural response, the unloading part of the curve, is worthy of study because in the case of some thin-walled structures it is easy to predict that collapse is sudden and without warning. The development of this theory is presented in Chapter 6, which also describes some of the methods available for determining the collapse loads of some classes of thin-walled structures. Thin-walled structures would never be used if designers were required to solve complicated non-linear partial differential equations for every design. There is available literature which presents the results of such analyses in more digestible forms and they can be used as guides for designers. Codes of practice and the Merrison Design Rules also contain much valuable information, and they are reviewed in Chapter 7. This chapter shows how some of the theory presented in earlier chapters has been developed into design rules and it also deals briefly with some practical problems encountered in the fabrication and erection of thin-walled structures.

1.2. INTRODUCTION TO THE CONCEPTS OF STRUCTURAL STABILITY

When a simple beam is loaded, its deflection does not have any influence upon the distribution of stresses, but when a strut carrying an axial load P deflects sideways through a distance δ, the bending stresses increase. This is the so-called P–δ effect, and it is the reason why struts buckle whereas beams with stocky cross-sections do not. In the case of thin-walled structures, the P–δ effect can be very localized and lead to local buckling.

The usual mild-steel pin-ended column buckles at the Euler load $P_E(=\pi^2 EI/L^2)$ if it remains elastic up to that load. However, a shorter column may start to yield soon after it deflects sideways. Its behaviour is then partly elastic and partly plastic (i.e. elasto-plastic) until the point is reached at which a plastic hinge is almost fully developed at its mid-point. Thereafter the column almost behaves as if it were made from a rigid-plastic material (Fig. 1.2.1). The elastic region and the rigid-plastic region can be analysed in a relatively simple manner using elastic and rigid-plastic theories, respectively, but elasto-plastic theory is much more

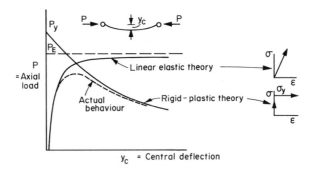

Fig. 1.2.1. Curves derived from linear elastic and rigid-plastic theories form a framework around the actual behaviour of a strut.

complicated and is seldom of direct practical use but it is used as a research tool.

The purpose of this section is to describe how the behaviour of a steel structure which buckles and undergoes some yielding may be analysed. In this section the elastic response and the rigid-plastic response of a solid cross-section are first analysed, and in later chapters these ideas are applied to thin-walled structures. Thus, this book is concerned only with the elastic theory of thin-walled structures in Chapters 2–5, and with rigid-plastic and failure theories in Chapter 6. In the case of a strut with a solid cross-section, the curves representing elastic and rigid-plastic response form a framework around the actual behaviour of the strut (Fig. 1.2.1). Tests recently conducted by Khoo (1979) indicate that the same approach can be used to study the behaviour of thin-walled structures.

To see how these ideas may be applied in practice, the case of a simple pin-ended steel strut is now considered; first its elastic and then its rigid-plastic response is analysed. A centrally loaded strut which was initially straight buckles as shown in Fig. 1.2.2(a). For small deflections its governing equation is

$$EIy^{iv} + P\ddot{y} = 0 \qquad (1.2.1)$$

where I is the second moment of area of the cross-section about the axis of bending.

The solution of this equation is

$$y = A \sin \alpha x + B \cos \alpha x + Cx + D \qquad (1.2.2)$$

where

$$\alpha = \sqrt{\left(\frac{P}{EI}\right)}. \qquad (1.2.3)$$

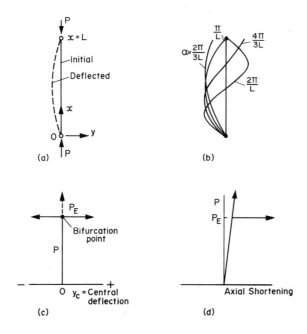

Fig. 1.2.2. Behaviour of a perfect pinned strut.

There are four boundary conditions to be satisfied, two at each end. At $x = 0$, $y = \ddot{y} = 0$, whence $B = D = 0$. At $x = L$, $y = \ddot{y} = 0$, whence $C = 0$ and $A \sin \alpha L = 0$. The last condition is satisfied when $A = 0$ in which case a complete solution of eqn (1.2.1) is

$$y = 0. \qquad (1.2.4)$$

This so-called trivial solution implies that a perfect strut is in equilibrium for all values of the axial load P if it remains straight. However, the boundary conditions are also satisfied when $\sin \alpha L$ is zero, in which case a complete solution of eqn (1.2.1) is

$$y = A \sin \alpha x \qquad (1.2.5)$$

and αL assumes discrete values, i.e.

$$\alpha L = L \sqrt{\left(\frac{P}{EI}\right)} = 0, \pi, 2\pi, \ldots, n\pi, \ldots \qquad (1.2.6)$$

Thus the buckling modes given by eqn (1.2.5) are also discrete and it is seen that A can have any small value when P has the values $n^2 \pi^2 \, EI/L^2$. Thus the trivial solution $y = 0$ is stable until P reaches the first critical load $P_E = \pi^2 \, EI/L^2$ after which the trivial solution is unstable. An alternative test for determining whether an equilibrium condition is stable or unstable is described in Section 1.3.1.

One usually thinks of natural phenomena as being smooth and continuous, as for example, the flow of water, sound vibrations, and so on. The fact that the strut buckles only at discrete loads and in a special manner peculiar only to that load is, therefore, somewhat curious. This kind of behaviour has been given a special name because it occurs in many branches of physics and engineering. The discrete values of α are called *eigenvalues,* and their associated functions (in this case $\sin n\pi x/L$) are called *eigenfunctions*. This kind of problem is called an *eigenvalue problem.*

This problem can be looked at in another way which brings out the importance of the boundary conditions. Let us imagine that we pin the end $(x = 0)$ of the strut so that the function which describes its deflection is $A \sin \alpha x$, as was seen in the derivation of eqn (1.2.5). If we now continuously vary α as shown in Fig. 1.2.2(b), it is seen that it is only for discrete values of α (in other words the axial load) that the boundary conditions at $x = L$ (that is, $y = \ddot{y} = 0$) are satisfied. Thus we see that not only is an eigenvalue problem associated with a governing equation, but also the discreteness of the solutions is associated with the boundary conditions.

Figure 1.2.2(c) is a plot of the load against lateral deflection at mid-height and Fig. 1.2.2(d) a plot of load against axial shortening. The branching at P_E seen in these graphs is called a *bifurcation*. So far it has been assumed that the strut was perfect. The shape of an imperfect strut (Fig. 1.2.3(a)) can be defined by a Fourier series, and it is easily shown

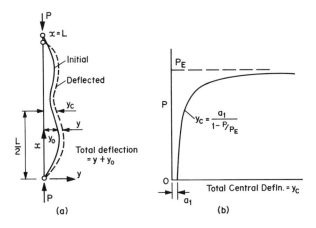

Fig. 1.2.3. Behaviour of an initially imperfect strut.

that the governing equation is

$$EIy^{iv} + P\ddot{y} = -P\ddot{y}_0 = P \sum a_n \frac{n^2\pi^2}{L^2} \sin \frac{n\pi x}{L} \quad (n = 1, 2, \ldots) \quad (1.2.7)$$

where general solution is the sum of the complementary functions and a particular solution, i.e. the form of the solution is

$$y = A \sin \alpha x + B \cos \alpha x + Cx + D + \sum b_n \sin \frac{n\pi x}{L}. \quad (1.2.8)$$

The boundary conditions are again $y = \ddot{y} = 0$ at both $x = 0$ and $x = L$, and hence the complete solution is

$$y = \sum_{n=1,2}^{\infty} \frac{Pa_n}{P_{En} - P} \sin \frac{n\pi x}{L} \quad (1.2.9)$$

where we define

$$P_{En} = \frac{n^2\pi^2 EI}{L^2}. \quad (1.2.10)$$

The total deflection is the sum of y and $\sum a_n \sin n\pi x/L$ i.e.

$$\text{total deflection} = \sum_{n=1,2}^{\infty} \frac{a_n}{1 - P/P_{En}} \sin \frac{n\pi x}{L}. \quad (1.2.11)$$

There are several points to notice about this solution.

(a) The solution is no longer just a discrete set of eigenvalues with deflections of indefinite magnitude (compare (1.2.6) et seq.), but for each value of P there is a definite deflection at every point in the strut.

(b) The deflection (1.2.11) becomes infinitely large at the same set of loads obtained from the eigenvalues of the perfect strut.

(c) The first term of the series (1.2.11) dominates in the vicinity of $P = P_E$. (Here it is assumed that all of the a_n values are of the same order of magnitude. In practice it is usual for $a_1 \gg a_2 \gg a_3 \gg a_4 \ldots$).

(d) The total deflection at $x = L/2$ can be approximated by a hyperbola

$$y_c = \frac{a_1}{1 - P/P_E} \quad (1.2.12)$$

which is plotted in Fig. 1.2.3(b).

The rigid-plastic behaviour of the same strut is now considered assuming that it is of rectangular cross-section $b \times d$ ($b > d$). Under axial load a plastic hinge forms at the central cross-section where the stress distribution is as shown in Fig. 1.2.4. The central portion can be thought of as

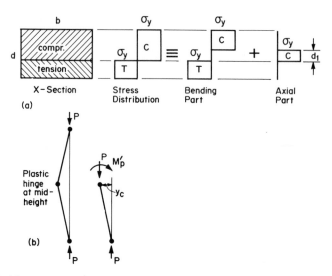

Fig. 1.2.4. (a) Stress distribution at a plastic hinge in a strut of rectangular cross-section; (b) plastic mechanism of a pin-ended strut as it collapses.

carrying the axial load P while the outer portions carry the reduced plastic moment M'_p.

Thus

$$P = \sigma_y b d_1 \quad \text{and} \quad M'_p = \sigma_y b \frac{(d-d_1)}{2} \frac{(d+d_1)}{2} = \sigma_y b \left(\frac{d^2 - d_1^2}{4}\right). \quad (1.2.13)$$

On eliminating d_1

$$M'_p = M_p \left(1 - \frac{P^2}{P_y^2}\right) \quad (1.2.14)$$

where

M_p = full plastic moment of the cross-section

$$= \sigma_y \frac{bd^2}{4}, \quad (1.2.15)$$

P_y = squash load of the cross-section

$$= \sigma_y bd. \quad (1.2.16)$$

Figure 1.2.5 is a plot of eqn (1.2.14) and shows the reduction in plastic moment with axial load.

For equilibrium of one half of the strut (Fig. 1.2.4(b))

$$M'_p = P y_c. \quad (1.2.17)$$

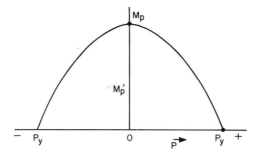

Fig. 1.2.5. Graph showing reduction in plastic moment capacity with axial load.

From (1.2.14) to (1.2.17) by eliminating M'_{p}

$$P = \sigma_y bd\left[\sqrt{\left(\frac{4y_c^2}{d^2}+1\right)} - \frac{2y_c}{d}\right].\qquad(1.2.18)$$

This equation gives the load-carrying capacity of the strut once a plastic hinge has formed. It is plotted in Fig. 1.2.6 which also contains the elastic curve for a strut with initial imperfection a_1. Figure 1.2.6 also shows the response of an actual strut and it is seen that it begins to depart from the elastic curve at point B when the first fibre reaches the yield stress. The load at which this occurs is easily calculated. The strut reaches its maximum load capacity at C, after which it declines and the curve approaches the theoretical rigid-plastic curve asymptotically.

In deriving the rigid-plastic curve (eqn (1.2.18)) it has been assumed that there is no strain-hardening, in which case the plastic hinges are each concentrated into a very small length of the strut. When strain-hardening exists each of the plastic hinges is forced to spread over a finite length of the strut and the rigid-plastic collapse curve is raised. However, this is

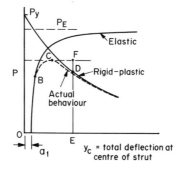

Fig. 1.2.6. After the maximum capacity of the strut is passed at C the out-of-balance force FD is an acceleration force.

only significant for large deflections or for very short struts, because in
these cases the simple model of the plastic mechanism (Fig. 1.2.4(b)) is
not accurate. In practice the region of interest is usually in the vicinity of
the maximum load where the deflections are small and strain hardening is
then ignored. Figure 1.2.6 shows that the elastic and rigid-plastic theories
complement one another. The elastic theory is able to define the deflections and stresses up to the point of first yield and to define the load at
which first yield occurs. The position of the rigid-plastic curve determines
the absolute limit of load-carrying capacity, for above it is a region in
which the structure cannot carry a load and remain in a state of equilibrium. It intersects the elastic line as if to say 'thus far and no further'.

The rigid-plastic curve is a graph of load-carrying capacity and therefore gives an idea of the suddenness of collapse. If the load is applied as a
dead load and the maximum capacity of the strut is reached at C
(Fig. 1.2.6), the load which it can sustain beyond C, for example at D, is
given by the ordinate ED. The actual load applied is, however, EF, so the
out-of-balance force, FD, appears as the force accelerating the load.
Thus, it is seen that if the rigid-plastic curve droops very steeply the
accelerating force increases rapidly and collapse is sudden. On the other
hand, a strut which has a less steeply drooping curve will collapse in a
more gradual manner. It is also easy to appreciate that a strut whose
rigid-plastic curve is steep will have a failure load whose value is very
sensitive to the magnitude of the initial imperfection a_1. This is because a
small increase in a_1 results in a horizontal shift to the right of the elastic
curve and there is a notable decrease in collapse load. If, on the other
hand, the rigid-plastic curve is nearly horizontal, such a shift has an
almost insignificant effect upon the collapse load.

The phenomenon which has just been described and analysed can be
looked at in another interesting way. Although the plastic zones in a strut
grow in extent in a continuous way the phenomenon will be described as
if it happens in a step-by-step manner. Because the slope of the plastic
part of the stress–strain curve of mild steel is zero it is seen that if a part
of a structure starts to yield then that part no longer contributes to the
stiffness of the structure. From the point of view of calculating the elastic
critical load one might as well cut out the parts of the structure which are
known to have become plastic. In the case of the imperfect strut considered above, yielding commences at mid-length on the concave side.
After a while this plastic compression zone penetrates some way into the
cross-section. Its removal from the point of view of analysis is equivalent
to replacing the strut by another (Fig. 1.2.7(a)) which has a notch cut into
it. This has two weakening effects, firstly, it reduces the critical load below
that of the original strut, and secondly, it increases the effective eccentricity of the axial load P.

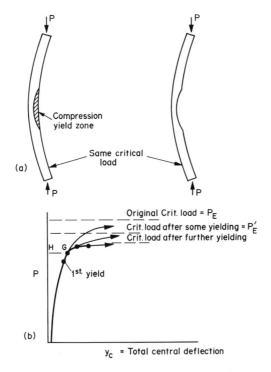

Fig. 1.2.7. After the onset of yielding the critical load deteriorates.

Thus, at point G (Fig. 1.2.7(b)) the weakened strut has a lower critical load indicated by the load P'_E, which becomes the new asymptote for the elastic curve, and an increased initial imperfection, which is greater than the length of GH. If this reasoning is continued in a step-by-step manner it is seen that finally the critical load deteriorates to the actual load being applied to the strut and this load can be defined as the failure load. This phenomenon first described by Wood (1958) is known as *deterioration of the critical load* or simply *deterioration of stability*. It will be seen that a similar phenomenon occurs in the case of thin-walled structures if they buckle locally.

All of the phenomena described so far for the case of a pin-ended column apply equally to braced and unbraced frameworks and other structures such as thin-walled structures. The only essential difference is that the load P_E must be replaced by the critical load of the structure, i.e. by P_{cr} Methods for evaluating P_{cr} are available in the literature (e.g. Matheson 1971) for frameworks and the evaluation of critical loads and stresses in thin-walled structures is one of the topics covered in this book (Chapters 4 and 5).

1.3. SOME GENERAL CONSIDERATIONS RELATING TO THE THEORETICAL ANALYSIS OF STRUCTURAL STABILITY

In the analysis of structural stability there are some considerations which need to be examined and understood before applying them to thin-walled structures. They include the order of analysis (first, second, or higher), imperfection sensitivity, interaction of critical loads, eigenfunctions, and the method of approximate analysis called the perturbation technique. Many of these concepts are illustrated by applying them to simple mechanical devices in order that a clear understanding of their meaning in physical terms can be obtained. Many of these ideas are given a fuller treatment in the excellent book by Croll and Walker (1972).

1.3.1. First-, second-, and higher-order analysis

The stability of a system can be studied by examining its total potential energy U. Figure 1.3.1 illustrates, by means of an analogue, the necessary and sufficient conditions for stable, neutral, and unstable equilibrium. They apply equally well to structures where the total potential energy U is the sum of the potential energy of the external loads (forces) and the energy absorbed by the structure during deformation (its strain energy). For equilibrium of a structure whose deflection is measured by the parameter θ

$$\frac{\mathrm{d}U}{\mathrm{d}\theta} = 0 \tag{1.3.1}$$

and if $\dfrac{\mathrm{d}^2U}{\mathrm{d}\theta^2} > 0$, the structure is in stable equilibrium;

if $\dfrac{\mathrm{d}^2U}{\mathrm{d}\theta^2} = 0$, neutral equilibrium, i.e. critical equilibrium $\tag{1.3.2}$

if $\dfrac{\mathrm{d}^2U}{\mathrm{d}\theta^2} < 0$, unstable equilibrium.

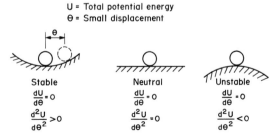

U = Total potential energy
θ = Small displacement

Stable	Neutral	Unstable
$\dfrac{\mathrm{d}U}{\mathrm{d}\theta} = 0$	$\dfrac{\mathrm{d}U}{\mathrm{d}\theta} = 0$	$\dfrac{\mathrm{d}U}{\mathrm{d}\theta} = 0$
$\dfrac{\mathrm{d}^2U}{\mathrm{d}\theta^2} > 0$	$\dfrac{\mathrm{d}^2U}{\mathrm{d}\theta^2} = 0$	$\dfrac{\mathrm{d}^2U}{\mathrm{d}\theta^2} < 0$

Fig. 1.3.1. Three types of equilibrium.

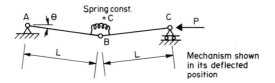

Fig. 1.3.2. Spring/link mechanism which illustrates instability.

These rules are now applied to the simple mechanism illustrated in Fig. 1.3.2, where the spring absorbs energy due to the angular changes at B. It is seen that total energy of the mechanism after deformation θ is

$U[\theta, P] = $ energy absorbed by spring $-$ work done by P

$$= 2C\theta^2 - 2PL(1 - \cos \theta). \tag{1.3.3}$$

For equilibrium

$$\frac{dU}{d\theta} = 4C\theta - 2PL \sin \theta = 0. \tag{1.3.4}$$

Equation (1.3.4) can be satisfied in two ways leading to two equilibrium

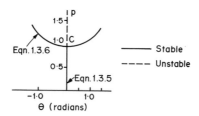

Fig. 1.3.3. Equilibrium paths of mechanism shown in Fig. 1.3.2.

curves (Fig. 1.3.3). They are

$$\theta = 0 \tag{1.3.5}$$

or

$$p = \frac{PL}{2C} = \frac{\theta}{\sin \theta} \tag{1.3.6}$$

where p is a non-dimensional load factor. The stability of points on these curves is examined by evaluating $d^2U/d\theta^2$

$$\frac{d^2U}{d\theta^2} = 4C - 2PL \cos \theta = 4C(1 - p \cos \theta). \tag{1.3.7}$$

Along the first curve ($\theta = 0$) it is seen that

$d^2U/d\theta^2 > 0$, i.e. stable for $p < 1$

$d^2U/\theta^2 = 0$, i.e. critical for $p = 1$ (1.3.8)

$d^2U/\theta^2 < 0$, i.e. unstable for $p > 1$

and for the second curve ($p = \theta/\sin \theta$)

$d^2U/d\theta^2$ is always > 0 for all practically realizable cases.   (1.3.9)

In Fig. 1.3.3 stable equilibrium curves are indicated by full lines and unstable equilibrium curves by interrupted lines. However, at the critical state itself, $d^2U/d\theta^2 = 0$, and we need to know whether U is a minimum (and thus stable) or not. The most convenient way of testing for this is to expand U about the critical point C in a Taylor's series.

$$U(\theta_c + \Delta\theta) = U(\theta_c) + \frac{\Delta\theta}{1!}\left(\frac{dU}{d\theta}\right)_c + \frac{(\Delta\theta)^2}{2!}\left(\frac{d^2U}{d\theta^2}\right)_c$$

$$+ \frac{(\Delta\theta)^3}{3!}\left(\frac{d^3U}{d\theta^3}\right)_c + \frac{(\Delta\theta)^4}{4!}\left(\frac{d^4U}{d\theta^4}\right)_c \cdots \qquad (1.3.10)$$

It is found by continued differentiation of U (eqn (1.3.3)) that all of the derivatives up to the third are zero and that

$$\frac{d^4U}{d\theta^4} = +4C \qquad (1.3.11)$$

at the critical point. Hence a small change in θ results in an increase in U, indicating that the mechanism is stable at the critical point. A small disturbance at this point will cause the mechanism to follow one of the lateral curves (Fig. 1.3.3). The critical point is a *bifurcation* point. So far the mechanism has been analysed by an exact method and the results are valid for all deflections. In real structures it is usually too difficult to do this, and besides, interest is usually centred only upon the vicinity of the critical point. Approximate methods are used extensively in stability analysis. To illustrate the technique the same strut is analysed, but the cosine term is replaced by the truncated series

$$\cos \theta = 1 - \frac{\theta^2}{2} + \frac{\theta^4}{24} \qquad (1.3.12)$$

and an approximation for U is U' where

$$U'[\theta, P] = 2C\theta^2 - PL\left(\theta^2 - \frac{\theta^4}{12}\right). \qquad (1.3.13)$$

The equilibrium condition becomes

$$\frac{dU'}{d\theta} = 4C\theta - 2PL\left(\theta - \frac{\theta^3}{6}\right) \tag{1.3.14}$$

which gives the two possible equilibrium curves

$$\theta = 0, \tag{1.3.15}$$

$$p = \frac{PL}{2C} = 1 + \frac{\theta^2}{6}. \tag{1.3.16}$$

The curve given by (1.3.16) is an excellent approximation to (1.3.6).
   If the cosine term is truncated after two terms, i.e.

$$\cos\theta = 1 - \frac{\theta^2}{2} \tag{1.3.17}$$

it is easily shown that the two equilibrium curves are

$$\theta = 0, \tag{1.3.18}$$

$$p = \frac{PL}{2C} = 1. \tag{1.3.19}$$

   This analysis is linear and yields the solution that non-zero values of θ can occur only when $p = 1$. This is the eigenvalue problem mentioned in Section 1.2. It illustrates that if a structural analysis is based upon a quadratic energy function it is only possible to obtain an eigenvalue problem, and its solution will only give the buckling load (or stress in the case of a plate) and the buckling shape (i.e. mode). However, many structures (such as plates) exhibit elastic buckling at a bifurcation, but their equilibrium curve rises as the deflections increase giving an increase in strength after buckling; such structures also give adequate warning of distress. Some other structures, however, have elastic post-buckling equilibrium curves which droop away from the bifurcation point (Fig. 1.3.4). Along such curves, and at the bifurcation point, the structure

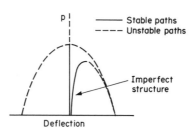

Fig. 1.3.4. Structures with drooping stability paths show great sensitivity to initial imperfection.

is unstable. Such structures exhibit great sensitivity to the magnitude of the initial imperfections (Croll and Walker 1972). A small increase in imperfection may result in a relatively large decrease in the maximum load. These important properties of a strut can only be studied and evaluated if a second-order analysis is performed.

The terms first- and second-order analysis are sometimes used in the literature in another sense. An analysis in which it is assumed that the deflections have no influence upon the stresses (e.g. in a simple beam) is sometimes referred to as a *first-order analysis* while one in which the deflections have an effect (e.g. the *P–δ* effect in a strut) is called a *second-order analysis*. It is more consistent mathematically to refer to a strut or plate analysis which only yields the buckling load and mode (i.e. the eigenvalue problem) as a *first-order stability analysis* and one which reveals information about the post-buckling responses as a *second-order stability analysis*. This terminology is adopted here.

1.3.2. Interaction of critical loads

In a later part of this book it will be shown that in the post-buckling range a flat plate loaded with in-plane forces exhibits an increase in its load-carrying capacity (Fig. 1.3.5). Designers of stiffened plates (Fig. 1.1.1(b)) with in-plane loading try to make use of this increase in strength after local buckling has occurred.

They also have to decide on the size and spacing of the transverse stiffeners and, of course, the temptation is to design for simultaneous local buckling and overall buckling of the panel as an Euler column. This can lead to a dangerous situation because two simultaneous critical loads can interact with one another and reduce the actual critical load of the structure. It is a phenomenon which occurs in many structures as most structures have more than one critical load and buckling mode. This

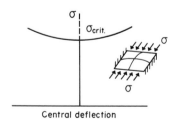

Fig. 1.3.5. Load-deflection curve of a flat plate.

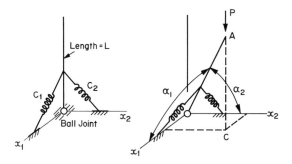

Fig. 1.3.6. Model used by Croll and Walker (1972) to illustrate interaction of critical loads.

phenomenon is also illustrated by Croll and Walker (1972) by considering the model illustrated in Fig. 1.3.6.

Letting

$$\theta_1 = \frac{\pi}{2} - \alpha_1 \quad \text{and} \quad \theta_2 = \frac{\pi}{2} - \alpha_2 \tag{1.3.20}$$

the vertical displacement Δ of the load P in the deformed state is the difference between L and AC. But, from Pythagoras' theorem,

$$\left(\frac{AC}{L}\right)^2 = 1 - \sin^2 \theta_1 - \sin^2 \theta_2$$

$$= 1 - \left(\theta_1 - \frac{\theta_1^3}{3!} + \frac{\theta_1^5}{5!} - \ldots\right)^2 - \left(\theta_2 - \frac{\theta_2^3}{3!} + \frac{\theta_2^5}{5!} - \ldots\right)^2$$

$$\simeq 1 - \left(\theta_1^2 - \frac{\theta_1^4}{3} + \theta_2^2 - \frac{\theta_2^4}{3}\right)$$

if we ignore terms of higher order than θ_1^4 and θ_2^4. By applying the binomial theorem and again truncating terms of higher order than θ_1^4, θ_2^4 we obtain

$$\Delta = L[\tfrac{1}{2}(\theta_1^2 + \theta_2^2) - \tfrac{1}{24}(\theta_1^4 - 6\theta_1^2\theta_2^2 + \theta_2^4)]. \tag{1.3.21}$$

Thus, the change in total potential energy of the model as it buckles is

$$U = \tfrac{1}{2}C_1\theta_1^2 + \tfrac{1}{2}C_2\theta_2^2 - P\Delta$$

$$= C_1\{\tfrac{1}{2}\theta_1^2 + \tfrac{1}{2}c\theta_2^2 - p[\tfrac{1}{2}(\theta_1^2 + \theta_2^2) - \tfrac{1}{24}(\theta_1^4 - 6\theta_1^2\theta_2^2 + \theta_2^4)]\} \tag{1.3.22}$$

where

$$c = \frac{C_2}{C_1} \quad \text{and} \quad p = \frac{PL}{C_1}.$$ (1.3.23)

Equilibrium requires that

$$\frac{\partial U}{\partial \theta_1} = C_1\{\theta_1 - p[\theta_1 - \tfrac{1}{24}(4\theta_1^3 - 12\theta_1\theta_2^2)]\} = 0,$$ (1.3.24)

$$\frac{\partial U}{\partial \theta_2} = C_1\{c\theta_2 - p[\theta_2 - \tfrac{1}{24}(-12\theta_1^2\theta_2 + 4\theta_2^3)]\} = 0.$$ (1.3.25)

These equations are satisfied by the following solutions

$$\theta_1 = 0, \qquad \theta_2 = 0 \quad \text{for all} \quad p,$$ (1.3.26)

$$\theta_2 = 0, \qquad 1 - p(1 - \tfrac{1}{6}\theta_1^2) = 0,$$ (1.3.27)

$$\theta_1 = 0, \qquad c - p(1 - \tfrac{1}{6}\theta_2^2) = 0,$$ (1.3.28)

$$1 - p(1 - \tfrac{1}{6}\theta_1^2 + \tfrac{1}{2}\theta_2^2) = 0, \ c - p(1 + \tfrac{1}{2}\theta_1^2 - \tfrac{1}{6}\theta_2^2) = 0.$$ (1.3.29)

Let us now assume that $c > 1$. The first three of these equations are represented by the paths S_1, S_2, and S_3 in Fig. 1.3.7, S_2 being stable in the region close to the p-axis (indicated by the full-line), and S_3 being unstable (indicated by the dotted line). However, the conditions described by eqn (1.3.29) are indicated by the space curves S_4. Thus there is another bifurcation at A after which the path is unstable. The coordinates of A are

$$\theta_1 = \pm\left[\frac{6(c-1)}{3+c}\right]^{\frac{1}{2}}; \quad p = \frac{3+c}{4}$$

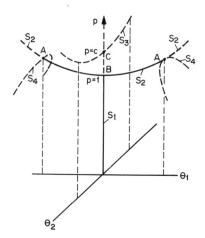

Fig. 1.3.7. Equilibrium paths of model shown in Fig. 1.3.6.

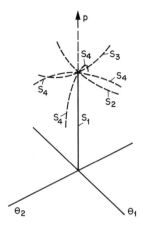

Fig. 1.3.8. Stability paths of model shown in Fig. 1.3.6 when $c = 1$.

showing that there are three critical loads, that is,

$$p_c = 1, \ 3 + c/4 \text{ and } c. \tag{1.3.30}$$

The reader should contemplate how the model would behave when it has initial imperfections and sketch the path on Fig. 1.3.7.

We now consider what happens when $c \to 1$. The three critical loads coalesce and the paths become the parabolas shown in Fig. 1.3.8. The inverted form of the S_4 parabolas suggest a marked increase in sensitivity of this model to initial imperfections. The maximum load-carrying capacity is rapidly reduced by small increases in initial imperfections.

However, a more alarming feature of this model is that it demonstrates that had we simply analysed the model with $c > 1$ for buckling in the x_1 direction (Fig. 1.3.6)—that direction obviously giving the lower critical load—we would have obtained the S_2 path and predicted that it has a load-stiffening characteristic. We would not have detected the presence of the unstable mode at A. In the design of a stiffened panel with in-plane loading this unsafe situation can be avoided quite simply by separating the critical loads for local and overall buckling. This is achieved by separating the transverse stiffeners, which lowers the critical stress for overall buckling, or by increasing the width-to-thickness ratio b/t of the plate, which lowers the critical stress for local buckling.

In this example the interaction of the critical loads results in a marked increase in the sensitivity of the structure to initial imperfections. In Section 3.5 it will be shown that in some other structures the critical loads interact and reduce the critical load.

1.3.3. Stability analysis by the perturbation technique

Another method which is frequently used to analyse the post-buckling behaviour of structures is called the *perturbation technique*. Its application to many difficult buckling problems has been described by Thompson and Hunt (1973).

The perturbation technique is used to solve that class of problem in which it is possible to obtain easily at least one point on the curve describing the behaviour of a body in some physical situation. For example it is often easy to determine the buckling load of a given structure when it has no initial imperfections. The perturbation technique is then used to determine the post-buckling behaviour of the structure, which is often a difficult non-linear problem. The parameters in the equations governing the behaviour of the structure are each expanded as an infinite power series in terms of a variable which is called the *perturbation parameter ε*. An essential feature of ε is that it vanishes at the point where the solution is known. The technique is illustrated by applying it to an example used by Sewell (1965).

The rigid bar structure ABC (Fig. 1.3.9) is supported at A and B by two springs of equal stiffness k and a load P is applied at C. For small values of P the only deflection is the vertical deflection of the springs but when P is sufficiently large there is an adjacent equilibrium position which involves a rotation θ. Later on this problem will be solved by the perturbation technique but in order to make the explanation clearer it is first solved by an analytical method. For vertical and rotational equilibrium, respectively,

$$P = F_A + F_B \qquad\qquad (1.3.31)$$

$$PL \sin\theta = (F_B - F_A)c \cos\theta \qquad\qquad (1.3.32)$$

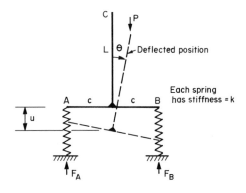

Fig. 1.3.9. Rigid bar and spring model.

After introducing the spring stiffness k into these equations

$$P = 2ku \tag{1.3.33}$$

$$\sin \theta \left[P - \frac{2kc^2}{L} \cos \theta \right] = 0 \tag{1.3.34}$$

The complete set of equilibrium paths is shown in Fig. 1.3.10 where it is seen that they intersect at the point $(P, u, \theta) = (2kc^2/L, c^2/L, 0)$. The loading–buckling path of this structure is that indicated by the arrows.

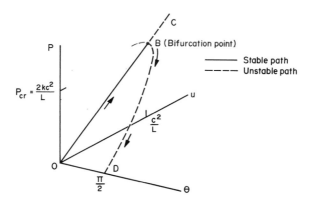

Fig. 1.3.10. Equilibrium paths of model shown in Fig. 1.3.9.

The same problem is now solved by the perturbation technique to show what information it will reveal. The parameters P, u, and θ are expanded as a Maclaurin series in terms of a perturbation parameter ε which will be chosen later on.

$$P(\varepsilon) = P(0) + \varepsilon \dot{P}(0) + \frac{\varepsilon^2}{2} \ddot{P}(0) + \ldots \tag{1.3.35}$$

$$u(\varepsilon) = u(0) + \varepsilon \dot{u}(0) + \frac{\varepsilon^2}{2} \ddot{u}(0) + \ldots \tag{1.3.36}$$

$$\theta(\varepsilon) = \theta(0) + \varepsilon \dot{\theta}(0) + \frac{\varepsilon^2}{2} \ddot{\theta}(0) + \ldots \tag{1.3.37}$$

where the dots denote differentiation with respect to ε and $P(0)$, $u(0)$, etc., are the values of P, u, etc., at $\varepsilon = 0$. The individual terms in eqns

(1.3.31) and (1.3.32) are now expanded as a Maclaurin series so that

$$0 = [P(0) - (F_A(0) + F_B(0))] + \varepsilon[\dot{P}(0) - (\dot{F}_A(0) + \dot{F}_B(0))]$$

$$+ \frac{\varepsilon^2}{2}[\ddot{P}(0) - (\ddot{F}_A(0) + \ddot{F}_B(0))] + \frac{\varepsilon^3}{6}[\dddot{P}(0) - (\dddot{F}_A(0) + \dddot{F}_B(0))] + \dots$$

$$(1.3.38)$$

$$0 = \frac{PL}{c}\left(\theta + \frac{\theta^3}{3} + \frac{2\theta^5}{15} + \dots\right) - (F_B - F_A)$$

$$= \frac{P(0)L}{c}\left[\theta(0) + \frac{\theta(0)^3}{3} + \frac{2\theta(0)^5}{15} + \dots\right] - (F_B(0) - F_A(0))$$

$$+ \varepsilon\left[\frac{\dot{P}(0)L\theta(0)}{c} + \frac{P(0)L\dot{\theta}(0)}{c} + \frac{\dot{P}(0)L\theta(0)^3}{3c}\right.$$

$$+ \frac{P(0)L\theta(0)^2\dot{\theta}(0)}{c} + \dots - (\dot{F}_B(0) - \dot{F}_A(0))\Bigg]$$

$$+ \frac{\varepsilon^2}{2}\left[\frac{2\dot{P}(0)L\dot{\theta}(0)}{c} + \frac{\ddot{P}(0)L\theta(0)}{c} + \frac{P(0)L\ddot{\theta}(0)}{c}\right.$$

$$+ \frac{\ddot{P}(0)L\theta(0)^3}{3c} + 2\dot{P}(0)L\theta(0)^2\dot{\theta}(0) + \frac{2P(0)L\theta(0)\dot{\theta}(0)^2}{c}$$

$$+ \frac{P(0)L\theta(0)^2\ddot{\theta}(0)}{c} + \dots - (\ddot{F}_B(0) - \ddot{F}_A(0))\Bigg] + \dots$$

$$(1.3.39)$$

When the expansions are carried out about a point which has $\theta(0) = 0$ (e.g. point B in Fig. 1.3.10) the last equation simplifies to

$$0 = -(F_B(0) - F_A(0)) + \varepsilon\left[\frac{P(0)L\dot{\theta}(0)}{c} - (\dot{F}_B(0) - \dot{F}_A(0))\right]$$

$$+ \frac{\varepsilon^2}{2}\left[\frac{2\dot{P}(0)L\dot{\theta}(0)}{c} + \frac{P(0)L\ddot{\theta}(0)}{c} - (\ddot{F}_B(0) - \ddot{F}_A(0))\right]$$

$$+ \frac{\varepsilon^3}{6}\left[\frac{3\ddot{P}(0)L\dot{\theta}(0)}{c} + \frac{3\dot{P}(0)L\ddot{\theta}(0)}{c}\right.$$

$$+ \frac{P(0)L(\dddot{\theta}(0) + 2\dot{\theta}(0)^3)}{c} - (\dddot{F}_B(0) - \dddot{F}_A(0))\Bigg] + \dots$$

$$(1.3.40)$$

Equations (1.3.38) and (1.3.40) together with the stiffness equation, $F = ke$, where $e = $ compression of a spring, are valid for all ε so the coefficients of ε must vanish. These coefficients are now examined.

Zero-order solution: P is arbitrary, $u = P/2k$, $\theta = 0$.

This is called the trivial solution and it is represented by any point on the straight line OBC in Fig. 1.3.10.

First-order solution: $\dot{P} = 2k\dot{u}$ and $\dot{\theta}(P - 2kc^2/L) = 0$.

There are two kinds of solution here. Firstly if P and $\dot{P}$ are given, then $\dot{u}$ and $\dot{\theta}$ will have the values $\dot{P}/2k$ and 0, respectively. This solution is again represented by any point on OBC in Fig. 1.3.10, except for the point $P = 2kc^2/L$. When P has this special value, $\dot{\theta}$ can have any value, meaning that there are infinitely many equilibrium paths through this special point B. To determine which of these paths is taken by the structure it is necessary to consider the second order solution.

Second-order solution: $\ddot{P} = 2k\ddot{u}$ and $\dot{P}\dot{\theta} = 0$.

Of all the infinitely many, possible paths permitted by the first-order solution at B this solution allows only two, that is, $\dot{P} \neq 0$ and $\dot{\theta} = 0$, whence $\theta(\varepsilon) \propto [P(\varepsilon) - P(0)]^2$ at most, and $\dot{\theta} \neq 0$ and $\dot{P} = 0$, whence $[P(\varepsilon) - P(0)] \propto [\theta(\varepsilon)]^2$ at most. To follow these paths in detail it is necessary to proceed to the third-order solution.

Third-order solution: $\dddot{P} = 2k\dddot{u}$ and $\dot{\theta}[\ddot{P} + 2kc^2\theta^2/L] + \dot{P}\ddot{\theta} = 0$.

For the first of the possible paths at B, that is, $\dot{P} \neq 0$, $\dot{\theta} = 0$ it is seen that $\ddot{\theta} = 0$, which shows that this solution is developing the trivial path OBC in Fig. 1.3.10. For the second of the above paths at B, that is, $\dot{\theta} \neq 0$, $\dot{P} = 0$, the amplitude of $\dot{\theta}$ is fixed by $\ddot{P}$ from the equation

$$\ddot{P} = 2kc^2\theta^2/L = 0$$

provided $\ddot{P} < 0$.

 Thus this path curves downwards along the line BD in Fig. 1.3.10. An approximation to this curve is obtained by truncating the series at this stage and letting $|\theta|$ be the perturbation parameter. For this case $|\dot{\theta}| = 1$. The equation of the path BD is approximately

$$P = 2kc^2\left(1 - \frac{\theta^2}{2}\right)\Big/L, \qquad u = c^2\left(1 - \frac{\theta^2}{2}\right)\Big/L \qquad (1.3.41)$$

which should be compared with eqns (1.3.33) and (1.3.34). It is shown in Section 5.3.5 that the perturbation technique can be applied to problems of plate stability. In these problems the governing equations are non-linear partial differential equations but the process of building up the equilibrium paths follows the same steps.

1.4. ENGINEERS' THEORY FOR SHEAR IN BEAMS AND SAINT VENANT TORSION OF THIN-WALLED CROSS-SECTIONS

It will be shown in Chapter 2 that generally when a beam is twisted direct and shear warping stresses (σ_w and τ_w) will exist. These stresses are additional to those direct and shear stresses which arise when a beam is subjected to combined bending and shear forces. These latter stresses are dealt with in texts on elementary strength of materials and structural mechanics, and this section on the so-called engineers' theory is included here for the sake of completeness.

Whenever a beam is twisted uniformly along its length the fibres must undergo a shear strain to accommodate this twist. Associated with these strains are shear stresses. This problem was first solved by Saint Venant (1797–1886) and the shear stresses are known as *Saint Venant shear stresses* τ_{st}. However, it will be shown in Chapter 2 that even when a beam is twisted in a non-uniform manner, Saint Venant shear stresses will occur and they will be additional to the warping stresses. In Section 1.4.2 the behaviour of a beam which is twisted uniformly along its length is studied. In this case, the warping stresses are zero, so they can be ignored until Chapter 2. The engineers' theory of beams with hollow profiles is presented in Section 1.4.3.

1.4.1. Shear in beams with an open profile—engineers' theory

A small element of a thin-walled beam, which is being bent about a principal axis *xx*, is considered (Fig. 1.4.1). In general, the bending moment changes along its length as indicated in the diagram. On the

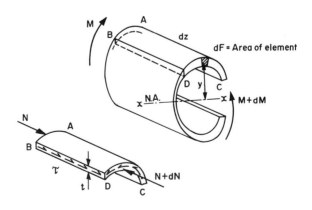

Fig. 1.4.1. Shear in beams by engineers' theory.

left-hand end the direct stress from engineers' bending theory is[†]

$$\sigma = \frac{My}{F_{yy}} \tag{1.4.1}$$

where F_{yy} is the second moment of area about the xx-axis ($= \int y^2\, dF$ where integration is taken over the whole of the cross-section). Hence the force acting on the element of area dF is

$$\sigma\, dF = \frac{My}{F_{yy}}\, dF. \tag{1.4.2}$$

The *excess* force in the corresponding element of area in the right-hand end is

$$\frac{d}{dz}(\sigma\, dF)\, dz = \frac{dMy\, dF}{F_{yy}} \tag{1.4.3}$$

The total excess force on the end CD is

$$dN = \int_C^D \frac{dMy\, dF}{F_{yy}} = \frac{dM}{F_{yy}} \int_C^D y\, dF \tag{1.4.4}$$

This excess force must be balanced by the shear stress τ along the surface BD. By equating forces

$$\tau t\, dz = \frac{dM}{F_{yy}} \int_C^D y\, dF \tag{1.4.5}$$

whence

$$\tau = \frac{1}{F_{yy}t}\left(\frac{dM}{dz}\right)\int_C^D y\, dF$$

$$= \frac{V(\bar{y}F)}{F_{yy}t} \tag{1.4.6}$$

where V is the transverse shear force and $(\bar{y}F)$ is the first moment of the area of cross-section between C and D about the neutral axis, $\bar{y}$ being the distance from the neutral axis to the centroid of the area F between C and D.

The shear stress τ acting along BD has an equal complementary shear stress acting on the surface CD, as shown in Fig. 1.4.1. It is easily seen from the above reasoning that in the case of cross-sections of constant

† *Footnote*: F_{yy} is usually given the symbol I_{xx} in textbooks written in English. The reason for this change is explained in Chapter 2.

thickness t, the shear stress has a maximum at the neutral axis because that is where $(\bar{y}F)$ has its maximum value. A more satisfactory way of writing eqn (1.4.6) is to express it in terms of the *shear flow*.

$$\text{Shear flow} = \tau t = \frac{V(\bar{y}F)}{F_{yy}} \qquad (1.4.7)$$

Now it is possible to make a more general statement which has no restrictions, that is, the shear flow is a maximum at the neutral axis. Figure 1.4.2 shows the shear flow through a number of cross-sections of practical interest. The length of the arrows is intended to show the relative magnitude of the shear flow. When the shear flow through each of these cross-sections is considered, one sees that the shear stress in each plate element gives a stress resultant which acts along the central plane of that plate. Thus the shear stresses give rise to a number of forces which are fixed in position and in magnitude relative to one another. (Apart from their dependence upon V the forces only depend upon the geometry of the cross-section—see eqn (1.4.6).) Such a system of forces has a resultant (see for example Fig. 1.4.3). If the shear force V is not applied exactly along the line of this resultant, the cross-section will be subjected to an applied twisting moment (torque). Every cross-section has a *shear-centre* M which is the point through which the shear force must be applied if there is to be no twisting.

When a cross-section has two planes of symmetry, the shear centre M and the centre of gravity S coincide (Fig. 1.4.2). If there is one plane of

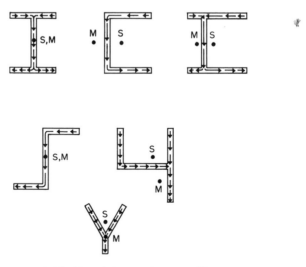

1.4.2. Shear in beams of open profile.

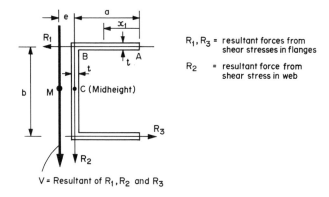

R_1, R_3 = resultant forces from shear stresses in flanges

R_2 = resultant force from shear stress in web

V = Resultant of R_1, R_2 and R_3

Fig. 1.4.3. Resultant of shear stress passes through the shear centre M.

symmetry (e.g. as there is for a channel (Fig. 1.4.2)) M and S are located on that plane and it is a relatively simple matter to establish their coordinates. If the section consists of a number of plates meeting along one line, it is obvious from the argument developed in this section that M lies on that line. S is, of course, found in the well-known way (i.e. by taking moments of the areas of the plate elements about any axis). However, when the cross-section is less regular (Fig. 1.4.4) a systematic method for locating M is required. The analysis in Chapter 2 treats the general case of both open and closed cross-sections which lack symmetry and it will be seen that they require a much lengthier analysis than those which have one or two axes of symmetry.

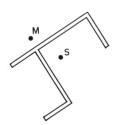

Fig. 1.4.4. Irregular profiles have M located in irregular places.

Example 1.4.1

Locate the shear centre M of the channel section illustrated in Fig. 1.4.3, and evaluate the shear stresses at the points B and C when the transverse shear force is V. The second moment of area F_{yy} is approximately

$$\frac{tb^3}{12} + 2at\left(\frac{b}{2}\right)^2 = \frac{tb^3}{12}\left(1 + \frac{6a}{b}\right).$$

From A to B (eqn 1.4.6)

$$\tau = \frac{Vx_1 b}{2F_{yy}}.$$

Hence

$$R_1 = \int_0^a \tau t \, dx_1 = \frac{Vbta^2}{4F_{yy}} = \frac{3Va^2}{b^2(1+6a/b)}.$$

For vertical equilibrium $R_2 = V$ if the small vertical shear force carried by the flange is ignored. For zero rotational effect

$$R_1 b = R_2 e$$

whence

$$e = \frac{R_1 b}{R_2}$$

$$= \frac{3a^2}{b(1+6a/b)}$$

$$\tau_B = \frac{Vab}{2F_{yy}} = \frac{6Va}{tb^2(1+6a/b)}$$

$$\tau_C = V/bt \text{ (approximately)}$$

or

$$\tau_C = [V/bt][1+b/6a]^{-1} \text{ when eqn (1.4.6) is used.}$$

In deriving eqn (1.4.1), it is assumed that plane sections remain plane (Bernoulli's hypothesis). However, the shear stresses τ (eqn 1.4.6) are accompanied by shear strains $(=\tau/G)$ which cause the plane of the cross-section to distort. In the box girder shown in Fig. 1.4.5 the element at B undergoes a distortion as shown and the sum effect of these for all such elements is that the originally flat plane of the cross-section distorts as shown. The distortions are easily calculated by simply integrating the shear strain along the cross-section. These distortions lead to changes in the simple stress distribution given by eqn (1.4.1), i.e. the direct stresses are no longer proportional to the distance from the neutral axis of bending. The stresses at the centre of the flanges are less than those near the web (Fig. 1.4.5). This phenomenon is known as *shear lag* and it has been investigated by Reissner (1946) and Hadji-Argyris (1944) using an energy method and, more recently in box-girders, by Malcolm and Redwood (1970) and by Moffatt and Dowling (1975) using finite elements. The effects of shear lag are considered in Chapters 4 and 7.

There is another way by which engineers' theory can be invalidated, that is, by distortion of the cross-section in its plane. This can occur, for example, in a thin-walled I-beam which has wide flanges. The outer edges

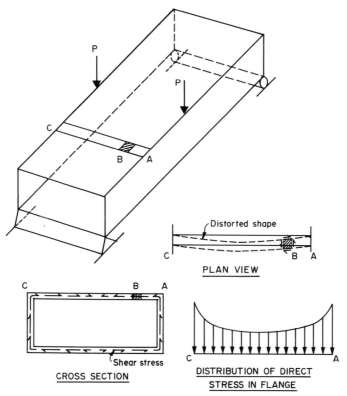

PLAN VIEW

CROSS SECTION

DISTRIBUTION OF DIRECT
STRESS IN FLANGE

Fig. 1.4.5. The cross-section of a simply supported box girder does not remain plane because of shear strains. The phenomenon which results in non-uniform flange stresses is called shear lag.

of the flange may distort as shown in Fig. 1.4.6 and thereby are incapable of carrying the stresses derived from engineers' theory. Thus, the theory based on Bernoulli's hypothesis is also invalidated if local buckling occurs, as it results in distortions of the cross-section. In structural design stiffeners are often used to try to reduce distortion of the cross-section.

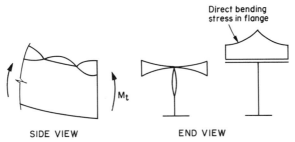

SIDE VIEW   END VIEW

Fig. 1.4.6. A wide-flanged beam exhibiting flange buckling resulting in non-uniform stresses in the flange.

1.4.2. Saint Venant torsion of thin-walled cross-sections

When a beam of uniform cross-section which is free to warp out of its plane (Fig. 1.1.2(b)) is twisted by torques M at each end, it is said to be in a state of uniform torque. Whatever the cross-section of the beam, it is found that for this type of loading the only stresses are shear stresses. Figure 1.4.7(a) shows how they act on an element on the surface of a circular shaft, and Fig. 1.4.7(c) shows how they are distributed through the cross-section of a thin plate which is twisted. The purpose of this section is to analyse beams of various cross-sections, i.e. open profiles,

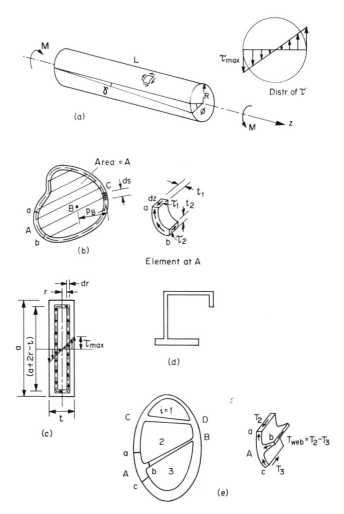

Fig. 1.4.7. Saint Venant torsion.

closed profiles with single cells or with more than one cell, and mixed profiles in which part is open and the other part is closed. Expressions for the stress distribution and the torsional stiffness of the beam are derived. The simplest case is a shaft of circular cross-section. Its solution is well-known, so the results are simply quoted here.

In the case of a cylindrical shaft (Fig. 1.4.7(a)) of radius R, length L, and applied torque M, the following formulae give the maximum shear stress τ_{max}, the maximum shear strain γ_{max} (i.e. at the outer surface), and the angle of rotation ϕ of one end relative to the other end.

$$\tau_{max} = \frac{MR}{J_p}; \qquad \gamma_{max} = \frac{\tau_{max}}{G}. \tag{1.4.8}$$

$$\phi = \frac{ML}{GJ_p} \tag{1.4.9}$$

where

$$J_p = \frac{\pi R^4}{2} \text{ for a solid shaft.} \tag{1.4.10}$$

J_p is the polar moment of inertia and GJ_p is called the torsional stiffness of the shaft.

In the case of a closed profile with one cell only (Fig. 1.4.7(b)), it is seen that for longitudinal equilibrium of the element A

$$\tau_1 t_1 \, dz = \tau_2 t_2 \, dz$$

i.e.

$$T_1 = T_2 \tag{1.4.11}$$

where T_1 and T_2 are the shear flow τt. That is, the shear flow around a closed profile is constant. The moment of the force on an element at C about an *arbitrary* point B is

$$dM = Tp_B \, ds. \tag{1.4.12}$$

By integrating around the section it is seen that

$$M = 2TA \tag{1.4.13}$$

where A is the area enclosed by the profile.

The angle of twist is found by equating the internal and external energies

$$\tfrac{1}{2}M\phi = \frac{L}{2}\oint \tau t \gamma \, ds = \frac{L}{2}\oint \frac{\tau^2 t}{G} \, ds.$$

Inserting eqn (1.4.13), we obtain

$$\frac{\phi}{L} = \frac{d\phi}{dz} = \frac{1}{2GA} \oint \frac{T}{t} ds = \frac{M \oint \frac{ds}{t}}{G \, 4A^2}. \tag{1.4.14}$$

Comparison with eqn (1.4.9) shows that for a closed profile with one cell the equivalent property of the cross section to the polar moment of inertia is

$$J_p = \frac{4A^2}{\oint \frac{ds}{t}}. \tag{1.4.15}$$

This property is called the Saint Venant torsion constant or simply the torsion constant and the same symbol J_p is used.

Equation (1.4.13) can be used to find the torsional properties of a thin rectangle (Fig. 1.4.7(c)), the element shown being treated as a closed profile. The shear strain distribution is assumed to be linear in r, so that, as in the case of a cylinder, compatibility of this element with its neighbouring elements will be satisfied. But the shear stress is proportional to the shear strain, whence

$$\tau(r) = \frac{2\tau_{max} r}{t}. \tag{1.4.16}$$

From eqn (1.4.13)

$$dM = 2\tau(r) \, dr \cdot 2r(a + 2r - t) = \frac{8\tau_{max}}{t} r^2(a + 2r - t) \, dr.$$

Hence

$$M = \frac{8\tau_{max}}{t} \int_0^{t/2} (ar^2 + 2r^3 - tr^2) \, dr = \tau_{max}\left(\frac{at^2}{3} - \frac{t^3}{12}\right).$$

For a thin plate the last term can be ignored, whence,

$$M = \tfrac{1}{3}\tau_{max} a t^2. \tag{1.4.17}$$

As in the case of the closed profile the angle of twist is found by equating internal and external energies.

$$\tfrac{1}{2} M\phi = \frac{L}{2} \int_0^{t/2} \oint \frac{\tau^2 \, dr}{G} \, ds.$$

On substituting from (1.4.15) and (1.4.17) and recognizing that the line integral around the element is simply $2a$, we obtain

$$M = \frac{\phi}{L} G \frac{1}{3} a t^3.$$ (1.4.18)

Hence the torsion constant of an isolated thin plate is

$$J_p = \frac{1}{3} a t^3.$$ (1.4.19)

This result will be used extensively in later work. For open cross-sections comprised of several thin rectangles joined together (Fig. 1.4.7(d)), the torsion constant is

$$J_p = k \sum \frac{1}{3} a_n t_n^3$$ (1.4.20)

where k is a factor which makes allowance for small fillets. k is 1 for thin sections bent from a sheet and 1.2 to 1.3 for rolled sections. The maximum shear stress is obtained by applying eqn (1.4.17) to each plate in the cross-section, it being understood that then M is simply the share of the applied torque carried by the plate being examined.

For mixed cross-sections, i.e. sections comprised of a closed profile and one or more open profiles attached to it, the above results can be added. The total torque M is the sum of the torques carried by the open profiles, i.e. $\sum M_{\text{open}}$ and that carried by the closed profile, i.e. M_{closed}. Hence

$$M = (\sum M)_{\text{open}} + M_{\text{closed}} = G \frac{d\phi}{dz} [(\sum J_p)_{\text{open}} + (J_p)_{\text{closed}}] = G \frac{d\phi}{dz} J_p$$ (1.4.21)

where J_p is the torsion constant of the entire cross-section.

The torque carried by any single plate or the closed profile is obtained by simple proportion.

$$M_{\text{open}} = M \frac{(\sum J_p)_{\text{open}}}{J_p}; \qquad M_{\text{closed}} = \frac{(J_p)_{\text{closed}}}{J_p}.$$ (1.4.22)

Having obtained the torque carried by, say, a plate of the cross-section, the stress can be found by applying the previous formulae. It will be found that in most cases the torsion constant of the closed part dominates and the open parts can usually be ignored.

When a cross-section consists of several closed profiles joined together (Fig. 1.4.7(e)), equilibrium considerations at a join, such as A, show that

$$T_{\text{web}} = T_i - T_{i-1}.$$ (1.4.23)

For the ith cell we have from eqn (1.4.13)

$$M_i = 2A_i T_i.$$

Hence, for the whole cross-section

$$M = \sum 2A_i T_i. \tag{1.4.24}$$

The angle of twist for the ith cell is obtained from eqn (1.4.14), thus

$$\frac{d\phi}{dz} = \frac{1}{2GA_i}\left[\oint \frac{T\,ds}{t}\right]_i$$

$$= \frac{1}{2GA_i}\left[-T_{i-1}\int_A^B \frac{ds}{t} + T_i\oint \frac{ds}{t} - T_{i+1}\int_C^D \frac{ds}{t}\right]. \tag{1.4.25}$$

Since $d\phi/dz$ is the same for all cells, it is seen that a set of simultaneous equations in the unknowns

$$\Psi_i = \frac{T_i}{G\dfrac{d\phi}{dz}} \tag{1.4.26}$$

can be established and solved. On substituting these Ψ values into eqn (1.4.24) the following relationship is obtained.

$$M = G\frac{d\phi}{dz}\sum 2\Psi_i A_i. \tag{1.4.27}$$

Hence the torsion constant of a multi-celled profile is

$$J_p = 2\sum \Psi_i A_i. \tag{1.4.28}$$

When the profile consists of a single cell only, eqns (1.4.13) to (1.4.15) show that

$$\Psi = \frac{T}{G\phi'} = \frac{2A}{\oint \dfrac{ds}{t}}. \tag{1.4.29}$$

But this simple formula cannot be used for multi-celled profiles; it is necessary to solve the set of eqns (1.4.25) with (1.4.26).

In this section we have considered only the problem of the twisting of a shaft under the action of a torque which is *uniform along its length* and warping is freely allowed. There is no longitudinal stress and this distinguishes this type of torsion from warping torsion treated in Chapter 2.

Example 1.4.2

Each of the cross-sections illustrated in Fig. 1.4.8 is subjected to a torque of 1000 N mm. Determine the angle of rotation per metre of length and the maximum shear stress according to Saint Venant's theory. $G = 80\,000$ MPa.

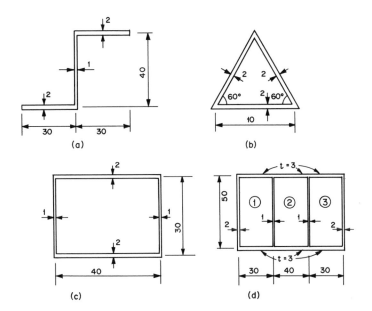

Fig. 1.4.8. Cross-sections of beams with Saint Venant torsion ($M = 1000$ N mm) in Example 1.4.2.

(a) Z section

From eqn (1.4.20) with $k = 1$

$$J_p = \tfrac{1}{3} \times 30 \times 2^3 + \tfrac{1}{3} \times 40 \times 1^3 + \tfrac{1}{3} \times 30 \times 2^3$$
$$= 80 + 13.333 + 80$$
$$= 173.333 \text{ mm}^4$$

$$\therefore \quad \phi = \frac{ML}{GJ_p} = \frac{1000 \times 1000}{80\,000 \times 173.333} = 0.0721 \text{ radian.}$$

Torque carried by one flange $= 1000 \times \dfrac{80}{173.333} = 461.5$ N mm

$$\therefore \quad \text{max. stress in flange (eqn 1.4.17)} = \frac{3 \times 461.5}{30 \times 2^2} = 11.54 \text{ MPa}$$

Torque carried by web $= \dfrac{1000 \times 13.333}{173.333} = 76.92$ N mm

$$\therefore \quad \text{max. stress in web} = \frac{3 \times 76.92}{40 \times 1^2} = 5.77 \text{ MPa}$$

i.e. max. shear stress occurs in flange (on its outer surface).

(b) *Equilateral triangle*

Area of profile $= A = 43.30$ mm^2.

From eqn (1.4.15) for a single-cell profile

$$J_p = \frac{4 \times 43.30^2}{10/2 + 10/2 + 10/2} = 499.97 \text{ mm}^4$$

$$\phi = \frac{1000 \times 1000}{80\,000 \times 499.97} = 0.025 \text{ radian.}$$

From eqn (1.4.13)

$$\tau = \frac{T}{t} = \frac{M}{2At} = \frac{1000}{2 \times 43.30 \times 2} = 5.77 \text{ MPa.}$$

This stress is uniformly distributed through the thickness.

(c) *Hollow rectangle*

$A = 1200$ mm^2.

From eqn (1.4.15)

$$J_p = \frac{4 \times 1200^2}{40/2 + 30/1 + 40/2 + 30/1} = 57\,600 \text{ mm}^4$$

$$\phi = \frac{1000 \times 1000}{80\,000 \times 57\,600} = 0.000\,217 \text{ radian.}$$

In the flanges

$$\tau = \frac{M}{2At} = \frac{1000}{2 \times 1200 \times 2} = 0.208 \text{ MPa.}$$

In the webs

$$\tau = \frac{1000}{2 \times 1200 \times 1} = 0.416 \text{ MPa.}$$

(d) *Rectangle with three cells*

From eqns (1.4.25) and (1.4.26)

Cell 1:

$$1 = \frac{1}{2 \times 50 \times 30} \left[\Psi_1 \left(\frac{30}{3} + \frac{50}{1} + \frac{30}{3} + \frac{50}{2} \right) - \Psi_2 \frac{50}{1} \right]$$

Cell 2:

$$1 = \frac{1}{2 \times 50 \times 40} \left[-\Psi_1 \frac{50}{1} + \Psi_2 \left(\frac{40}{3} + \frac{50}{1} + \frac{40}{3} + \frac{50}{1} \right) - \Psi_3 \frac{50}{1} \right]$$

Cell 3:

$$1 = \frac{1}{2 \times 50 \times 30} \left[-\Psi_2 \frac{50}{1} + \Psi_3 \left(\frac{30}{3} + \frac{50}{2} + \frac{30}{3} + \frac{50}{1} \right) \right]$$

The solution of these three equations is

$$\Psi_1 = \Psi_3 = 82.464\ 46 \qquad \Psi_2 = 96.682\ 46.$$

From eqn (1.4.28)

$$J_p = 2[82.464\ 46 \times 50 \times 30 \times 2 + 96.682\ 46 \times 50 \times 40]$$
$$= 881\ 516.6\ \text{mm}^4$$

$$\phi = \frac{1000 \times 1000}{80\ 000 \times 881\ 516.6} = 0.000\ 014\ 18\ \text{radian.}$$

From eqn (1.4.26)

$$T_1 = \Psi_1 G\phi/L = 82.464\ 46 \times 80\ 000 \times 0.000\ 014\ 18/1000 = 0.093\ 55$$

i.e. in upper and lower flanges $\tau = 0.093\ 55/3 = 0.031\ 18$ MPa
in outer web $\tau = 0.093\ 55/2 = 0.046\ 77$ MPa
$T_2\Psi_2 G\phi/L = 96.682\ 46 \times 80\ 000 \times 0.000\ 014\ 18/1000 = 0.109\ 68$
$T_2 = \Psi_2 G\phi/L = 96.682\ 46 \times 80\ 000 \times 0.000\ 014\ 18/1000 = 0.109\ 68$
i.e. in upper and lower flanges $\tau = 0.109\ 68/3 = 0.036\ 56$ MPa.
Shear flow in inner webs $= T_2 - T_1 = 0.016\ 13\ \text{N mm}^{-1}$
i.e. $\tau = 0.016\ 13/1 = 0.016\ 13$ MPa.

The maximum shear stress occurs in the outer web. The distribution of shear stresses is shown in Fig. 1.4.9.

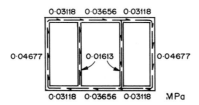

Fig. 1.4.9. Distribution of Saint Venant shear stresses in beam shown in Fig. 1.4.8(d) for clockwise torque at 1000 N mm.

1.4.3. Shear in beams with a closed profile—engineers' theory

An element of a beam with a single-celled profile bent about its neutral axis *xx*, and carrying a transverse shear force *V* acting through its shear

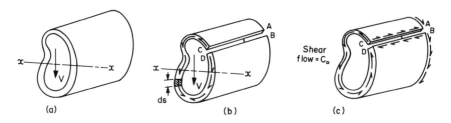

Fig. 1.4.10. The shear stress in a hollow profile due to a transverse shear force V acting through its shear centre is the sum of those in an open profile due to V and a shear flow C_0.

centre is illustrated in Fig. 1.4.10(a). It can be changed into an open profile by introducing a longitudinal cut (Fig. 1.4.10(b)) and the theory developed in Section 1.4.1 can then be applied to it. The shear stress τ at any point in this profile is given by eqn (1.4.6). At this point the shear strain is τ/G and when an element of width ds is considered it is seen that the relative movement in the axial direction between its two faces is $\tau\,ds/G$. Therefore, the total relative displacement in the axial direction between D and C is

$$\Delta_{DC} = \oint \tau/G \; ds. \tag{1.4.30}$$

This displacement or so-called dislocation can be eliminated by introducing the shear flow C_0 (Fig. 1.4.10(c)) which was released when the longitudinal cut was made. As seen in Section 1.4.2 a shear flow such as C_0 is constant around the profile (see eqn (1.4.11). Hence if there is to be no dislocation at D and C

$$\oint (\tau + C_0/t)/G \; ds = 0 \tag{1.4.31}$$

i.e.

$$C_0 = -\oint \tau \; ds / \oint ds/t. \tag{1.4.32}$$

When the profile has n cells (Fig. 1.4.11) a similar technique is employed. After reducing the profile to an open one by longitudinal cuts a constant shear flow C_i is introduced into the ith cell ($i = 1, \ldots, n$). The dislocation at the cut in the ith cell is equated to zero as follows (G has

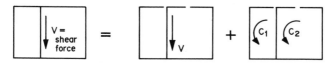

Fig. 1.4.11. A profile with n cells $(n>1)$ has n unknown shear flows C_1 to C_n plus the shear stress due to V.

been eliminated).

$$\oint (\tau + C_i/t)\,ds - \sum \int_{\text{webs}} C_j/t\,ds = 0 \tag{1.4.33}$$

where the last terms are the contributions from the neighbouring cells to the distortion of the web which it has in common with the ith cell. Equation (1.4.33) leads to n simultaneous equations in the unknown shear flows C_1 to C_n. The following examples show how eqns (1.4.32) and (1.4.33) may be used for hollow profiles.

Example 1.4.3

A transverse shear force $V = 96$ kN acts through the shear centre vertically downwards on the hollow profile illustrated in Fig. 1.4.12(a). Determine the distribution of shear stress.

Introduce a cut at point **1** to convert the profile into an open profile.

$$F_{yy} = 2 \times 4 \times 200 \times 200^2 + 4 \times 400^3/12 + 2 \times 400^3/12 = 96 \times 10^6 \text{ mm}^4.$$

$$\therefore \quad V/F_{yy} = 10^{-3} \text{ N mm}^{-4}.$$

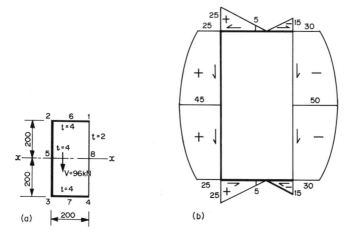

(a)

(b)

Fig. 1.4.12. (a) Profile analysed in Example 1.4.3; (b) shear stress distribution.

For the open profile, using eqn (1.4.6)

$$\tau_1 \qquad\qquad\qquad\qquad\qquad = 0,$$

$$\tau_6 = \frac{100 \times 4 \times 200}{4} \times 10^{-3} \qquad\qquad = 20 \text{ MPa},$$

$$\tau_2 = \frac{200 \times 4 \times 200}{4} \times 10^{-3} \qquad\qquad = 40 \text{ MPa},$$

$$\tau_5 = \left(\frac{200 \times 4 \times 200}{4} + \frac{200 \times 4 \times 100}{4}\right) \times 10^{-3} = 60 \text{ MPa},$$

$$\tau_3 = \frac{200 \times 4 \times 200}{4} \times 10^{-3} \qquad\qquad = 40 \text{ MPa},$$

$$\tau_7 = \left(\frac{200 \times 4 \times 200}{4} - \frac{100 \times 4 \times 200}{4}\right) \times 10^{-3} = 20 \text{ MPa},$$

$$\tau_4 \qquad\qquad\qquad\qquad\qquad = 0,$$

$$\tau_8 = -\frac{200 \times 2 \times 100}{2} \times 10^{-3} \qquad\qquad = -20 \text{ MPa}.$$

$\oint \tau \, ds$ (by Simpson's Rule):

1 to **2**: $\dfrac{200}{6}(0 + 4 \times 20 + 40) \quad = \quad 4000$

2 to **3**: $\dfrac{400}{6}(40 + 4 \times 60 + 40) = 21333$

3 to **4**: $\dfrac{200}{6}(40 + 4 \times 20 + 0) \quad = \quad 4000$

4 to **1**: $\dfrac{400}{6}(0 - 4 \times 20 + 0) \quad = \underline{-5333}$

$$\oint \tau \, ds = \underline{24000}$$

$$\oint ds/t = \frac{800}{4} + \frac{400}{2} = 400$$

$$\therefore \quad C_0 = -24\,000/400 = -60$$

Shear stress

At **1** (flange) $\tau = \quad 0 - 60/4 = -15 \text{ MPa}$

At **6** $\qquad\quad \tau = \quad 20 - 60/4 = \quad 5 \text{ MPa}$

At **2** $\qquad\quad \tau = \quad 40 - 60/4 = \quad 25 \text{ MPa}$

At **5** $\qquad\quad \tau = \quad 60 - 60/4 = \quad 45 \text{ MPa}$

At **3** $\tau = 40 - 60/4 = 25$ MPa

At **7** $\tau = 20 - 60/4 = 5$ MPa

At **4** (flange) $\tau = 0 - 60/4 = -15$ MPa

At **4** (web) $\tau = 0 - 60/2 = -30$ MPa

At **8** $\tau = -20 - 60/2 = -50$ MPa

At **1** (web) $\tau = 0 - 60/2 = -30$ MPa

These stresses are sketched in Fig. 1.4.12(b).

Example 1.4.4

A transverse shear force $V = 128.61$ kN acts vertically downwards through the shear centre of the profile illustrated in Fig. 1.4.13(a). Sketch the distribution of shear stress.

From the usual analysis of section properties the distance $\bar{y}$ from the upper flange to the neutral axis is 209.72 mm and the second moment of area about the neutral axis $F_{yy} = 0.973\,266 \times 10^9$ mm^4. In this cross-section cuts are introduced at points **1** and **2** as indicated in Fig. 1.4.13(b).

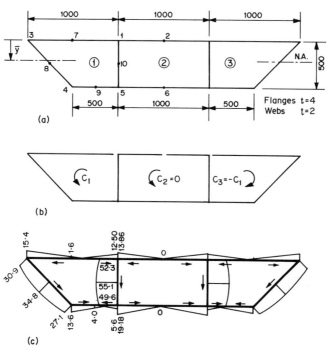

Fig. 1.4.13. (a) Profile analysed in Example 1.4.4; (b) opened profile with unknown shear flows; (c) total shear stress due to transverse shear force; $V = 10$ MN acting through the shear centre.

Because of symmetry the shear flow C_2 in cell 2 is zero when the cut is made at point **2**. Thus it is only necessary to close up the dislocation in cell 1 by introducing a shear flow C_1. With $V/F_{yy} = 0.000\,132\,143\,1\,\text{N mm}^{-4}$ the use of eqn (1.4.6) gives:

Cell 1:

$$\tau_1 \qquad\qquad\qquad\qquad\qquad\qquad\qquad\qquad\qquad\qquad = 0$$

$$\tau_7 \quad = \frac{500 \times 4 \times 209.72}{4} \times 0.132\,143\,1 \times 10^{-3} \qquad = 13.857\,\text{MPa}$$

$$\tau_3 \quad = \frac{1000 \times 4 \times 209.72}{4} \times 0.132\,143\,1 \times 10^{-3} \quad = 27.713\,\text{MPa}$$

$$\tau_8 \quad = \frac{(1000 \times 4 \times 209.72 + 250 \times 2\sqrt{2} \times 84.72)}{2}$$
$$\times 0.132\,143\,1 \times 10^{-3} \qquad\qquad = 59.384\,\text{MPa}$$

$$\tau_4(\text{web}) = \frac{(1000 \times 4 \times 209.72 - 500 \times 2\sqrt{2} \times 40.28)}{2}$$
$$\times 0.132\,143\,1 \times 10^{-3} \qquad\qquad = 51.662\,\text{MPa}$$

$$\tau_4\,(\text{flange}) \qquad\qquad\qquad\qquad\qquad\qquad\qquad = 25.831\,\text{MPa}$$

$$\tau_9 \quad = \frac{(1000 \times 4 \times 209.72 - 500 \times 2\sqrt{2} \times 40.28 - 250 \times 4 \times 290.28)}{4}$$
$$\times 0.132\,143\,1 \times 10^{-3} \qquad\qquad = 16.241\,\text{MPa}$$

$$\tau_5\,(\text{left flange}) = \frac{(1000 \times 4 \times 209.72 - 500 \times 2\sqrt{2} \times 40.28 - 500 \times 4 \times 290.28)}{4}$$
$$\times 0.132\,143\,1 \times 10^{-3} \qquad\qquad = 6.652\,\text{MPa}$$

Cell 2:

$$\tau_2 \qquad\qquad\qquad\qquad\qquad\qquad\qquad\qquad\qquad\qquad = 0$$

$$\tau_1(\text{flange}) = \frac{(500 \times 4 \times 209.72)}{4} \times 0.132\,143\,1 \times 10^{-3} \quad = 13.857\,\text{MPa}$$

$$\tau_1\,(\text{web}) \qquad\qquad\qquad\qquad\qquad\qquad\qquad = 27.713\,\text{MPa}$$

$$\tau_{10} \quad = \frac{(500 \times 4 \times 209.72 + 250 \times 2 \times 84.72)}{2}$$
$$\times 0.132\,143\,1 \times 10^{-3} \qquad\qquad = 30.512\,\text{MPa}$$

$$\tau_5\,(\text{web}) = \frac{(500 \times 4 \times 209.72 - 500 \times 2 \times 40.28)}{2}$$
$$\times 0.132\,143\,1 \times 10^{-3} \qquad\qquad = 25.052\,\text{MPa}$$

$$\tau_6 = \frac{(1500 \times 4 \times 209.72 - 500 \times (2 + 2\sqrt{2}) \times 40.28 \\ - 1000 \times 4 \times 290.28)}{4}$$

$$\times 0.132\ 143\ 1 \times 10^{-3} \qquad\qquad = 0$$

In cell 1 $\oint \tau\ ds$ is found by Simpson's Rule (which checks)

1 to 3: $\dfrac{1000}{6}(0 + 4 \times 13.857 + 27.713)$ $\qquad = \quad 13857$

3 to 4: $\dfrac{500\sqrt{2}}{6}(55.426 + 4 \times 59.384 + 51.662)$ $\qquad = \quad 40614$

4 to 5: $\dfrac{500}{6}(25.831 + 4 \times 16.241 + 6.652)$ $\qquad = \quad 8121$

5 to 1: $-\dfrac{500}{6}(25.052 + 4 \times 30.512 + 27.713)$ $\qquad = -14568$

$$\oint \tau\ ds = \quad 48024$$

$$\oint \frac{ds}{t} = \frac{1000}{4} + \frac{500\sqrt{(2)}}{2} + \frac{500}{4} + \frac{500}{2} = 978.55$$

Hence $C_1 = -48\ 024/978.55 = -49.1\ \text{N mm}^{-1}$.

The total shear stress at every point in cell 1 is now found by subtracting $49.1/t$ from the values calculated on the basis of an open profile. For example at point 8 the shear stress is $59.384 - 49.1/2 = 34.8\ \text{MPa}$. Final results of this calculation are plotted in Fig. 1.4.13(c).

EXERCISES

1.1 The rigid strut AB is pinned at each end and supported by a spring AC. For the given geometry show that the total potential energy of the system when AB rotates through a clockwise angle θ is

$$U = \frac{kL^2}{2}\{1 - [(1 - 2\sin\theta)^2 + 4(1 - \cos\theta)^2]^{\frac{1}{2}}\}^2 - 2PL(1 - \cos\theta)$$

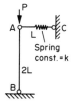

Ex. 1.1

Expand the sine and cosine functions and use the binomial theorem to expand the square root, retaining terms of the order θ^4. Show that the equilibrium paths are $\theta = 0$ and $P = 2kL + 31kL\,\theta^2$.

1.2   A pin-ended strut of length 1000 mm has the cross-section illustrated. If the yield stress is 250 MPa what is the squash load? Calculate the lateral deflection at a central plastic hinge at the following axial loads, $P = 100, 60, 30$ kN when bending occurs about the xx-axis.
(Answer: Squash   load $= 120$ kN;   central   deflections $= 4.7$,   22.5, 64.50 mm.)

Ex. 1.2

1.3   A beam whose cross-section is illustrated carries a transverse shear force $V = 10$ kN. Evaluate the shear stresses at the points A, B and C. Take $F_{yy} = 165\,888$ mm^4.
(Answer: 31.5, 62.9, 72.7 MPa.)

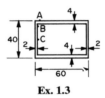

Ex. 1.3

1.4   If the transverse load in the last exercise is applied through one of the webs what *additional* shear stresses occur? Also what is the angle of twist per unit length? ($G = 80\,000$ MPa.)
(Answer: In   flanges   17.36,   in   webs   34.72 MPa, $\phi/L = 13.51 \times 10^{-6}$ rad mm^{-1}.)

1.5   In the cross-section illustrated all of the angles are 60° and all plates

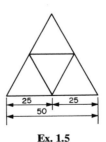

Ex. 1.5

have a thickness of 1. Determine the angle of twist per unit length and the shear stress distribution if the torque is 1000 N mm.
(Answer: $\phi/L = 0.355\ 56 \times 10^{-6}$ rad mm^{-1}. $\tau_{outer} = 0.4106$ MPa, $\tau_{inner} = 0.2052$ MPa.)

1.6   Evaluate J_p for the cross-section illustrated. If the internal webs are removed what is the value of J_p and what conclusion can be drawn?
(Answer: $\Psi_1 = 150.96$, $\Psi_2 = 177.65$, $J_p = 2.54 \times 10^6$ mm^4, $J_p = 2.48 \times 10^6$ mm^4.)

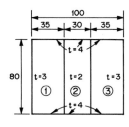

Ex. 1.6

2

FIRST-ORDER THEORY FOR BENDING AND TORSION OF THIN-WALLED BEAMS

2.1. INTRODUCTION

This chapter is concerned with the behaviour of a straight, thin-walled, prismatic beam which is loaded with distributed loads p_x, p_y, and p_z, with distributed torque m_z, and with concentrated loads and torques at arbitrary points z_i (Fig. 2.1.1). It is desired to determine the relationship between the applied loads, stresses and deformations using a first-order theory. As stated previously the influence of deformations on the stresses is ignored in the first-order theory of beams. Therefore this theory as it is developed here cannot deal with the problem of buckling but is concerned with the distribution of stresses and deflections in the beam. In Chapter 3 the theory is modified so that it can be used to solve certain

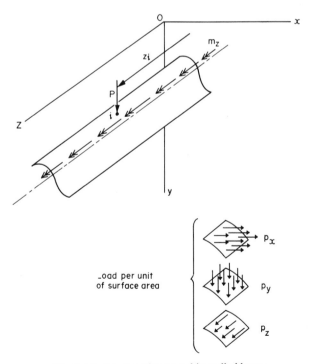

Fig. 2.1.1. Loads acting on a thin-walled beam.

first-order buckling problems, i.e. eigenvalue problems. In Fig. 1.1.2 it is seen that when a thin-walled beam is restrained from warping additional stresses arise in the longitudinal and transverse directions. These stresses do not appear in the case of uniform (Saint Venant) torsion. It was C. Bach (1909) who first noted these stresses by considering a channel for which the centre of gravity S and the shear centre M do not coincide (see Fig. 1.4.2). He observed that a transverse load P acting at the end of a cantilever through S produced vertical deflection and rotation of the cross-section about a longitudinal axis. He also noticed that plane sections did not remain plane and that a warping out of the plane occurred. By systematically changing the position of P he found the position of M and noted that when P acted through M, Bernoulli's hypothesis was satisfied. The additional stresses also disappeared.

2.2. ASSUMPTIONS USED AND SOME REMARKS ON THE NOTATION

Thus, in a thin-walled beam which is twisted only, there can be two sets of shear stresses (and of course shear strains), that is, the Saint Venant shear stresses τ_s and the warping shear stresses τ_w. These are quite different from shear stresses associated with bending (see Sections 1.4.1 and 1.4.3).

There are no longitudinal direct stresses associated with Saint Venant (uniform) torsion, but there are with warping (non-uniform) torsion and they, together with the associated warping shear stresses τ_w, are called the *warping stresses*. The problem analysed in this chapter is the derivation of formulae for these warping stresses, the amount of twisting and the longitudinal displacements or *warping* of the section.

For open and closed profiles the warping of the cross-section due to twisting must be obtained by slightly different routes. In Chapter 1 it was shown that for *open* profiles the Saint Venant shear stresses τ_s (and therefore the associated shear strains γ_s) along the centreline of the profile are zero (see Fig. 1.4.7(c)). For open profiles undergoing non-uniform torsion the warping shear strains are considered to be of a secondary nature so it is assumed that $\gamma_w = 0$, i.e. the total shear strain γ at the centre line is

$$\gamma = \gamma_s + \gamma_w = 0 + 0 = 0. \tag{2.2.1}$$

This means that the in-plane warping shear stresses cannot be derived directly from the shear strains γ_w, i.e. from the stress–strain equation $\tau_w = \gamma_w G$, because they too would have to be zero. This difficulty is overcome by first integrating the equation for warping shear strain, as a function of two displacement components u and v, that is:

$$\gamma_w = \gamma_{zs} = \frac{\partial u}{\partial s} + \frac{\partial v}{\partial z} = 0 \qquad (2.2.2)$$

to obtain the out-of-plane displacement u (warping) of the profile as a function of the other displacements (Section 2.3.3). The longitudinal warping stress σ_w is then expressed as a function of the displacement u through the stress–strain relationship (Section 2.3.4) and finally the governing equations are obtained from the laws of equilibrium, but for convenience they too are expressed in terms of the displacements (Section 2.3.5).

Much of the theory developed for open profiles may be used for *closed* profiles. For a closed profile the Saint Venant shear stresses τ_s are uniformly distributed across the walls of each cell of the profile and their associated strains γ_s cannot be ignored. If they were ignored it would be found that no warping stresses could develop. For *closed* profiles many theories have been developed but here only two are considered, the first due to von Karman and Christensen (1944) is referred to as the approximate theory, while the second presented by Benscoter (1954) is referred to as the more accurate theory. The approximate theory for closed profiles relates very closely to the theory for open profiles and, where it is logical to do so (in Sections 2.3–2.5) the approximate theory for closed profiles is introduced alongside of the theory for open profiles. Later, in Section 2.6.1, the whole of the approximate theory for closed profiles is summarized and illustrated by examples and in Section 2.6.2 the more accurate theory is developed and illustrated.

In the approximate theory for closed profiles the out-of-plane displacement u is obtained as a function of the other displacements by integrating the equation for the shear strain of a small element

$$\gamma_{zs} = \frac{\partial u}{\partial s} + \frac{\partial v}{\partial z} = \frac{T_i}{Gt} \qquad (2.2.3)$$

for each cell where T_i is the shear flow in the ith cell arising from Saint Venant torsion. Implicit in this step is the assumption that the shear strain due to warping γ_w is negligible compared to the Saint Venant shear strains, i.e. $\gamma_w = 0$, or, the total shear strain

$$\gamma = \gamma_s + \gamma_w = \gamma_s. \qquad (2.2.4)$$

This integration is carried around each cell of the profile which also has to satisfy the requirement that there are no dislocations in the out-of-plane displacements. It will be seen later that, in the final analysis, this approximate theory gives the same governing equations for open profiles, for closed single-cell profiles and for closed multi-cell profiles. The only difference in the theory of these three classes of profile is that some of the section properties (yet to be introduced) are not the same.

Some hollow sections undergo large warping strains γ_w and their effect cannot be ignored. Obviously if γ_w is large the warping shear stresses τ_w will also be large and the approximate theory gives unsatisfactory results. For such cases the more accurate theory should be used (Section 2.6.2). Here the approach is to assume a mathematical form for the tangential displacement v and the longitudinal displacement u of an arbitrary point on the profile. This form introduces a new deflection parameter $\theta(z)$ which is found to be a function of the angle of rotation ϕ of the profile. The governing equations are found via the stress–strain relationship from equilibrium conditions and it is revealed that they have similar forms to those obtained for open profiles and for closed profiles by the approximate theory. The differences again arise only through the interpretation of the coefficients of the terms in the equations.

In the theory developed in Sections 2.3–2.5 the following is a summary of the assumptions made.

(1) The material is elastic, isotropic and obeys Hooke's Law.

(2) The beam is straight before loading and there is no local buckling or other deformation of the cross-section.

(3) The longitudinal stress σ_z and the in-plane shear stress τ_{zs} in an element of the beam (Fig. 2.2.1) are the only significant stresses. The following stresses and strains are neglected:

σ_x, σ_y (hence the in-plane stress normal to the z-axis),

$\varepsilon_x, \varepsilon_y$,

and, in the case of open profiles,

$\gamma_{xy}, \gamma_{xz}, \gamma_{yz}$ (hence the in-plane shear strain of the element).

A consequence of the assumption that the warping strain $\gamma_w = 0$ is that corresponding shear stresses should, by Hooke's Law, also be zero. But this will not be so except perhaps at a few points along the beam. In other words, it is anticipated now that expressions for the in-plane shear

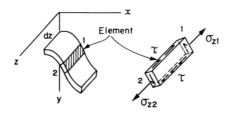

Fig. 2.2.1. The only significant stresses in thin-walled beam theory are σ_z and τ.

stresses cannot be derived directly from γ_w and they must be obtained by other means. In the engineers' theory for shear stresses due to the bending of beams (Section 1.4 1) this problem was overcome in a similar manner by using an equation of equilibrium to evaluate the shear stresses (eqn (1.4.5)); it being tacitly assumed that the in-plane shear stresses and strains do not change the distribution of bending stresses given by eqn (1.4.1) from the Bernoulli hypothesis. However, when there is twisting there will be additional longitudinal and shear stresses (the warping stresses) arising from the bimoment which may exist in the beam (Fig. 1.1.2) and which are ignored in the engineers' theory.

For closed profiles there is an additional requirement that there shall be no longitudinal dislocation in *each* cell. For closed profiles the in-plane shear strains cannot be neglected (see Section 2.3).

(4) The loads are conservative, i.e. they act in their original direction and position throughout the loading.

(5) The deformations are sufficiently small that the usual assumption about curvature can be made, that is,

$$\frac{1}{R_x} = \xi'' \quad \text{and} \quad \frac{1}{R_y} = \eta'', \qquad (2.2.5)$$

where ξ and η are displacements which are defined later.

The following remarks give an explanation of the notation used. All of the notation is listed elsewhere and the intention here is to present a more detailed explanation of some of the symbols used in this chapter.

As seen in Chapter 1 the symbol A is used for the enclosed area of a cell in a single or multi-cell profile. The area of actual material in a cross-section is denoted by the symbol F (from the German word, Fläche = area) to avoid confusion with the area A of a hole in the profile,

$\int_F \Phi \, dF$ means the function Φ is integrated over the entire cross-section, i.e. over the area occupied by material.

In this chapter the governing equations are derived by first establishing an original coordinate system, then an intermediate coordinate system and finally a principal coordinate system. The coordinates of a point in these three systems have to be denoted by different symbols (see below). Furthermore it is found later that a so-called warping function, ω, has to be defined with respect to a pole, denoted by B in the first two systems and the shear centre M in the principal coordinate system, and with respect to a starting point for the integration, denoted by V.

To summarize, three sets of coordinates axes are used:

(1) Original coordinate system (origin A) $\bar{x}$, $\bar{y}$, $\bar{z}$. The pole B and the

starting point V for calculating the warping function are arbitrary. Usually

$$\int_F \bar{\omega}_B \, dF = F_{\bar{\omega}_B} \neq 0,$$

(2) The intermediate coordinate system (origin S, the centroid) $\tilde{x}, \tilde{y}, \tilde{z}$. These axes are parallel to $\bar{x}, \bar{y}, \bar{z}$. The pole B is retained in its original position but the starting point V is changed so that

$$\int_F \tilde{\omega}_B \, dF = F_{\tilde{\omega}_B} = 0$$

(3) The principal coordinate system (origin S) x, y, z. The x- and y-axes are inclined at an angle ψ to the $\tilde{x}$ and $\tilde{y}$ axes and the pole is shifted to M.

s = distance measured around the profile to point i from a starting point V.

s_E = distance measured around profile from edge of profile.

$\bar{\alpha}(s)$ = angle between the tangent to the profile middle line and the $\bar{x}$ axis, (Fig. 2.3.7).

ψ = angle between $\tilde{x}$ and x axes (Fig. 2.4.1).

$\alpha(s) = \bar{\alpha}(s) - \psi$ = angle between tangent to the profile middle line and the x-axis.

The following displacements are used:

$\bar{\zeta}(\bar{z}, s), \bar{\xi}(\bar{z}, s), \bar{\eta}(\bar{z}, s)$ = displacement components in the direction of the $\bar{z}, \bar{x}$, and $\bar{y}$ axes,

$\tilde{\zeta}(\tilde{z}, s), \tilde{\xi}(\tilde{z}, s), \tilde{\eta}(\tilde{z}, s)$ = displacement components in the direction of the $\tilde{z}, \tilde{x}$, and $\tilde{y}$ axes,

$\zeta(z, s), \xi(z, s), \eta(z, s)$ = displacement components in the direction of the z, x, and y axes,

$\phi_z(\bar{z}) = \phi_z(\tilde{z}) = \phi_z(z) = \phi(z)$ = angle of twist,

$u(\bar{z}, s), v(\bar{z}, s), w(\bar{z}, s)$ = displacement components in the $\bar{z}$, tangential and normal directions.

The following letters indicate points:

A = origin of original coordinate system,

C = centre of rotation of profile,

S = origin of intermediate and principal coordinate systems (these origins are located at the centroid of the profile),

M = shear centre = origin (i.e. pole) for calculating ω in the principal coordinate system,

B = pole for calculating ω in the original and intermediate coordinate system,

V = starting point = point from which s is measured,

i = general point on the middle line of the profile,

E = edge point on the middle line of the profile.

Warping:

$\bar{\omega}_B(s) = \bar{\omega}_B$ = warping function with pole B in the coordinate system $\bar{x}, \bar{y}, \bar{z}$; generally, $\int_F \bar{\omega} \, dF \neq 0$,

$\tilde{\omega}_B(s) = \tilde{\omega}_B$ = warping function with pole B in the coordinate system $\tilde{x}, \tilde{y}, \tilde{z}$ but starting point V arranged so that $\int_F \tilde{\omega} \, dF = 0$,

$\omega_M(s) = \omega_M = \omega$ = warping function with pole at shear centre M in the x, y, z (principal) coordinate system.

Component of radius:

$\bar{p}_B(s) = \bar{p}_B$ = length of perpendicular from pole B to the tangent to the middle line of the profile at i in $\bar{x}, \bar{y}, \bar{z}$ coordinate system (Fig. 2.3.8).

Material constants:

E = Young's modulus of elasticity,

G = modulus of shear rigidity,

ν = Poisson's ratio,

Thickness:

$t(\bar{z}, s)$ = thickness of profile,

$t(s)$ = thickness of profile when it is independent of $\bar{z}$.

Sectional properties:

F = area of cross-section;

(a) In original coordinate $\bar{x}, \bar{y}, \bar{z}$:

$$F_{\bar{x}} = \int_F \bar{x}(s)dF = \text{first moment of area about } \bar{y}\text{-axis},$$

$$F_{\bar{y}} = \int_F \bar{y}(s)dF = \text{first moment of area about } \bar{x}\text{-axis},$$

$$F_{\bar{\omega}_B} = F_{\bar{\omega}} = \int_F \bar{\omega}_B(s) \, dF = \text{first moment sectorial area about pole B},$$

$$F_{\bar{x}\bar{x}} = \int_F \bar{x}^2(s) \, dF = \text{second moment of area of profile about } \bar{y}\text{-axis},$$

$$F_{\bar{y}\bar{y}} = \int_F \bar{y}^2(s) \, dF = \text{second moment of area of profile about } \bar{x}\text{-axis},$$

$$F_{\bar{x}\bar{y}} = \int_F \bar{x}(s)\bar{y}(s) \, dF = \text{product moment of area of profile in } \bar{x}, \bar{y}, \bar{z}$$
coordinate system,

$$F_{\bar{\omega}_B \bar{x}} = F_{\bar{\omega}\bar{x}} = \int_F \bar{\omega}_B(s)\bar{x}(s)\,dF = \text{sectorial product of area,}$$

$$F_{\bar{\omega}_B \bar{y}} = F_{\bar{\omega}\bar{y}} = \int_F \bar{\omega}_B(s)\bar{y}(s)\,dF = \text{sectorial product of area,}$$

$$F_{\bar{\omega}_B \bar{\omega}_B} = F_{\bar{\omega}\bar{\omega}} = \int_F \bar{\omega}_B^2(s)\,dF = \text{warping constant with pole B,}$$

$$J_p = \text{torsion (Saint Venant) constant;}$$

(b) After the first step of orthogonalization (i.e. $\tilde{x}, \tilde{y}, \tilde{z}$ coordinate):

$$F_{\tilde{x}} = \int_F \tilde{x}(s)\,dF = \text{first moment of area about } \tilde{y}\text{-axis,}$$

$$F_{\tilde{y}} = \int_F \tilde{y}(s)\,dF = \text{first moment of area about } \tilde{x}\text{-axis,}$$

$$F_{\tilde{\omega}_B} = F_{\tilde{\omega}} = \int_F \tilde{\omega}_B(s)\,dF = \text{first moment of sectorial area with pole at}$$
B,

$$F_{\tilde{x}\tilde{x}} = \int_F \tilde{x}^2(s)\,dF = \text{second moment of area about } \tilde{y}\text{-axis,}$$

$$F_{\tilde{y}\tilde{y}} = \int_F \tilde{y}^2(s)\,dF = \text{second moment of area about } \tilde{x}\text{-axis,}$$

$$F_{\tilde{x}\tilde{y}} = \int_F \tilde{x}(s)\tilde{y}(s)\,dF = \text{product moment of area in } \tilde{x}, \tilde{y}, \tilde{z} \text{ coordinates,}$$

$$F_{\tilde{\omega}_B \tilde{x}} = F_{\tilde{\omega}\tilde{x}} = \int_F \tilde{\omega}_B(s)\tilde{x}(s)\,dF = \text{sectorial product of area,}$$

$$F_{\tilde{\omega}_B \tilde{y}} = F_{\tilde{\omega}\tilde{y}} = \int_F \tilde{\omega}_B(s)\tilde{y}(s)\,dF = \text{sectorial product of area,}$$

$$F_{\tilde{\omega}_B \tilde{\omega}_B} = F_{\tilde{\omega}\tilde{\omega}} = \int_F \tilde{\omega}_B^2(s)\,dF = \text{warping constant with pole at B;}$$

(c) After the second step of orthogonalization (i.e. x, y, z coordinate):

$$F_x = \int_F x(s)\,dF = \text{first moment of area about } y\text{-axis,}$$

$$F_y = \int_F y(s)\,dF = \text{first moment of area about } x\text{-axis,}$$

$$F_{\omega_M} = F_\omega = \int_F \omega_M(s)\,dF = \text{first moment of sectorial area with pole at}$$
M,

$$F_{xx} = \int_F x^2(s)\, dF = \text{second moment of area about } y\text{-axis},$$

$$F_{yy} = \int_F y^2(s)\, dF = \text{second moment of area about } x\text{-axis},$$

$$F_{xy} = \int_F x(s)y(s)\, dF = \text{product moment of area in } x, y, z \text{ coordinates},$$

$$F_{\omega_M x} = F_{\omega x} = \int_F \omega_M(s)x(s)\, dF = \text{sectorial product of area},$$

$$F_{\omega_M y} = F_{\omega y} = \int_F \omega_M(s)y(s)\, dF = \text{sectorial product of area},$$

$$F_{\omega_M \omega_M} = F_{\omega \omega} = \int_F \omega_M^2(s)\, dF = \text{warping constant with pole at M}.$$

Strains:

$\varepsilon_{\bar{z}}(\bar{z}, s) = $ longitudinal strain in $\bar{z}$ direction,

$\gamma_{\bar{z}s} = $ shear strain in the $\bar{z}-s$ plane.

Stresses and shear flows:

$\sigma_{\bar{z}}(\bar{z}, s) = $ normal stress in $\bar{z}$ direction,

$\tau_{\bar{z}s}(\bar{z}, s) = $ shear stress in the $\bar{z}-s$ plane,

$T(\bar{z}, s) = t(s)\tau_{\bar{z}s}(\bar{z}, s) = $ shear flow.

External loads:

$p_{\bar{x}}(\bar{z}, s); p_{\bar{y}}(\bar{z}, s); p_{\bar{z}}(\bar{z}, s) = $ surface loads per unit area of profile in the $\bar{x}$, $\bar{y}$, and $\bar{z}$ directions, respectively,

$q_{\bar{x}}(\bar{z}); q_{\bar{y}}(\bar{z}); q_{\bar{z}}(\bar{z}) = $ line loads per unit length (i.e. $dz = 1$) in the $\bar{x}$, $\bar{y}$, and $\bar{z}$ directions, respectively,

$m_{\bar{z}}(\bar{z}) = $ externally applied torsional moment per unit length of beam.

After the second step of orthogonalisation these quantities become:

$p_x(z, s); p_y(z, s); p_z(z, s) = $ surface loads per unit of area of profile in the x, y, and z directions, respectively,

$q_x(z, s); q_y(z, s); q_z(z, s) = $ line loads per unit length (i.e. $dz = 1$) in the x, y, and z directions, respectively,

$m_z(z) = $ externally applied torsional moment per unit length of beam.

Stress resultants in the $\bar{x}, \bar{y}, \bar{z}$ coordinates system:

$N_{\bar{z}}(\bar{z}) = $ normal force (tensile) in the $\bar{z}$ direction,

$V_{\bar{x}}(\bar{z}), V_{\bar{y}}(\bar{z}) = $ shear force in the $\bar{x}$ and $\bar{y}$ directions,

$M_{\bar{x}}(\bar{z}), M_{\bar{y}}(\bar{z}) = $ bending moment about the $\bar{x}$ and $\bar{y}$ areas,

$M_{\bar{\omega}}(z) = $ bimoment,

$M_{DS}(\bar{z}) = $ warping torsion moment,

(The stress resultants in the $\tilde{x}, \tilde{y}, \tilde{z}$ and x, y, z coordinate systems are obtained by replacing $\bar{x}$ by $\tilde{x}$ and so on.)

$$M_{st} = \text{Saint Venant torsion moment.}$$

The following quantities are obtained by integrating from an edge of a profile to a point which is at a distance s_E from that edge:

$$F(s_E) = \int_0^{s_E} 1 \, dF = \text{area of profile from edge to point,}$$

$$F_{\bar{x}}(s_E) = \int_0^{s_E} \bar{x} \, dF = \text{first moment of area about } \bar{y}\text{-axis,}$$

$$F_{\bar{y}}(s_E) = \int_0^{s_E} \bar{y} \, dF = \text{first moment of area about } \bar{x}\text{-axis,}$$

$$F_{\bar{\omega}_B}(s_E) = \int_0^{s_E} \bar{\omega}_B \, dF = \text{statical warping function moment.}$$

In these four formulae $\bar{x}$ may be changed to $\tilde{x}$ or x depending upon which coordinate system is being used at the time and similar changes can be made to $\bar{y}$ and $\bar{\omega}$.

2.3. GOVERNING EQUATIONS OBTAINED BY THE EQUILIBRIUM/COMPATIBILITY METHOD

In this section the governing equations are obtained by considering

—equilibrium of an element,
—compatibility of an element,
—the stress–strain relationship (i.e., Hooke's Law),

and then by eliminating many of the variables so that finally the governing equations, in which the unknowns are the components of deflections of an element, are derived. An element has six degrees of freedom and therefore six parameters (three of displacement ξ, η, ζ, and three rotations ϕ_x, ϕ_y, ϕ_z) are required to define its movement. However, there are two simple relationships which enable ϕ_x and ϕ_y to be eliminated immediately. When a small fibre of length dz and originally lying parallel to the z-axis is displaced its rotation about the y-axis is given by the rate at which the x-component of the displacement, i.e. ξ varies with respect to z. Thus

$$\phi_y = \frac{d\xi}{dz} = \xi'.$$

Similarly the fibre rotates about the x-axis through

$$\phi_x = -\frac{d\eta}{dz} = -\eta'.$$

In arriving at these results the sense of the rotations is obtained by using the left-hand screw rule. Thus, if the left-hand grips a coordinate axis with the thumb indicating the positive direction of that axis the fingers of the left hand indicate the positive sense of a rotation about that axis. The same convention is used later for torques and bending moments.

Thus the above result means that it is only necessary to consider the unknowns ξ, η, ζ, and ϕ_z. Therefore the aim is to derive four governing equations in these variables. For ease of writing, the suffix z on ϕ is dropped since it is no longer required and hence $\phi = \phi_z$.

In the following derivation there are five sets of equations:

—the equilibrium equations (Section 2.3.1) in stresses, loads, etc.;
—the stress–strain relations (Section 2.3.2);
—the compatibility equations (Section 2.3.3) which express the geometrical relationships between deformations and displacements;
—expressions for the stress, strains and shear flow in terms of the displacements (Section 2.3.4);
—the governing equations (Section 2.3.5) which are four simultaneous differential equations in ξ, η, ζ, and ϕ.

2.3.1. The first set of equations—the equilibrium equations

In this section it is first shown how a set of distributed loads $p_{\bar{x}}$, $p_{\bar{y}}$, and $p_{\bar{z}}$ may be replaced by a set of line loads $q_{\bar{x}}$, $q_{\bar{y}}$, and $q_{\bar{z}}$ which are statically equivalent. The laws of equilibrium are then used to derive equations which relate these external line loads to the internal stresses.

If the beam carries loads distributed around its cross-section (Fig. 2.3.1(a)), that is $p_{\bar{x}}(\bar{z}, s)$, $p_{\bar{y}}(\bar{z}, s)$, and $p_{\bar{z}}(\bar{z}, s)$, (N mm^{-2}) they can be replaced by statically equivalent line loads (N mm^{-1}).

$$q_{\bar{x}}(\bar{z}) = \int_F p_{\bar{x}}(\bar{z}, s) \, ds$$

$$q_{\bar{y}}(\bar{z}) = \int_F p_{\bar{y}}(\bar{z}, s) \, ds \qquad\qquad (2.3.1)$$

$$q_{\bar{z}}(\bar{z}) = \int_F p_{\bar{z}}(\bar{z}, s) \, ds$$

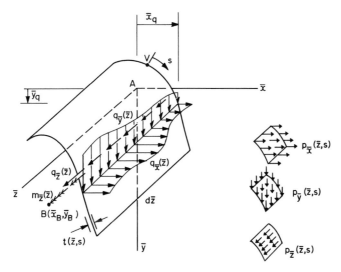

(a) Distributed surface loads can be
replaced by line loads

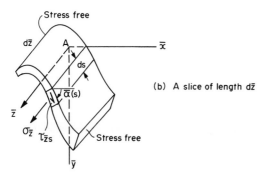

(b) A slice of length $d\bar{z}$

Fig. 2.3.1. (a) Distributed surface loads can be replaced by line loads; (b) a slice of length $d\bar{z}$.

where the locations of the line loads are given by

$$q_{\bar{x}}(\bar{z})\bar{y}_q = \int_F \bar{y}_p(s)p_{\bar{x}}(\bar{z}, s)\, ds$$

$$q_{\bar{y}}(\bar{z})\bar{x}_q = \int_F \bar{x}_p(s)p_{\bar{y}}(\bar{z}, s)\, ds$$

$$q_{\bar{z}}(\bar{z})\bar{x}_z = \int_F \bar{x}_p(s)p_{\bar{z}}(\bar{z}, s)\, ds \qquad\qquad (2.3.2)$$

$$q_{\bar{z}}(\bar{z})\bar{y}_z = \int_F \bar{y}_p(s)p_{\bar{z}}(\bar{z}, s)\, ds$$

In the above equations the symbol $\int_F$ implies that the integration is carried out around the whole profile. Equilibrium of the forces acting on the element shown in Fig. 2.3.1(b) is now considered. In the $\bar{z}$-direction

$$\int_F \frac{\partial[\sigma_{\bar{z}}(\bar{z}, s)t(\bar{z}, s)]}{\partial \bar{z}} \, d\bar{z} \, ds + q_{\bar{z}}(\bar{z}) \, d\bar{z} = 0. \tag{2.3.3}$$

In the $\bar{x}$-direction

$$\int_F \frac{\partial[\tau_{\bar{z}s}(\bar{z}, s)t(\bar{z}, s)]}{\partial \bar{z}} \cos \bar{\alpha}(s) \, d\bar{z} \, ds + q_{\bar{x}}(\bar{z}) \, d\bar{z} = 0. \tag{2.3.4}$$

In the $\bar{y}$-direction

$$\int_F \frac{\partial[\tau_{\bar{z}s}(\bar{z}, s)t(\bar{z}, s)]}{\partial \bar{z}} \sin \bar{\alpha}(s) \, d\bar{z} \, ds + q_{\bar{y}}(\bar{z}) \, d\bar{z} = 0. \tag{2.3.5}$$

The moment of the forces acting on the element about a line parallel to the $\bar{z}$-axis through an arbitrary point B is zero, whence

$$\int_F \frac{\partial[\tau_{\bar{z}s}(\bar{z}, s)t(\bar{z}, s)]}{\partial \bar{z}} \bar{p}_B(s) \, d\bar{z} \, ds + \frac{\partial M_{st}}{\partial \bar{z}} \, d\bar{z}$$
$$+ m_{\bar{z}}(\bar{z}) \, d\bar{z} - q_{\bar{x}}(\bar{z})[\bar{y}_q - \bar{y}_B] \, d\bar{z} + q_{\bar{y}}(\bar{z})[\bar{x}_q - \bar{x}_B] \, d\bar{z} = 0 \tag{2.3.6}$$

where the first term is the difference between the torsional effects of the shear stress $\tau_{\bar{z}s}$ on each end of the element. Because the element is being twisted some of the torque will be that obtained from the theory of Saint Venant, that is, M_{st} ($= GJ_p\phi'$), as explained in Section 1.4.2. The second term in eqn (2.3.6) is the difference in M_{st} between the two ends of the element. The last three terms are the torsional effects of the externally applied torque per unit length and the line loads derived in eqns (2.3.1) and (2.3.2).

Since the element has six degrees of freedom it should be possible to write down two more equations of rotational equilibrium, one about the $\bar{x}$-axis and one about the $\bar{y}$-axis. However, this results in trivial equations. By projecting the length ds of the element onto the axes

$$\cos \bar{\alpha}(s) \, ds = d\bar{x} \tag{2.3.7}$$

$$\sin \bar{\alpha}(s) \, ds = d\bar{y} \tag{2.3.8}$$

and for convenience a new variable $\bar{\omega}_B$ is introduced, thus,

$$\bar{p}_B(s) \, ds = d\bar{\omega}_B. \tag{2.3.9}$$

By substituting into eqns (2.3.3) to (2.3.6)

$$\int_F \frac{\partial[\sigma_{\bar{z}}(\bar{z}, s)t(\bar{z}, s)]}{\partial \bar{z}} \, ds + q_{\bar{z}}(\bar{z}) = 0, \tag{2.3.10}$$

$$\int_F \frac{\partial[\tau_{\bar{z}s}(\bar{z}, s)t(\bar{z}, s)]}{\partial \bar{z}} \, d\bar{x} + q_{\bar{x}}(\bar{z}) = 0, \tag{2.3.11}$$

$$\int_F \frac{\partial[\tau_{\bar{z}s}(\bar{z}, s)t(\bar{z}, s)]}{\partial \bar{z}} \, d\bar{y} + q_{\bar{y}}(\bar{z}) = 0, \tag{2.3.12}$$

$$\int_F \frac{\partial[\tau_{\bar{z}s}(\bar{z}, s)t(\bar{z}, s)]}{\partial \bar{z}} \, d\bar{\omega}_B + \frac{\partial M_{st}}{\partial \bar{z}} + m_{\bar{z}}(\bar{z})$$

$$- q_{\bar{x}}(\bar{z})[\bar{y}_q - \bar{y}_B] + q_{\bar{y}}(\bar{z})[\bar{x}_q - \bar{x}_B] = 0. \tag{2.3.13}$$

These four equilibrium equations are expressed in terms of the displacements in Section 2.3.5.

2.3.2. The second set of equations—the relationship between stresses and strains

In Section 2.1 it was stated that the in-plane stresses normal to the z-axis are assumed to be zero. Hence the longitudinal stress is obtained from the longitudinal strain as follows:

$$\sigma_{\bar{z}}(\bar{z}, s) = E\varepsilon_{\bar{z}}(\bar{z}, s). \tag{2.3.14}$$

The in-plane shear stress $\tau_{\bar{z},s}(\bar{z}, s)$ cannot be obtained by simply applying Hooke's Law. This problem was anticipated in Section 2.1 in view of the assumption that warping strain $\gamma_w = 0$. As in engineers' theory an equilibrium equation is used. By considering equilibrium of the shaded element shown in Fig. 2.3.2 in the $\bar{z}$ direction, it is easily shown that if second-order terms are neglected then

$$\frac{\partial(\sigma_{\bar{z}}t)}{\partial \bar{z}} + \frac{\partial(\tau_{\bar{z}s}t)}{\partial s} + p_{\bar{z}}(\bar{z}, s) = 0$$

and after integration from an edge to an arbitrary point which is s_E from the edge (Fig. 2.3.2)

$$\tau_{\bar{z}s}(\bar{z}, s_E) = \frac{1}{t(\bar{z}, s_E)} \left\{ -\int_0^{s_E} \frac{\partial[\sigma_{\bar{z}}(\bar{z}, s)t(\bar{z}, s)]}{\partial \bar{z}} \, ds - \int_0^{s_E} p_{\bar{z}}(\bar{z}, s) \, ds + C(\bar{z}) \right\}$$

$$\tag{2.3.15}$$

where $C(\bar{z})$ is a function of $\bar{z}$ only.

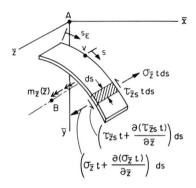

Fig. 2.3.2. Internal stresses in an element.

From eqns (2.3.14) and (2.3.15)

$$\tau_{\bar{z}s}(\bar{z}, s_E) = \frac{1}{t(\bar{z}, s_E)}\left\{-E\int_0^{s_E} \frac{\partial[\varepsilon_{\bar{z}}(\bar{z}, s)t(\bar{z}, s)]}{\partial \bar{z}} - \int_0^{s_E} p_{\bar{z}}(\bar{z}, s)\,\mathrm{d}s + C(\bar{z})\right\}.$$

$$(2.3.16)$$

2.3.3. The third set of equations—the compatibility conditions

It is assumed (Section 2.1) that the cross-section of the beam retains its shape while it is bent and twisted. Thus the cross-section moves as a rigid body and such a movement can be considered as a rotation about a point.

This section describes the geometrical relationships which exist between displacements when a cross-section is rotated about an arbitrary point C but retains its shape.

The in-plane displacements

The displacements $\bar{\xi}_i(\bar{z})$ and $\bar{\eta}_i(\bar{z})$ of the point i due to a rigid-body rotation in the first coordinate system are derived in terms of the displacements $\bar{\xi}_A(\bar{z})$ and $\bar{\eta}_A(\bar{z})$ of the origin A (Fig. 2.3.3). It is seen from this diagram that

$$\bar{\xi}_i = \bar{\xi}_A - \bar{x}_i - \bar{y}_i \sin\phi + \bar{x}_i \cos\phi \tag{2.3.17}$$

$$\bar{\eta}_i = \bar{\eta}_A - \bar{y}_i + \bar{y}_i \cos\phi + \bar{x}_i \sin\phi. \tag{2.3.18}$$

If the angle of rotation ϕ is small, first-order theory can be used, i.e.

$$\sin\phi \simeq \phi, \qquad \cos\phi \simeq 1 \tag{2.3.19}$$

whence

$$\bar{\xi}_i = \bar{\xi}_A - \bar{y}_i\phi \tag{2.3.20}$$

$$\bar{\eta}_i = \bar{\eta}_A + \bar{x}_i\phi. \tag{2.3.21}$$

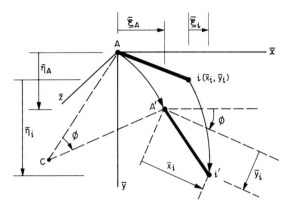

Fig. 2.3.3. The displacements $\bar{\xi}_i$, $\bar{\eta}_i$ of an arbitrary point i can be expressed in terms of the displacement of the origin A when the vector Ai is rotated about C.

The physical meaning of these expressions should be studied by the reader because it is an approximation which often occurs in the analysis of small displacements. In second order theory the approximations given in equation (2.3.19) are not accurate enough. It is also required to be able to express $\bar{\xi}_i$ and $\bar{\eta}_i$ in terms of the displacement components $\bar{\xi}_B$ and $\bar{\eta}_B$ of an arbitrary point B (Fig. 2.3.4). From eqns (2.3.20) and (2.3.21)

$$\bar{\xi}_B = \bar{\xi}_A - \bar{y}_B\phi \tag{2.3.22}$$

$$\bar{\eta}_B = \bar{\eta}_A + \bar{x}_B\phi \tag{2.3.23}$$

$\bar{\xi}_A$ and $\bar{\eta}_A$ can be eliminated from eqns (2.3.20)–(2.3.23) whence

$$\bar{\xi}_i = \bar{\xi}_B - [\bar{y}_i - \bar{y}_B]\phi \tag{2.3.24}$$

$$\bar{\eta}_i = \bar{\eta}_B + [\bar{x}_i - \bar{x}_B]\phi. \tag{2.3.25}$$

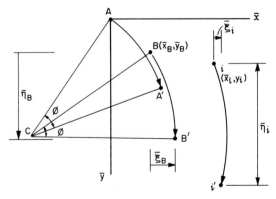

Fig. 2.3.4. The displacements $\bar{\xi}_i$, $\bar{\eta}_i$ can be expressed in terms of $\bar{\xi}_B$, $\bar{\eta}_B$, and ϕ.

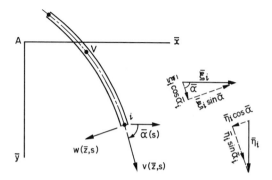

Fig. 2.3.5. Components of $\bar{\xi}_i$, $\bar{\eta}_i$ in tangential and normal directions are v, w, respectively.

Again, the reader should interpret these expressions in a geometrical way. In the case of a thin-walled section (Fig. 2.3.5) rotating as a rigid body through angle ϕ, interest centres around the displacements $v(\bar{z}, s)$ and $w(\bar{z}, s)$ which lie in the plane and normal to the plane of the sheet, respectively. By considering the components of $\bar{\xi}_i(\bar{z}, s)$ and $\bar{\eta}_i(\bar{z}, s)$ as shown and using eqns (2.3.20) and (2.3.21)

$$v(\bar{z}, s) = \bar{\xi}_A \cos \bar{\alpha} + \bar{\eta}_A \sin \bar{\alpha} + [\bar{x}_i \sin \bar{\alpha} - \bar{y}_i \cos \bar{\alpha}]\phi \qquad (2.3.26)$$

$$w(\bar{z}, s) = -\bar{\xi}_A \sin \bar{\alpha} + \bar{\eta}_A \cos \bar{\alpha} + [\bar{x}_i \cos \bar{\alpha} + \bar{y}_i \sin \bar{\alpha}]\phi. \qquad (2.3.27)$$

The terms in the square brackets are simply components of the radius vector Ai (Fig. 2.3.6). Hence eqns (2.3.26) and (2.3.27) become

$$v(\bar{z}, s) = \bar{\xi}_A \cos \bar{\alpha} + \bar{\eta}_A \sin \bar{\alpha} + \bar{p}_A\phi \qquad (2.3.28)$$

$$w(\bar{z}, s) = -\bar{\xi}_A \sin \bar{\alpha} + \bar{\eta}_A \cos \bar{\alpha} + \bar{q}_A\phi. \qquad (2.3.29)$$

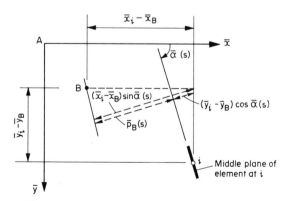

Fig. 2.3.6. Physical meaning of sectorial radius $\bar{p}_B(s)$.

These displacements can also be referred to the arbitrary point B. $v(\bar{z}, s)$ and $w(\bar{z}, s)$ are the components of $\bar{\xi}_i$ and $\bar{\eta}_i$ in the tangential and normal directions. Hence from eqns (2.3.24) and (2.3.25)

$$v(\bar{z}, s) = \bar{\xi}_B \cos \bar{\alpha} + \bar{\eta}_B \sin \bar{\alpha} + \bar{p}_B \phi \qquad (2.3.30)$$

$$w(\bar{z}, s) = -\bar{\xi}_B \sin \bar{\alpha} + \bar{\eta}_B \cos \bar{\alpha} + \bar{q}_B \phi \qquad (2.3.31)$$

where the terms $\bar{p}_B$ and $\bar{q}_B$ are the components of a radius vector, i.e.

$$\bar{p}_B = (\bar{x}_i - \bar{x}_B) \sin \bar{\alpha} - (\bar{y}_i - \bar{y}_B) \cos \bar{\alpha} \qquad (2.3.32)$$

$$\bar{q}_B = (\bar{x}_i - \bar{x}_B) \cos \bar{\alpha} + (\bar{y}_i - \bar{y}_B) \sin \bar{\alpha}. \qquad (2.3.33)$$

Figure 2.3.6 shows that $\bar{p}_B$ is the perpendicular from B to the tangent to the middle line of the profile at the point i.

The out-of-plane displacement

The remaining component of deflection is that in the $\bar{z}$-direction. In the first instance an element from a beam with *open profile* is considered. As discussed in Section 2.1 the shear strain at the middle surface (Fig. 2.3.7) is assumed to be zero. Hence

$$\gamma_{\bar{z}s} = \frac{\partial u(\bar{z}, s)}{\partial s} + \frac{\partial v(\bar{z}, s)}{\partial \bar{z}} = 0. \qquad (2.3.34)$$

It is also assumed that the shear strain normal to the surface, $\gamma_{\bar{z}n}$, is zero. By integrating eqn (2.3.34) between an arbitrary starting point V and point i on the profile distance s from V, and substituting from eqn (2.3.30) the following expression for the displacement in the longitudinal,

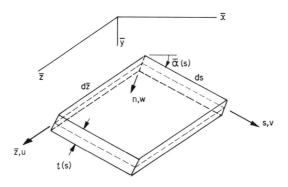

Fig. 2.3.7. An element from the beam.

i.e. $\bar{z}$, direction is obtained.

$$u(\bar{z}, s) = \bar{\zeta}(\bar{z}) - \bar{\xi}'_B(\bar{z}) \int_0^s \cos \bar{\alpha}(s) \, ds - \bar{\eta}'_B(\bar{z}) \int_0^s \sin \bar{\alpha}(s) \, ds$$

$$- \phi'(\bar{z}) \int_0^s \bar{p}_B(s) \, ds \quad (2.3.35)$$

where $\bar{\zeta}(\bar{z})$ is a function of $\bar{z}$ only and arises from the integration and the primes indicate differentiation with respect to $\bar{z}$. It is noted that

$$d\bar{x} = ds \cos \bar{\alpha}(s); \qquad d\bar{y} = ds \sin \bar{\alpha}(s) \qquad (2.3.36)$$

and the following function is defined

$$\bar{\omega}_B(s) = \int_0^s \bar{p}_B(s) \, ds. \qquad (2.3.37)$$

This term is called the *warping function*. Its properties will be considered presently. On substitution from eqns (2.3.37) and (2.3.36) into eqn (2.3.35) the following expression for $u(\bar{z}, s)$ is obtained

$$u(\bar{z}, s) = \bar{\zeta}(\bar{z}) - \bar{\xi}'_B(\bar{z})\bar{x}(s) - \bar{\eta}'_B(\bar{z})\bar{y}(s) - \phi'(\bar{z})\bar{\omega}_B(s). \qquad (2.3.38)$$

The first term represents a longitudinal extension uniform over the cross-section, the second term being linear in $\bar{x}$ represents a rotation about the $\bar{y}$-axis while the third term represents a rotation about the $\bar{x}$-axis. The fourth term is the distortion, or so-called *warping*, of the cross-section out of its plane due to twisting. This equation can be written conveniently in the following matrix form

$$u(\bar{z}, s) = -\{v\}'[w] \qquad (2.3.39)$$

where $\{\}$ and $[\,]$ indicate row and column vectors, respectively.

$$\{v\} = \left\{ -\int \bar{\zeta}(\bar{z}) \, d\bar{z}, \bar{\xi}_B(\bar{z}), \bar{\eta}_B(z), \phi(\bar{z}) \right\} \qquad (2.3.40)$$

$$[w] = \int_0^s \begin{bmatrix} 0 \\ \cos \bar{\alpha}(s) \\ \sin \bar{\alpha}(s) \\ \bar{p}_B(s) \end{bmatrix} ds = \begin{bmatrix} 1 \\ \bar{x}(s) \\ \bar{y}(s) \\ \bar{\omega}_B(s) \end{bmatrix}. \qquad (2.3.41)$$

An element from the ith cell of a beam with a *closed profile* is now considered using the approximate theory referred to in Section 2.1. The shear strain is not zero in this case but is given by the stress–strain relationship expressed in terms of the Saint Venant shear flow T_i in the ith cell

$$\frac{\partial u(\bar{z}, s)}{\partial s} + \frac{\partial v(\bar{z}, s)}{\partial \bar{z}} = \gamma_{\bar{z}s} = \frac{T_i(\bar{z})}{Gt(s)} \qquad (2.3.42)$$

i.e. only the Saint Venant shear strain is considered.
From eqns (1.4.26) and (1.4.29)

$$\Psi_i(\bar{z}) = \frac{T_i(\bar{z})}{G\phi'(\bar{z})} \quad \text{and} \quad \Psi(\bar{z}) = \frac{T(\bar{z})}{G\phi'(\bar{z})} \left(= \frac{2A}{\oint \frac{ds}{t}} \right) \qquad (2.3.43)$$

for multi-cell and single cell profiles, respectively. Hence

$$\gamma_{\bar{z}s} = \frac{\Psi_i(\bar{z})}{t(s)} \phi'(\bar{z}). \qquad (2.3.44)$$

From eqns (2.3.42), (2.3.30), and (2.3.44)

$$u(\bar{z}, s) = \bar{\zeta}(\bar{z}) - \bar{\xi}_B'(\bar{z}) \int_0^s \cos \bar{\alpha}(s) \, ds - \bar{\eta}_B'(\bar{z}) \int_0^s \sin \bar{\alpha}(s) \, ds$$

$$- \phi'(\bar{z}) \int_0^s \left(\bar{p}_B(s) - \frac{\Psi_i(\bar{z})}{t(s)} \right) ds. \qquad (2.3.45)$$

This equation has the same form as eqn (2.3.38) except that

$$[w] = \int_0^s \begin{bmatrix} 0 \\ \cos \bar{\alpha}(s) \\ \sin \bar{\alpha}(s) \\ \bar{p}_B(s) - \dfrac{\Psi_i(\bar{z})}{t(s)} \end{bmatrix} ds \qquad = \begin{bmatrix} 1 \\ \bar{x}(s) \\ \bar{y}(s) \\ \bar{\omega}_B(s) \end{bmatrix} \qquad (2.3.46)$$

Thus eqn (2.3.39) is valid for both open and closed profiles provided $\bar{\omega}_B(s)$ is correctly evaluated from eqns (2.3.37) or (2.3.43) and (2.3.46). The warping function is now considered to see how it may be evaluated and to study its properties. From eqn (2.3.37) and Fig. 2.3.8, it is seen that the warping function is an area, only an element of which is shown in Fig. 2.3.8. It is self-evident that the magnitude of the warping function depends upon the location of B (called the *pole*) and of the point V in the profile from which the integration is started (called the *starting point*).[†] Figure 2.3.9 shows the plot of $\bar{\omega}_B(s)$ for some simple profiles. A convenient formula for calculating warping functions is that due to Leibnitz

---

† Vlasov calls V the sectorial origin.

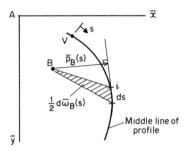

Fig. 2.3.8. The physical meaning of the warping function as an area.

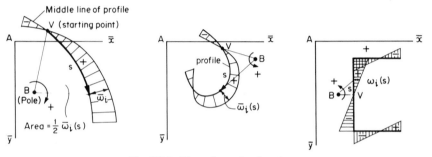

Fig. 2.3.9. Plots of warping functions.

(1646–1716). In Fig. 2.3.10 it is seen that an expression for the area BV_1V_2 is required.

$$\text{Area } (BV_1V_2) = \tfrac{1}{2}[(\bar{x} - \bar{x}_B) + d\bar{x}][(\bar{y} - \bar{y}_B) + d\bar{y}]$$
$$- (\bar{y} - \bar{y}_B + \tfrac{1}{2}d\bar{y})\, d\bar{x} - \tfrac{1}{2}(\bar{x} - \bar{x}_B)(\bar{y} - \bar{y}_B).$$

Alternatively,

$$\text{Area } (BV_1V_2) = -\tfrac{1}{2}[(\bar{x} - \bar{x}_B) + d\bar{x}][(\bar{y} - \bar{y}_B) + d\bar{y}]$$
$$+ (\bar{x} - \bar{x}_B + \tfrac{1}{2}d\bar{x})\, d\bar{y} + \tfrac{1}{2}(\bar{x} - \bar{x}_B)(\bar{y} - \bar{y}_B).$$

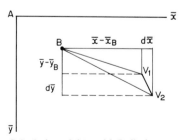

Fig. 2.3.10. Calculation of $d\bar{\omega}_B$ with Leibnitz sectorial formula.

On adding these two expressions the *Leibnitz sectorial formula* is obtained, that is,

$$2 \times \text{area } (\text{BV}_1\text{V}_2) = d\bar{\omega}_B(s) = (\bar{x} - \bar{x}_B)\,d\bar{y} - (\bar{y} - \bar{y}_B)\,d\bar{x} \qquad (2.3.47)$$

It is convenient to have available a general formula which enables transformations from one pole to another. Two poles J and K (Fig. 2.3.11) are considered and the suffix B in eqn (2.3.47) is replaced by J and K, thus

$$d\bar{\omega}_J(s) = (\bar{x} - \bar{x}_J)\,d\bar{y} - (\bar{y} - \bar{y}_J)\,d\bar{x} \qquad (2.3.48)$$

$$d\bar{\omega}_K(s) = (\bar{x} - \bar{x}_K)\,d\bar{y} - (\bar{y} - \bar{y}_K)\,d\bar{x} \qquad (2.3.49)$$

Hence

$$d[\bar{\omega}_J(s) - \bar{\omega}_K(s)] = (\bar{x}_K - \bar{x}_J)\,d\bar{y} + (\bar{y}_J - \bar{y}_K)\,d\bar{x}.$$

On integration

$$\bar{\omega}_J(s) = \bar{\omega}_K(s) + (\bar{y}_J - \bar{y}_K)\bar{x} - (\bar{x}_J - \bar{x}_K)\bar{y} + \Delta\bar{\omega}_{JK} \qquad (2.3.50)$$

where $\Delta\bar{\omega}_{JK}$ is an integration constant. If $\bar{\omega}_J(s)$ and $\bar{\omega}_K(s)$ are equated at the starting point $V(\bar{x}_0, \bar{y}_0)$ from eqn (2.3.50) the integration constant is

$$\Delta\bar{\omega}_{JK} = -(\bar{y}_J - \bar{y}_K)\bar{x}_0 + (\bar{x}_J - \bar{x}_K)\bar{y}_0.$$

Hence the transformation formula is

$$\bar{\omega}_J(s) = \bar{\omega}_K(s) + (\bar{y}_J - \bar{y}_K)(\bar{x} - \bar{x}_0) - (\bar{x}_J - \bar{x}_K)(\bar{y} - \bar{y}_0). \qquad (2.3.51)$$

Finally, a further physical meaning is given to the warping function $\bar{\omega}(s)$. Equation (2.3.38) applies to both open and closed profiles except that their warping functions have to be calculated in different ways.

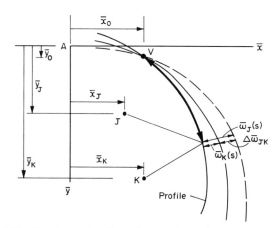

Fig. 2.3.11. Transformation of warping function $\bar{\omega}$ from pole K to pole J.

Suppose that the beam is simply twisted about the axis through the pole B without bending in either the $\bar{x}$ or the $\bar{y}$ directions and without longitudinal extension. Under these circumstances

$$\bar{\zeta}(\bar{z}) = \bar{\xi}'_B(\bar{z}) = \bar{\eta}'_B(\bar{z}) = 0.$$

Furthermore, if the angle of twist $\phi'(\bar{z})$ could be made -1 rad per unit length eqn (2.3.38) becomes

$$u(\bar{z}, s) = \bar{\omega}_B(s). \tag{2.3.52}$$

Thus it is seen that the warping function is the out-of-plane displacement of the cross-section under the circumstances just described. The following examples are intended to show the effect of changing the location of the pole B and of the starting point V on the warping function $\omega(s)$ or, as has just been demonstrated, upon the out-of-plane displacement of the cross-section.

Example   2.3.1

Derive the warping functions illustrated in Fig. 2.3.12 for the given locations of B and V. For the channel the flanges and web are 30 and 40 mm long, respectively, and the box section measures 40×30 mm. The clockwise sense for calculating $\omega(s)$ is positive. $t = 1$ for all profiles.

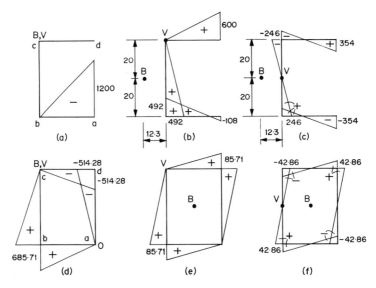

Fig. 2.3.12. Example 2.3.1 shows the effect of changing the pole and starting point locations on the warping function $\omega(s)$.

Figure 2.3.12(a): For an open profile such as this, eqn (2.3.37) is used to evaluate $\omega(s)$ with $s = 0$ at the starting point V. Since p_B is zero for both cd and cb the warping function $\omega(s)$ is zero from b to d. From b to a point lying a distance s_1 to the right of b

$$\omega(s_1) = -\int_0^{s_1} 40 \, ds = -40s_1$$

where the minus sign appears because from b towards a the sense of ω is anticlockwise, i.e. negative. Thus from b to a $\omega(s)$ is a linear function and at a $\omega = -1200$.

Figure 2.3.12(b): By reasoning similar to that used above, from c to d $\omega(s)$ is a linear function and at d it is equal to $20 \times 30 = +600$. From c to b it is again linear and at b its value is $12.3 \times 40 = +492$. When the integration is continued to a the value is $492 - 20 \times 30 = -108$.

Figure 2.3.12(c): With $\omega(s) = 0$ at V, integration from V to c involves an anticlockwise rotation, whence $\omega(s)$ at c is $-12.3 \times 20 = -246$. Integration from c to d involves a clockwise rotation and hence $\omega(s)$ at d is $-246 + 20 \times 30 = +354$. The remainder of the calculation is straightforward.

Figure 2.3.12(d): For a single-cell profile Ψ is found from eqn (1.4.29)

$$\Psi = \frac{2A}{\oint \dfrac{ds}{t}} = \frac{2 \times 30 \times 40}{30/1 + 40/1 + 30/1 + 40/1} = 17.1429$$

Integrating from c to d to find ω_d from eqn (2.3.46)

$$\omega_d = \int_0^{30} \left(0 - \frac{17.1429}{1}\right) ds = -17.1429 \times 30 = -514.28$$

Integrating from d to a

$$\omega_a = -514.28 + \int_0^{40} (30 - 17.1429/1) \, ds = 0$$

Integrating from a to b

$$\omega_b = 0 + \int_0^{30} (40 - 17.1429/1) \, ds = 685.71$$

Integrating from b to c

$$\omega_c = 685 + \int_0^{40} (0 - 17.1429/1) \, ds = 0$$

(which checks)

Figure 2.3.12(e): As above $\Psi = 17.1429$

$$\omega_d = \int_0^{30} (20 - 17.1429/1)\, ds = 30 \times 2.8571 = 85.714$$

$$\omega_a = 85 + \int_0^{40} (15 - 17.1429/1)\, ds = 0$$

$$\omega_b = 0 + \int_0^{30} (20 - 17.1429/1)\, ds = 85.714$$

$$\omega_c = 85.714 + \int_0^{40} (15 - 17.1429/1)\, ds = 0 \qquad \text{(which checks)}$$

Figure 2.3.12(f): As above $\Psi = 17.1429$

$$\omega_c = \int_0^{20} (15 - 17.1429/1)\, ds = -42.857$$

$$\omega_d = -42.857 + \int_0^{30} (20 - 17.1429/1)\, ds = 42.857$$

$$\omega_a = 42.857 + \int_0^{40} (15 - 17.1429/1)\, ds = -42.857$$

$$\omega_b = -42.857 + \int_0^{30} (20 - 17.1429/1)\, ds = 42.857$$

$$\omega_v = 42.857 + \int_0^{20} (15 - 17.1429/1)\, ds = 0 \qquad \text{(which checks)}$$

From the discussion around eqn (2.3.52) it is seen that these values represent a scaled value of the warping or out-of-plane distortion of the cross-section. When they are plotted on the cross-section in the direction of the z-axis as in Fig. 2.3.13 they give a pictorial representation of the warping. Figure 2.3.13(b)–(c) show that a shift in the position of the starting point V is simply equivalent to shifting the whole diagram in the direction of the longitudinal axis. The same point is also illustrated in Fig. 2.3.13(e)–(f). It is shown later that it is usually necessary to shift V to a point so that the average warping is zero, i.e. the positive and negative effects balance. This kind of pictorial representation helps one to understand the mathematical operations involved in the analysis of a beam with torsional loads.

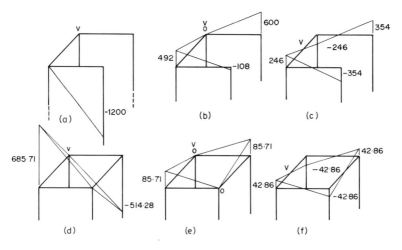

Fig. 2.3.13. When the warping functions of Fig. 2.3.12 are plotted in the direction of the z-axis they represent a scaled picture of the warping of the cross-section.

2.3.4. The fourth set of equations—the stresses and stress resultants expressed as functions of the displacements

The stresses and stress resultants are obtained from the longitudinal strain

$$\varepsilon_{\bar{z}}(\bar{z}, s) = \frac{\partial u(\bar{z}, s)}{\partial \bar{z}} \tag{2.3.53}$$

From eqn (2.3.38)

$$\varepsilon_{\bar{z}}(\bar{z}, s) = \bar{\zeta}'(\bar{z}) - \bar{\xi}_{\rm B}''(\bar{z})\bar{x}(s) - \bar{\eta}_{\rm B}''(\bar{z})\bar{y}(s) - \phi''(\bar{z})\bar{\omega}_{\rm B}(s) \tag{2.3.54}$$

or in matrix form

$$\varepsilon_{\bar{z}}(\bar{z}, s) = -\{v''\}[w]. \tag{2.3.55}$$

The physical meaning of the terms in (2.3.54) are first, a longitudinal strain, second and third are bending terms (from Bernoulli's hypothesis) and the last is a torsional term. The longitudinal stress can be obtained from the simple stress–strain relationship, i.e.

$$\sigma_{\bar{z}}(\bar{z}, s) = E\varepsilon_{\bar{z}}(\bar{z}, s) = E\bar{\zeta}'(\bar{z}) - E\bar{\xi}_{\rm B}''(\bar{z})\bar{x}(s) - E\bar{\eta}_{\rm B}''(\bar{z})\bar{y}(s) - E\phi''(\bar{z})\bar{\omega}_{\rm B}(s). \tag{2.3.56}$$

This is because the in-plane stress normal to the $\bar{z}$-axis is zero (see Section 2.1) and therefore the Poisson's ratio effect is zero. The first term is the longitudinal stress if there is no bending $(\bar{\xi}_{\rm B}''(\bar{z}) = \bar{\eta}_{\rm B}''(\bar{z}) = 0)$ or twisting $(\phi''(\bar{z}) = 0)$. The second and third terms are those obtained from

Bernoulli's bending theory while the fourth term is the longitudinal warping stress. In matrix form this equation is

$$\sigma_{\bar{z}}(\bar{z}, s) = -E\{v''\}[w]. \tag{2.3.57}$$

The shear flow $(= \tau_{\bar{z}s}(\bar{z}, s))$ is now derived from eqns (2.3.15) and (2.3.56), it being assumed that $t(\bar{z}, s)$ does not change rapidly with $\bar{z}$, i.e. $t'(\bar{z}, s) \approx 0$.

$$T(\bar{z}, s_E) = C(\bar{z}) - \int_0^{s_E} t(\bar{z}, s) \frac{\partial \sigma_{\bar{z}}(\bar{z}, s)}{\partial \bar{z}} \, ds - \int_0^{s_E} p_{\bar{z}}(\bar{z}, s) \, ds$$

$$= C(\bar{z}) - E\bar{\zeta}''(\bar{z}) \int_0^{s_E} t(\bar{z}, s) \, ds + E\bar{\xi}_B'''(\bar{z}) \int_0^{s_E} t(\bar{z}, s)\bar{x}(s) \, ds$$

$$+ E\bar{\eta}_B'''(\bar{z}) \int_0^{s_E} t(\bar{z}, s)\bar{y}(s) \, ds + E\phi'''(\bar{z}) \int_0^{s_E} t(\bar{z}, s)\bar{\omega}_B(s) \, ds$$

$$- \int_0^{s_E} p_{\bar{z}}(\bar{z}, s) \, ds. \tag{2.3.58}$$

The following sectional properties are defined as

$$F(s_E) = \int_0^{s_E} 1 \, dF \qquad (= \text{area})$$

$$\left.\begin{array}{l} F_{\bar{x}}(s_E) = \displaystyle\int_0^{s_E} \bar{x} \, dF \\[2ex] F_{\bar{y}}(s_E) = \displaystyle\int_0^{s_E} \bar{y} \, dF \end{array}\right\} \qquad (= \text{first moment of area}) \tag{2.3.59}$$

$$F_{\bar{\omega}_B}(s_E) = \int_0^{s_E} \bar{\omega}_B \, dF \qquad (\text{statical warping function moment})$$

and the abbreviation

$$q_{\bar{z}}(\bar{z}, s) = \int_0^{s_E} p_{\bar{z}}(\bar{z}, s) \, ds. \tag{2.3.60}$$

In these formulae the integrations must be started at an edge of the profile. In the case of an open profile this can be any edge and in closed profiles it is necessary to create an edge by cutting in the longitudinal direction.

Hence eqn (2.3.58) becomes

$$T(\bar{z}, s_E) = C(\bar{z}) - q_{\bar{z}}(\bar{z}, s_E) - EF(s_E)\bar{\zeta}''(\bar{z}) + E\bar{\xi}_B'''(\bar{z})F_{\bar{x}}(s_E)$$
$$+ E\bar{\eta}_B'''(\bar{z})F_{\bar{y}}(s_E) + E\phi'''(\bar{z})F_{\bar{\omega}_B}(s_E). \tag{2.3.61}$$

This equation is first considered as it applies to an *open profile*. It is assumed that the shear stress vanishes at the edge $s_E = 0$ of the profile and furthermore it is obvious that

$$F(0) = F_{\bar{y}}(0) = F_{\bar{x}}(0) = F_{\bar{\omega}_B}(0) = q_{\bar{z}}(\bar{z}, 0) = 0. \tag{2.3.62}$$

Hence

$$C(\bar{z}) = 0. \tag{2.3.63}$$

It should be noted that if the axes are chosen such that $F_{\bar{x}}(s_E^*) = F_{\bar{y}}(s_E^*) = F_{\bar{\omega}_B}(s_E^*) = 0$, where s_E^* is the s_E coordinate of the other end of the profile, and if the shear stress vanishes on that boundary the equation has the special form

$$q_{\bar{z}}(\bar{z}, s_E^*) + EF(s_E^*)\bar{\zeta}''(\bar{z}) = 0.$$

This is the case of a member elongating under a load $q_{\bar{z}}$ distributed along its length.

If there is then no distributed load along the beam in the direction of $\bar{z}$ (i.e. $p_{\bar{z}}(\bar{z}, s) = q_{\bar{z}}(\bar{z}) = 0$—see eqn (2.3.1)) then $\bar{\zeta}''(\bar{z}) = 0$, i.e. $\zeta(\bar{z}) = A\bar{z} + B$ where A and B are constants, which corresponds to the case of direct tension in the beam.

Equation (2.3.61) can be written in the matrix form

$$T(\bar{z}, s_E) = -q_{\bar{z}}(\bar{z}, s_E) + E\{v_B'''\}[F_{\bar{\omega}_B}] \tag{2.3.64}$$

where

$$[F_{\bar{\omega}_B}] = \begin{bmatrix} F(s_E) \\ F_{\bar{x}}(s_E) \\ F_{\bar{y}}(s_E) \\ F_{\bar{\omega}_B}(s_E) \end{bmatrix} \tag{2.3.65}$$

In eqn (2.3.61) or (2.3.64) it is seen that the terms $E\bar{\xi}_B'''(\bar{z})F_{\bar{x}}(s_E)$ and $E\bar{\eta}_B'''(\bar{z})F_{\bar{y}}(s_E)$ are the same shear flows that one obtains from engineers' theory. They arise from bending about the $\bar{y}$ and $\bar{x}$ axes, respectively. The last term is that arising from the *warping shear stress*.

Again turning briefly to the approximate theory for *closed profiles* referred to in Section 2.1, in the case of a closed profile with one cell an imaginary longitudinal cut may be introduced so as to convert it to an open profile. The cut edges are no longer stress free so, in general, $C(\bar{z}) \neq 0$. Hence for this profile

$$T(\bar{z}, s) = t\tau_{\bar{z}s} = C(\bar{z}) + T_0(\bar{z}, s) \tag{2.3.66}$$

where $T_0(\bar{z}, s)$ is the shear flow in the opened section given by eqn (2.3.64) and $C(\bar{z})$ is a statically indeterminate shear flow whose value is constant around the profile. A compatibility condition, that is, that there shall be no dislocations in the longitudinal direction at the cut, enables $C(\bar{z})$ to be found in terms of T_0.

$$\oint \gamma \, ds = \frac{1}{G} \oint \tau \, ds = 0 \qquad (2.3.67)$$

becomes

$$C(\bar{z}) \oint \frac{ds}{t} + \oint \frac{T_0(\bar{z}, s)}{t} \, ds = 0$$

i.e.

$$C(\bar{z}) = -\frac{\oint \dfrac{T_0(\bar{z}, s)}{t}}{\oint \dfrac{ds}{t}} \, ds. \qquad (2.3.68)$$

This expression enables the shear stress distribution to be found from eqn (2.3.66).

For closed profiles with n cells $(n > 1)$ it is necessary to use eqn (2.3.67) n times. Each cell is given an imaginary longitudinal cut and for the ith cell

$$C_i(\bar{z}) \oint_i \frac{ds}{t} + \oint_i \frac{T_{0i}(\bar{z}, s)}{t} \, ds - \sum_{k=1}^{m} C_k \int \frac{ds}{t} = 0 \qquad (2.3.68(a))$$

where the last term is the sum of the effects of the shear flows in the m cells which have a common join with the ith cell. Hence it is necessary to solve n simultaneous equations of the form (2.3.68(a)) to obtain the $C_i(\bar{z})$ values which can then be substituted into eqn (2.3.66) for each cell. This procedure is similar to that used in Section 1.4.2 to evaluate the Saint Venant torsion stresses in a closed profile in n cells $(n > 1)$. An example of the use of these equations is presented in Section 2.6.1. In using these equations it is important to realise that the axis of rotation of the cut profile should coincide with that of the uncut profile. (Generally the position of the shear centre M is changed by cutting a profile.)

2.3.5. The fifth set of equations—simultaneous differential equations of equilibrium in $\bar{\xi}$, $\bar{\eta}$, $\bar{\zeta}$, and ϕ

In this section the equations of equilibrium (eqns (2.3.10) to (2.3.13)) are expressed in terms of the displacements $\bar{\xi}$, $\bar{\eta}$, $\bar{\zeta}$, and ϕ. The special case of a beam whose cross-section is independent of $\bar{z}$, i.e. $t(\bar{z}, s) = t(s)$, is now analysed. Using the notation $t(s) \, ds = dF$ for an element of area eqns (2.3.10)–(2.3.13) are integrated by parts over the whole cross-section, whence

$$\int_F \frac{\partial \sigma_{\bar{z}}(\bar{z}, s)}{\partial \bar{z}} \, dF + q_{\bar{z}}(\bar{z}) = 0 \tag{2.3.69}$$

$$\left[\frac{\partial \tau_{\bar{z}s}(\bar{z}, s)}{\partial \bar{z}} t(s)\bar{x}\right]_{\text{edge 1}}^{\text{edge 2}} - \int_F \bar{x} \frac{\partial}{\partial s}\left[t(s)\frac{\partial \tau_{\bar{z}s}(\bar{z}, s)}{\partial \bar{z}}\right] ds + q_{\bar{x}}(\bar{z}) = 0 \tag{2.3.70}$$

$$\left[\frac{\partial \tau_{\bar{z}s}(\bar{z}, s)}{\partial \bar{z}} t(s)\bar{y}\right]_{\text{edge 1}}^{\text{edge 2}} - \int_F \bar{y} \frac{\partial}{\partial s}\left[t(s)\frac{\partial \tau_{\bar{z}s}(\bar{z}, s)}{\partial \bar{z}}\right] ds + q_{\bar{y}}(\bar{z}) = 0 \tag{2.3.71}$$

$$\left[\frac{\partial \tau_{\bar{z}s}(\bar{z}, s)}{\partial \bar{z}} t(s)\bar{\omega}_B\right]_{\text{edge 1}}^{\text{edge 2}} - \int_F \bar{\omega}_B \frac{\partial}{\partial s}\left[t(s)\frac{\partial \tau_{\bar{z}s}(\bar{z}, s)}{\partial \bar{z}}\right] ds$$

$$+ \frac{\partial M_{\text{st}}}{\partial \bar{z}} + m_{\bar{z}}(\bar{z}) - q_{\bar{x}}(\bar{z})[\bar{y}_q - \bar{y}_B] + q_{\bar{y}}(\bar{z})[\bar{x}_q - \bar{x}_B] = 0. \tag{2.3.72}$$

In an open profile the edges are free of stress so the first terms in eqns (2.3.70)–(2.3.72) are zero. In the case of a closed profile the edges, which are obtained by cutting longitudinally, carry equal stresses so these terms are again zero. From eqn (2.3.56)

$$\frac{\partial \sigma_{\bar{z}}(\bar{z}, s)}{\partial \bar{z}} \, dF = E[\bar{\zeta}''(\bar{z}) - \bar{\xi}_B'''(\bar{z})\bar{x}(s) - \bar{\eta}_B'''(\bar{z})\bar{y}(s) - \phi'''(\bar{z})\bar{\omega}_B(s)] \, dF \tag{2.3.73}$$

from (2.3.61)

$$\frac{\partial}{\partial \bar{z}}\left[\frac{\partial \tau_{\bar{z}s}(\bar{z}, s)t(s)}{\partial s}\right] ds = -\frac{\partial p_{\bar{z}}(\bar{z}, s)}{\partial \bar{z}} \, ds$$

$$- E[\bar{\zeta}'''(\bar{z}) - \bar{\xi}_B^{\text{iv}}(\bar{z})\bar{x}(s) - \bar{\eta}_B^{\text{iv}}(\bar{z})\bar{y}(s) - \phi^{\text{iv}}(\bar{z})\bar{\omega}_B(s)] \, dF \tag{2.3.74}$$

and from equations similar to (1.4.9)

$$\frac{\partial M_{\text{st}}}{\partial \bar{z}} = GJ_p\phi''(\bar{z}). \tag{2.3.75}$$

Therefore eqns (2.3.69)–(2.3.72) can be expressed in terms of the displacement as follows, where the integrals are all taken over the whole profile

$$\bar{\zeta}''(\bar{z})E\int dF - \bar{\xi}'''(\bar{z})E\int \bar{x}(s) \, dF - \bar{\eta}_B'''(\bar{z})E\int \bar{y}(s) \, dF - \phi'''(\bar{z})E\int \bar{\omega}_B(s) \, dF$$

$$+ q_{\bar{z}}(\bar{z}) = 0 \tag{2.3.76}$$

$$\bar{\zeta}'''(\bar{z})E\int \bar{x}(s) \, dF - \bar{\xi}_B^{\text{iv}}(\bar{z})E\int \bar{x}^2(s) \, dF - \bar{\eta}_B^{\text{iv}}(\bar{z})E\int \bar{x}(s)\bar{y}(s) \, dF$$

$$+ \phi^{\text{iv}}(\bar{z})E\int \bar{x}(s)\bar{\omega}_B(s) \, dF + \int \frac{\partial p_{\bar{z}}(\bar{z}, s)}{\partial \bar{z}} \bar{x}(s) \, ds + q_{\bar{x}}(\bar{z}) = 0 \tag{2.3.77}$$

$$\bar{\zeta}'''(\bar{z})E\int \bar{y}(s)\,\mathrm{d}F - \bar{\xi}_{\mathrm{B}}^{\mathrm{iv}}(\bar{z})E\int \bar{x}(s)\bar{y}(s)\,\mathrm{d}F - \bar{\eta}_{\mathrm{B}}^{\mathrm{iv}}(z)E\int \bar{y}^2(s)\,\mathrm{d}F$$

$$-\phi^{\mathrm{iv}}(\bar{z})E\int \bar{y}(s)\bar{\omega}_{\mathrm{B}}(s)\,\mathrm{d}F + \int \frac{\partial p_{\bar{z}}(\bar{z},s)}{\partial \bar{z}}\,\bar{y}(s)\,\mathrm{d}s + q_{\bar{y}}(\bar{z}) = 0 \quad (2.3.78)$$

$$\bar{\zeta}'''(\bar{z})E\int \bar{\omega}_{\mathrm{B}}(s)\,\mathrm{d}F - \bar{\xi}_{\mathrm{B}}^{\mathrm{iv}}(\bar{z})E\int \bar{x}(s)\bar{\omega}_{\mathrm{B}}(s)\,\mathrm{d}F$$

$$-\bar{\eta}_{\mathrm{B}}^{\mathrm{iv}}(\bar{z})E\int \bar{y}(s)\bar{\omega}_{\mathrm{B}}(s)\,\mathrm{d}F - \phi^{\mathrm{iv}}(\bar{z})E\int \bar{\omega}_{\mathrm{B}}^2(s)\,\mathrm{d}F + \int \frac{\partial p_{\bar{z}}(\bar{z},s)}{\partial \bar{z}}\,\bar{\omega}_{\mathrm{B}}(s)\,\mathrm{d}s$$

$$+ m_{\bar{z}}(\bar{z}) + GJ_{\mathrm{p}}\phi''(\bar{z}) - q_{\bar{x}}(\bar{z})[\bar{y}_q - \bar{y}_{\mathrm{B}}] + q_{\bar{y}}(\bar{z})[\bar{x}_q - \bar{x}_{\mathrm{B}}] = 0. \quad (2.3.79)$$

These four simultaneous equations can be abbreviated by introducing the following section properties:

$$F = \int_F 1^2\,\mathrm{d}F \cdot F_{\bar{x}} = \int_F 1 \cdot \bar{x}(s)\,\mathrm{d}F \cdot F_{\bar{y}} = \int_F 1 \cdot \bar{y}(s)\,\mathrm{d}F$$

$$F_{\bar{x}\bar{x}} = \int_F \bar{x}^2(s)\,\mathrm{d}F \cdot \quad F_{\bar{x}\bar{y}} = \int_F \bar{x}(s)\bar{y}(s)\,\mathrm{d}F$$

$$F_{\bar{y}\bar{y}} = \int_F \bar{y}^2(s)\,\mathrm{d}F$$

$$F_{\bar{\omega}} = \int_F 1 \cdot \bar{\omega}_{\mathrm{B}}(s)\,\mathrm{d}F \qquad\qquad (2.3.80)$$

$$F_{\bar{x}\bar{\omega}} = \int_F \bar{x}(s)\bar{\omega}_{\mathrm{B}}(s)\,\mathrm{d}F$$

$$F_{\bar{y}\bar{\omega}} = \int_F \bar{y}(s)\bar{\omega}_{\mathrm{B}}(s)\,\mathrm{d}F$$

$$F_{\bar{\omega}\bar{\omega}} = \int_F \bar{\omega}_{\mathrm{B}}^2(s)\,\mathrm{d}F.$$

It should be noted that $F_{\bar{x}\bar{x}}$ is the second moment of area about the $\bar{y}$-axis according to engineers' definition. The following matrices are defined.

$$[F] = \begin{bmatrix} F & F_{\bar{x}} & F_{\bar{y}} & F_{\bar{\omega}} \\ F_{\bar{x}} & F_{\bar{x}\bar{x}} & F_{\bar{x}\bar{y}} & F_{\bar{x}\bar{\omega}} \\ F_{\bar{y}} & F_{\bar{x}\bar{y}} & F_{\bar{y}\bar{y}} & F_{\bar{y}\bar{\omega}} \\ F_{\bar{\omega}} & F_{\bar{x}\bar{\omega}} & F_{\bar{y}\bar{\omega}} & F_{\bar{\omega}\bar{\omega}} \end{bmatrix} \qquad (2.3.81)$$

$$[J] = \begin{bmatrix} 0 & 0 & 0 & 0 \\ 0 & 0 & 0 & 0 \\ 0 & 0 & 0 & 0 \\ 0 & 0 & 0 & J_{\mathrm{p}} \end{bmatrix} \qquad (2.3.82)$$

$$[q] = \begin{bmatrix} q_{\bar{z}}(\bar{z}) \\ q_{\bar{x}}(\bar{z}) \\ q_{\bar{y}}(\bar{z}) \\ m_{\bar{z}}(\bar{z}) - q_{\bar{x}}(\bar{z})[\bar{y}_q - \bar{y}_B] + q_{\bar{y}}(z)[\bar{x}_q - \bar{x}_B] \end{bmatrix} \qquad (2.3.83)$$

$$\int_F \frac{\partial p_{\bar{z}}(\bar{z}, s)}{\partial \bar{z}}[w]\,ds = \int_F \begin{bmatrix} 0 \\ \dfrac{\partial p_{\bar{z}}(\bar{z}, s)}{\partial \bar{z}}\bar{x} \\ \dfrac{\partial p_{\bar{z}}(\bar{z}, s)}{\partial \bar{z}}\bar{y} \\ \dfrac{\partial p_{\bar{z}}(\bar{z}, s)}{\partial \bar{z}}\bar{\omega}_B \end{bmatrix} ds. \qquad (2.3.84)$$

Thus, eqns (2.3.76)–(2.3.79) can be written in the following matrix form

$$E[F][v]^{iv} - G[J][v]'' = [q] + \int \frac{\partial p_{\bar{z}}(\bar{z}, s)}{\partial \bar{z}}[w]\,ds. \qquad (2.3.85)$$

In eqn (2.3.85) the displacement vector $[v]$ (see eqn (2.3.40)) has elements which describe the displacement of the arbitrary point B and the integral in the last term is taken over the whole cross-section of the beam.

2.4. SIMPLIFICATION OF GOVERNING EQUATIONS BY CHANGE OF REFERENCE AXES

One way of solving the four simultaneous equations represented by eqn (2.3.85) would be to use a process of elimination. This is found to be an unsatisfactory way to proceed. In this section the governing equations [(2.3.76)–(2.3.79) or their matrix equivalent (2.3.85)] of a thin-walled beam are simplified by changing the axes of reference both in position and orientation. This process is known as *orthogonalization* and it is possible because the $[F]$ matrix is symmetrical. It should be noted that all variables $\bar{\zeta}$, $\bar{\xi}$, $\bar{\eta}$, and ϕ appear in all four equations, i.e. the equations are coupled. After orthogonalization each variable will appear in one and only one equation, i.e. the equations are then uncoupled. The uncoupling is arranged in two steps. In the first step the reference axes are shifted without rotation to the centroid S and the starting point V is shifted so as to make a certain integral equal zero. In the second step the reference axes are rotated to the principal axes of the cross-section and it will be shown that the warping function $\omega(s)$ is taken relative to the shear centre, M, i.e. the pole is moved from the arbitrary point B to the shear centre M. The set of axes has been represented so far by the symbols $\bar{x}$, $\bar{y}$, $\bar{z}$.

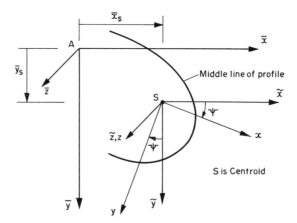

Fig. 2.4.1. The original axes $\bar{x}$, $\bar{y}$, $\bar{z}$, the intermediate axes $\tilde{x}$, $\tilde{y}$, $\tilde{z}$ and the principal axes x, y, z.

After the first step the new reference axes are given the symbols $\tilde{x}$, $\tilde{y}$, $\tilde{z}$ (Fig. 2.4.1) and after the second step the reference axes are simply x, y, z. In Section 2.4.1 eqn (2.3.85) is simplified and in later sections the section properties and expressions for the stress resultants are derived in terms of displacements.

2.4.1. Simplification of eqn (2.3.85)

Equation (2.3.85) is simplified by changing the coordinate axes from an arbitrarily chosen set $\bar{x}$, $\bar{y}$, $\bar{z}$, to ones which uncouple the equations, i.e. make them independent of one another. In matrix notation this means that the matrix $[F]$ becomes a diagonal matrix. To start this process the section properties contained in eqn (2.3.81) are first evaluated in the $\bar{x}$, $\bar{y}$, $\bar{z}$ coordinate system with an arbitrary pole B and an arbitrary starting point V. In the first step the origin is moved from A to a point S by a parallel shift of axes. The location of B is retained but the coordinates of S and the starting point V (Fig. 2.3.9) are chosen in a way which makes

$$F_{\bar{x}} = \int_F 1 \cdot \bar{x}(s)\, \mathrm{d}F = 0$$

$$F_{\bar{y}} = \int_F 1 \cdot \bar{y}(s)\, \mathrm{d}F = 0 \qquad\qquad (2.4.1)$$

$$F_{\bar{\omega}_B} = \int_F 1 \cdot \bar{\omega}(s)\, \mathrm{d}F = 0$$

i.e. the matrix $[F]$ becomes (see eqn (2.3.81))

$$[F] = \begin{bmatrix} F & 0 & 0 & 0 \\ 0 & F_{\bar{x}\bar{x}} & F_{\bar{x}\bar{y}} & F_{\bar{x}\bar{\omega}} \\ 0 & F_{\bar{x}\bar{y}} & F_{\bar{y}\bar{y}} & F_{\bar{y}\bar{\omega}} \\ 0 & F_{\bar{x}\bar{\omega}} & F_{\bar{y}\bar{\omega}} & F_{\bar{\omega}\bar{\omega}} \end{bmatrix}.$$ (2.4.2)

In the first part of the second step the axes are rotated through an angle ψ, the magnitude of which is chosen so that

$$F_{\bar{x}\bar{y}} = \int_F \bar{x}(s)\bar{y}(s)\,dF = 0.$$ (2.4.3)

In order that the remaining off-diagonal terms in eqn (2.4.2), that is, $F_{\bar{x}\bar{\omega}}$ and $F_{\bar{y}\bar{\omega}}$, can be made zero, i.e.,

$$F_{\bar{x}\bar{\omega}} = \int_F \bar{x}(s)\bar{\omega}(s)\,dF = 0$$

$$F_{\bar{y}\bar{\omega}} = \int_F \bar{y}(s)\bar{\omega}(s)\,dF = 0$$ (2.4.4)

the coordinates of the sectorial pole M or shear centre are found and this enables this last step to be taken. The final set of coordinates is now called x, y, z.

The transformation of axes just described does not affect the matrix $[J]$ (eqn (2.3.82)) and it is simple to show that the matrices $[q]$ and

$$\oint_F \frac{\partial p_{\bar{z}}(\bar{z}, s)}{\partial \bar{z}} [w]\,ds$$

(eqns (2.3.83) and (2.3.84)) are transformed by simply replacing $\bar{z}$ by z. Thus eqns (2.3.76)–(2.3.79) can be transformed to

$$EF\zeta''(z) = -q_z(z)$$
$$EF_{xx}\xi^{iv}(z) = q_x(z) + xq_z'(z)$$
$$EF_{yy}\eta^{iv}(z) = q_y(z) + yq_z'(z)$$ (2.4.5)
$$EF_{\omega\omega}\phi^{iv}(z) - GJ_p\phi''(z) = m_z(z) + \omega_M q_z'(z) - q_x(z)(y_q - y_M)$$
$$+ q_y(z)(x_q - x_M)$$

and eqn (2.3.81) to

$$[F] = \begin{bmatrix} F & & & \\ 0 & F_{xx} & \text{symmetrical} & \\ 0 & 0 & F_{yy} & \\ 0 & 0 & 0 & F_{\omega\omega} \end{bmatrix}.$$ (2.4.6)

In deriving the second and third of eqns (2.4.5) use has been made of eqn (2.3.2) and by analogy a similar step can be used for the fourth equation. The first, third, and fourth terms on the right-hand side of the last of eqns (2.4.5) are clearly externally applied torques per unit length. It is shown in Chapter 3 (see Theorem 3.2.2) that the second term is the first derivative of a bimoment and hence it also is a torque per unit length. Thus the right-hand side of the last of eqns (2.4.5) could be written as $m(z)$ where it is understood that it is the sum of these four external effects.

Although matrix eqn (2.3.85) remains as it is after the transformations, the individual equations are now uncoupled because $[F]$ is given by eqn (2.4.6). This uncoupling means that by suitable choice of axes the effects of the various loads can be treated separately. There are four effects, one corresponding to each of the eqns (2.4.5), and they are referred to the following axes:

longitudinal loads act through the centroid S in the z-direction;
bending loads about the y-axis;
bending loads about the x-axis;
torsional loads about a longitudinal axis through the shear centre M.

Thus the longitudinal axis through the shear centre is often referred to as the natural axis of rotation because a pure torsion about this axis does not involve the other three effects listed above.

2.4.2. Section properties and stress resultants in the principal coordinate system

In this section the requirements which must be fulfilled when the axes are shifted and rotated so that $[F]$ can be transformed to a diagonal matrix are considered. There are six off-diagonal terms which are to be made equal to zero and hence there will be six equations. The unknowns are as follows:

coordinates of S $(\bar{x}_s, \bar{y}_s)$	= 2 unknowns
angle of inclination of principal axes (ψ)	= 1 unknown
distance of starting point V from edge of profile (s_v)	= 1 unknown
coordinates of M $(\tilde{x}_M, \tilde{y}_M)$	= 2 unknowns
Total	= 6 unknowns

Expressions for the coordinates of S are first derived. The coordinates referred to the origin A are transformed to the origin at S in the first step by the following transformation formulae (Fig. 2.4.1)

$$\bar{x} = \tilde{x} + \bar{x}_s, \qquad \bar{y} = \tilde{y} + \bar{y}_s, \qquad \bar{z} = \tilde{z}. \tag{2.4.7}$$

Thus the expression for $F_{\tilde{x}}$ is transformed as follows

$$F_{\tilde{x}} = \int_F \tilde{x} \, dF = \int_F (\bar{x} + \bar{x}_s) \, dF = F_{\bar{x}} + \bar{x}_s F.$$

But the position of S is chosen so that $F_{\bar{x}} = 0$ (see eqn (2.4.1)). Hence

$$\bar{x}_s = \frac{F_{\tilde{x}}}{F} = \int_F \frac{\tilde{x} \, dF}{F} . \tag{2.4.8}$$

Similarly

$$\bar{y}_s = \frac{F_{\tilde{y}}}{F} = \int_F \frac{\tilde{y} \, dF}{F} . \tag{2.4.9}$$

These are the well-known equations for the coordinates of the centroid of a profile, and hence S is the centroid of the cross-section. The magnitude of $F_{\bar{\omega}_B}$ can be adjusted to zero by choosing (for a given pole B) the starting point V (Fig. 2.3.9). This point will be chosen shortly. The axes $\tilde{x}$ and $\tilde{y}$ must now be rotated through angle ψ so that

$$F_{xy} = \int_F x(s) y(s) \, dF = 0. \tag{2.4.10}$$

The coordinates $\tilde{x}$, $\tilde{y}$, $\tilde{z}$ are transformed to x, y, z as follows

$$x = \tilde{x} \cos \psi + \tilde{y} \sin \psi, \qquad y = -\tilde{x} \sin \psi + \tilde{y} \cos \psi, \qquad z = \tilde{z} \tag{2.4.11}$$

or alternatively

$$\tilde{x} = x \cos \psi - y \sin \psi, \qquad \tilde{y} = x \sin \psi + y \cos \psi, \qquad \tilde{z} = z. \tag{2.4.12}$$

Therefore from eqns (2.4.10) and (2.4.11)

$$F_{xy} = 0 = \int_F x(s) y(s) \, dF$$

$$= \int (\tilde{x} \cos \psi + \tilde{y} \sin \psi)(-\tilde{x} \sin \psi + \tilde{y} \cos \psi) \, dF$$

$$= -F_{\tilde{x}\tilde{x}} \sin \psi \cos \psi + F_{\tilde{x}\tilde{y}}(\cos^2 \psi - \sin^2 \psi) + F_{\tilde{y}\tilde{y}} \sin \psi \cos \psi$$

Hence,

$$\tan 2\psi = \frac{2F_{\tilde{x}\tilde{y}}}{-F_{\tilde{y}\tilde{y}} + F_{\tilde{x}\tilde{x}}} \tag{2.4.13}$$

The coordinates of the principal pole M are now evaluated.

Since the warping function depends only upon a sectorial area it is self-evident that provided the same pole B and starting point V are used

$$\tilde{\omega}_B(s) = \omega_B(s). \tag{2.4.14}$$

From eqn (2.3.50)

$$\omega_M(s) = \tilde{\omega}_M(s) = \tilde{\omega}_B(s) + (\tilde{y}_M - \tilde{y}_B)\tilde{x} - (\tilde{x}_M - \tilde{x}_B)\tilde{y} + \Delta\tilde{\omega}_{BM} \qquad (2.4.15)$$

$$F_{\omega_M x} = 0 = \int_F \omega_M(s) x(s)\, dF$$

$$= \int_F [\tilde{\omega}_B(s) + (\tilde{y}_M - \tilde{y}_B)\tilde{x} - (\tilde{x}_M - \tilde{x}_B)\tilde{y} + \Delta\tilde{\omega}_{BM}]$$

$$\times [\tilde{x} \cos\psi + \tilde{y} \sin\psi]\, dF \qquad (2.4.16)$$

However, from eqn (2.4.1)

$$F_{\tilde{x}} = \int_F \tilde{x}\, dF = 0 \quad \text{and} \quad F_{\tilde{y}} = \int_F \tilde{y}\, dF = 0,$$

whence

$$F_{\tilde{\omega}_B \tilde{x}} \cos\psi + F_{\tilde{\omega}_B \tilde{y}} \sin\psi - (\tilde{x}_M - \tilde{x}_B)F_{\tilde{y}\tilde{y}} \sin\psi + (\tilde{y}_M - \tilde{y}_B)F_{\tilde{x}\tilde{x}} \cos\psi$$
$$+ (\tilde{y}_M - \tilde{y}_B)F_{\tilde{x}\tilde{y}} \sin\psi - (\tilde{x}_M - \tilde{x}_B)F_{\tilde{x}\tilde{y}} \cos\psi = 0. \qquad (2.4.17)$$

Similarly, by equating $F_{\omega_M y}$ to zero

$$-F_{\tilde{\omega}_B \tilde{x}} \sin\psi + F_{\tilde{\omega}_B \tilde{y}} \cos\psi - (\tilde{x}_M - \tilde{x}_B)F_{\tilde{y}\tilde{y}} \cos\psi - (\tilde{y}_M - \tilde{y}_B)F_{\tilde{x}\tilde{x}} \sin\psi$$
$$+ (\tilde{y}_M - \tilde{y}_B)F_{\tilde{x}\tilde{y}} \cos\psi + (\tilde{x}_M - \tilde{x}_B)F_{\tilde{x}\tilde{y}} \sin\psi = 0. \qquad (2.4.18)$$

Multiplying eqn (2.4.17) by $(\cos\psi)$, eqn (2.4.18) by $(-\sin\psi)$ and adding

$$(\tilde{x}_M - \tilde{x}_B)F_{\tilde{x}\tilde{y}} - (\tilde{y}_M - \tilde{y}_B)F_{\tilde{x}\tilde{x}} = F_{\tilde{\omega}_B \tilde{x}}. \qquad (2.4.19)$$

On multiplying eqn (2.4.17) by $(\sin\psi)$, eqn (2.4.18) by $(\cos\psi)$ and adding.

$$(\tilde{x}_M - \tilde{x}_B)F_{\tilde{y}\tilde{y}} - (\tilde{y}_M - \tilde{y}_B)F_{\tilde{x}\tilde{y}} = F_{\tilde{\omega}_B \tilde{y}} \qquad (2.4.20)$$

After elimination the coordinates of M are found.

$$\tilde{x}_M - \tilde{x}_B = \frac{F_{\tilde{\omega}_B \tilde{y}}F_{\tilde{x}\tilde{x}} - F_{\tilde{\omega}_B \tilde{x}}F_{\tilde{x}\tilde{y}}}{F_{\tilde{x}\tilde{x}}F_{\tilde{y}\tilde{y}} - F_{\tilde{x}\tilde{y}}^2} \qquad (2.4.21)$$

$$\tilde{y}_M - \tilde{y}_B = \frac{F_{\tilde{\omega}_B \tilde{y}}F_{\tilde{x}\tilde{y}} - F_{\tilde{\omega}_B \tilde{x}}F_{\tilde{y}\tilde{y}}}{F_{\tilde{x}\tilde{x}}F_{\tilde{y}\tilde{y}} - F_{\tilde{x}\tilde{y}}^2} \qquad (2.4.22)$$

In arriving at these coordinates the location of the starting point V has not been considered. It is easily proved that the values of $F_{x\omega_B}$ and $F_{y\omega_B}$ (where B is any pole) are unchanged when the position of V is changed, provided that the origin of coordinates is located at the centroid S. If two starting points V_1 and V_2 (Fig. 2.4.2) and a general point i located on the middle line of the profile are considered the sectorial areas are related by the equation

$$\omega_B(s_1) = \omega_B(s_2) + 2 \times \text{area } (V_1 V_2 B). \qquad (2.4.23)$$

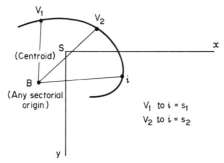

Fig. 2.4.2. When the origin is located at the centroid the position of V has no influence over the value of $F_{x\omega_B}$ and $F_{y\omega_B}$ where B is any sectorial origin.

Hence

$$\int_F x\omega_B(s_1)\, dF = \int_F x\omega_B(s_2)\, dF + [2 \times \text{area } (V_1 V_2 B)] \int_F x\, dF$$

$$= \int_F x\omega_B(s_2)\, dF \qquad (2.4.24)$$

since S is the centroid. Similarly

$$\int_F y\omega_B(s_1)\, dF = \int_F y\omega_B(s_2)\, dF. \qquad (2.4.25)$$

Thus it is seen that the location of V affects only one of the section properties, that is, $F_\omega \; (= \int_F \omega\, dF)$, once the origin of the $\tilde{x}, \tilde{y}$ coordinates is moved to the centre of gravity of the profile. If F_ω is to be zero it is self-evident that V should be located so that the positive and negative effects are equal. Obviously if the profile has an axis of symmetry V should be located at its intersection with the profile.

From the previous analysis it is seen that the origin of coordinates for $\tilde{x}, \tilde{y}, \tilde{z}$ (used in the calculation of $F_{\tilde{x}}$, $F_{\tilde{x}\tilde{x}}$, etc.) is located at the centroid S and the pole for calculating ω_M should be at the shear centre M. It is convenient to calculate ω_M in the following way which ensures that $\int_F \omega_M\, dF = 0$.

(1) Locate the origin A, the pole B, and the starting point V at convenient points, call the coordinates $\bar{x}, \bar{y}, \bar{z}$ and evaluate F, $\bar{\omega}_B$, and $F_{\bar{\omega}_B}$.

(2) Calculate the position of the centroid S (eqns (2.4.8) and (2.4.9)) and by a parallel shift of axes move the origin to S, calling the new coordinates $\tilde{x}, \tilde{y}$, and $\tilde{z}$. This transformation does not affect the warping function since it is determined from the sectorial area which remains unchanged for a given pole B and starting point V.

(3) To obtain $\tilde{\omega}_B(s)$ subtract from $\bar{\omega}_B(s)$ the value $\bar{\omega}_0$, where

$$\bar{\omega}_0 = F_{\bar{\omega}_B}/F \qquad (2.4.26)$$

(which is the 'average' value of $\bar{\omega}_B(s)$). This step then ensures that $\int_F \bar{\omega}_B(s)\, dF = 0$. Equation (2.4.26) is equivalent to eqns (2.4.8) and (2.4.9).

(4) Evaluate ψ from eqn (2.4.13) and the coordinates of the shear centre M from eqns (2.4.21) and (2.4.22). Now move the pole from B to M and transform $\bar{\omega}_B(s)$ to the value $\omega_M(s)$ so that $\int_F \omega_M(s)\, ds$ is also zero. This is achieved as follows by using eqn (2.3.51).

$$\omega_M(s) = \bar{\omega}_B(s) + (\bar{y}_M - \bar{y}_B)\bar{x} - (\bar{x}_M - \bar{x}_B)\bar{y} \tag{2.4.27}$$

To show that the shift of pole from B to M does not change F_ω eqn (2.4.27) can be integrated over the profile, remembering that $\int_F \bar{x}\, dF = \int_F \bar{y}\, dF = 0$. It will be noted that there is an automatic adjustment to the location of V in steps 3 and 4.

In the numerical evaluation of section properties the following trapezoidal formulae may be useful.

$$F = \int_F dF = \sum_{j=1}^n F_j \qquad (n = \text{number of segments})$$

$$F_x = \int_F x\, dF = \sum_{j=1}^n (x_L + x_R)_j \frac{F_j}{2} e$$

$$F_y = \int_F y\, dF = \sum_{j=1}^n (y_L + y_R)_j \frac{F_j}{2} \tag{2.4.28}$$

$$F_\omega = \int_F \omega\, dF = \sum_{j=1}^n (\omega_L + \omega_R)_j \frac{F_j}{2}$$

and the product integrals

$$F_{xy} = \int_F xy\, dF = \tfrac{1}{6} \sum_{j=1}^n (2x_L y_L + 2x_R y_R + x_L y_R + x_R y_L)_j F_j$$

$$F_{x\omega} = \int_F x\omega\, dF = \tfrac{1}{6} \sum_{j=1}^n (2x_L \omega_L + 2x_R \omega_R + x_L \omega_R + x_R \omega_L)_j F_j$$

$$F_{y\omega} = \int_F y\omega\, dF = \tfrac{1}{6} \sum_{j=1}^n (2y_L \omega_L + 2y_R \omega_R + y_L \omega_R + y_R \omega_L)_j F_j$$

$$f_{xx} = \int_F x^2\, dF = \tfrac{1}{3} \sum_{j=1}^n (x_L^2 + x_R^2 + x_L x_R)_j F_j \tag{2.4.29}$$

$$F_{yy} = \int_F y^2\, dF = \tfrac{1}{3} \sum_{j=1}^n (y_L^2 + y_R^2 + y_L y_R)_j F_j$$

$$F_{\omega\omega} = \int_F \omega^2\, dF = \tfrac{1}{3} \sum_{j=1}^n (\omega_L^2 + \omega_R^2 + \omega_L \omega_R)_j F_j$$

Table 2.4.1. Table of $\displaystyle\int_0^L F_1(x)F_2(x)\,\mathrm{d}x$

$F_1(x)$ / $F_2(x)$	$a \Box\ \ L$	$a \triangleright\ \ L$	$a \Box b\ \ L$
$c \Box\ \ L$	Lac	$\dfrac{Lac}{2}$	$\dfrac{Lc(a+b)}{2}$
$c \triangleright\ \ L$	$\dfrac{Lac}{2}$	$\dfrac{Lac}{3}$	$\dfrac{Lc(2a+b)}{6}$
$\triangleleft c\ \ L$	$\dfrac{Lac}{2}$	$\dfrac{Lac}{6}$	$\dfrac{Lc(a+2b)}{6}$
$c \Box d\ \ L$	$\dfrac{La(c+d)}{2}$	$\dfrac{La(2c+d)}{6}$	$\dfrac{La(2c+d)+Lb(c+2d)}{6}$
Parabolic $c\ d\ e$ $\dfrac{L}{2}\ \dfrac{L}{2}$	$\dfrac{La(c+4d+e)}{6}$	$\dfrac{La(c+2d)}{6}$	$\dfrac{La(c+2d)+Lb(2d+e)}{6}$

Table 2.4.1 contains a set of well-known formulae (Matheson 1971) for evaluating volume integrals. Experience suggests that they are probably easier to use than eqns (2.4.28) and (2.4.29).

Another useful set of formulae enable the product integrals to be transformed from the $\bar{x}\bar{y}$ axes to the $\tilde{x}\tilde{y}$ axes (Fig. 2.4.1). Thus from eqns (2.4.3), (2.4.7), (2.4.8), and (2.4.9)

$$F_{\tilde{x}\tilde{y}} = \int_F \tilde{x}\tilde{y}\,\mathrm{d}F = \int_F (\bar{x}-x_s)(\bar{y}-y_s)\,\mathrm{d}F = F_{\bar{x}\bar{y}} - x_s y_s F = F_{\bar{x}\bar{y}} - \frac{F_{\bar{x}}F_{\bar{y}}}{F}$$

Similarly

$$F_{\tilde{x}\tilde{\omega}} = \qquad F_{\bar{x}\bar{\omega}} - \frac{F_{\bar{x}}F_{\bar{\omega}}}{F}$$

$$F_{\tilde{y}\tilde{\omega}} = \qquad F_{\bar{y}\bar{\omega}} - \frac{F_{\bar{y}}F_{\bar{\omega}}}{F}$$

$$F_{\tilde{x}\tilde{x}} = \qquad F_{\bar{x}\bar{x}} - \frac{F_{\bar{x}}^2}{F} \qquad\qquad (2.4.30)$$

$$F_{\tilde{y}\tilde{y}} = \qquad F_{\bar{y}\bar{y}} - \frac{F_{\bar{y}}^2}{F}$$

$$F_{\tilde{\omega}\tilde{\omega}} = \qquad F_{\bar{\omega}\bar{\omega}} - \frac{F_{\bar{\omega}}^2}{F}$$

The formulae which involve $\tilde{\omega}$ have been transformed by $\tilde{\omega} = \bar{\omega} - \bar{\omega}_0 = \bar{\omega} - F_{\bar{\omega}}/F$. In this transformation the pole B is not changed but there is a change in the starting point V so as to make $\int_F \tilde{\omega}_B(s)\,dF = 0$ as will be seen from the previous argument on this point. The second moment of area about the principal axis y is obtained by substituting eqns (2.4.11) in the following manner.

$$F_{xx} = \int_F x^2\,dF$$

$$= \int_F (\tilde{x}\cos\psi + \tilde{y}\sin\psi)^2\,dF$$

$$= F_{\tilde{x}\tilde{x}}\cos^2\psi + F_{\tilde{y}\tilde{y}}\sin^2\psi + F_{\tilde{x}\tilde{y}}\sin 2\psi$$

$$= \tfrac{1}{2}F_{\tilde{x}\tilde{x}}(1+\cos 2\psi) + \tfrac{1}{2}F_{\tilde{y}\tilde{y}}(1-\cos 2\psi) + F_{\tilde{x}\tilde{y}}\sin 2\psi$$

$$= \tfrac{1}{2}(F_{\tilde{x}\tilde{x}} + F_{\tilde{y}\tilde{y}}) \pm \tfrac{1}{2}\sqrt{[(F_{\tilde{x}\tilde{x}} - F_{\tilde{y}\tilde{y}})^2 + 4F_{\tilde{x}\tilde{y}}{}^2]} \tag{2.4.31}$$

where eqn (2.4.13) has been used in the last step. Similarly

$$F_{yy} = \tfrac{1}{2}(F_{\tilde{x}\tilde{x}} + F_{\tilde{y}\tilde{y}}) \mp \tfrac{1}{2}\sqrt{[(F_{\tilde{x}\tilde{x}} - F_{\tilde{y}\tilde{y}})^2 + 4F_{\tilde{x}\tilde{y}}{}^2]}. \tag{2.4.32}$$

In the last two formulae one axis is associated with the plus-sign and the other with the minus-sign. The former axis is the major principal axis for bending and the other the minor principal axis. The transformation formula for the warping constant $F_{\omega_M\omega_M}$ is obtained from eqn (2.4.27)

$$F_{\omega_M\omega_M} = \int_F \omega_M^2\,dF$$

$$= \int_F (\tilde{\omega}_B(s) + (\tilde{y}_M - \tilde{y}_B)\tilde{x} - (\tilde{x}_M - \tilde{x}_B)\tilde{y})^2\,dF$$

$$= \int_F \tilde{\omega}_B^2(s)\,dF + (\tilde{y}_M - \tilde{y}_B)^2\int_F \tilde{x}^2\,dF + (\tilde{x}_M - \tilde{x}_B)^2\int_F \tilde{y}^2\,dF$$

$$\quad + 2(\tilde{y}_M - \tilde{y}_B)\int_F \tilde{\omega}_B(s)\tilde{x}\,dF - 2(\tilde{x}_M - \tilde{x}_B)\int_F \tilde{\omega}_B(s)\tilde{y}\,dF$$

$$\quad - 2(\tilde{y}_M - \tilde{y}_B)(\tilde{x}_M - \tilde{x}_B)\int_F \tilde{x}\tilde{y}\,dF.$$

Because the product integrals $F_{\omega_M x}$ and $F_{\omega_M y}$ are zero

$$F_{\omega_M x} = 0 = \int_F [\tilde{\omega}_B + (\tilde{y}_M - \tilde{y}_B)\tilde{x} - (\tilde{x}_M - \tilde{x}_B)\tilde{y}][\tilde{x}\cos\psi + \tilde{y}\sin\psi]\,dF$$

$$F_{\omega_M y} = 0 = \int_F [\tilde{\omega}_B + (\tilde{y}_M - \tilde{y}_B)\tilde{x} - (\tilde{x}_M - \tilde{x}_B)\tilde{y}][-\tilde{x}\sin\psi + \tilde{y}\cos\psi]\,dF.$$

By multiplying in turn by $\sin \psi$ and $\cos \psi$ and by adding and subtracting these equations it is easily shown that

$$\int_F [\tilde{\omega}_B \tilde{x} + (\tilde{y}_M - \tilde{y}_B)\tilde{x}^2 - (\tilde{x}_M - \tilde{x}_B)\tilde{x}\tilde{y}]\,dF = 0$$

$$\int_F (\tilde{\omega}_B \tilde{y} + (\tilde{y}_M - \tilde{y}_B)\tilde{x}\tilde{y} - (\tilde{x}_M - \tilde{x}_B)\tilde{y}^2]\,dF = 0.$$

When these relationships are used in the expression above for $F_{\omega_M \omega_M}$ the following formula is obtained

$$F_{\omega_M \omega_M} = F_{\tilde{\omega}_B \tilde{\omega}_B} + (\tilde{y}_M - \tilde{y}_B)F_{\tilde{\omega}_B \tilde{x}} - (\tilde{x}_M - \tilde{x}_B)F_{\tilde{\omega}_B \tilde{y}}. \tag{2.4.33}$$

This formula allows $F_{\omega\omega}$ to be transformed from the $\tilde{x}, \tilde{y}, \tilde{z}$ axes to the principal axes x, y, z. By combining eqns (2.4.21), (2.4.22), and (2.4.33) $F_{\omega\omega}$ can be expressed as

$$F_{\omega\omega} = F_{\tilde{\omega}_B \tilde{\omega}_B} + \frac{(F_{\tilde{\omega}_B \tilde{y}}F_{\tilde{x}\tilde{y}} - F_{\tilde{\omega}_B \tilde{x}}F_{\tilde{y}\tilde{y}})F_{\tilde{\omega}_B \tilde{x}} - (F_{\tilde{\omega}_B \tilde{y}}F_{\tilde{x}\tilde{x}} - F_{\tilde{\omega}_B \tilde{x}}F_{\tilde{x}\tilde{y}})F_{\tilde{\omega}_B \tilde{y}}}{F_{\tilde{x}\tilde{x}}F_{\tilde{y}\tilde{y}} - F_{\tilde{x}\tilde{y}}^2}$$

$$\tag{2.4.34}$$

2.4.3. Examples of evaluation of section properties for open profiles

The first two of the following examples show how the steps described above may be carried out for open profiles. In the first example it is obvious that many short cuts could be used because of symmetry but advantage is not taken of them here so that it can be seen how the method may be applied to a more irregular cross-section. In Example 2.4.3 a mixed profile is treated.

Example 2.4.1

Find all of the section properties of the channel section illustrated in Fig. 2.4.3(a) starting with the axes $\tilde{x}, \tilde{y}$ as indicated.

The section properties are first calculated in the $\tilde{x}\tilde{y}$ coordinate system and then transformed to the second and third coordinate systems. The $\tilde{x}$ and $\tilde{y}$ coordinates of points on the profile are shown in Fig. 2.4.3(b) and (c). The $\tilde{\omega}_B$ coordinate was calculated in Example 2.3.1 and is reproduced here in Fig. 2.4.3(d). Table 2.4.1 is now used to calculate the section properties.

F = area of section = 100 mm^2.

$$F_{\tilde{x}} = \int_F 1\tilde{x}\,dF = \frac{30 \times 1 \times 30}{2} + \frac{30 \times 1 \times 30}{2} = 900 \text{ mm}^3$$

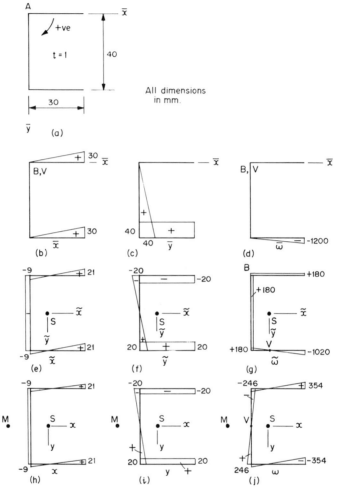

Fig. 2.4.3. Evaluation of coordinates in three coordinate systems for channel profile of Example 2.4.1.

where the formula in the third row and first column of Table 2.4.1 has been applied once to the upper flange and once to the lower flange.

$$F_{\bar{y}} = \int_F 1\bar{y} \, dF = \frac{40 \times 1 \times 40}{2} + 30 \times 1 \times 40 = 2000 \text{ mm}^3$$

$$F_{\bar{\omega}} = \int_F 1\bar{\omega} \, dF = \frac{-30 \times 1 \times 1200}{2} = -18\,000 \text{ mm}^4$$

$$F_{\bar{x}\bar{x}} = \int_F \bar{x}^2 \, dF = \frac{30 \times 30 \times 30}{3} + \frac{30 \times 30 \times 30}{3} = 18\,000 \text{ mm}^4$$

$$F_{\bar{y}\bar{y}} = \int_F \bar{y}^2 \, dF = \frac{40 \times 40 \times 40}{3} + 30 \times 40 \times 40 = 69\,333.3 \text{ mm}^4$$

$$F_{\bar{\omega}\bar{\omega}} = \int_F \bar{\omega}^2 \, dF = \frac{30(-1200)(-1200)}{3} = 14\,400\,000 \text{ mm}^6$$

$$F_{\bar{x}\bar{y}} = \int_F \bar{x}\bar{y} \, dF = \frac{30 \times 30 \times 40}{2} = 18\,000 \text{ mm}^4$$

$$F_{\bar{x}\bar{\omega}} = \int_F \bar{x}\bar{\omega} \, dF = \frac{30 \times 30(-1200)}{3} = -360\,000 \text{ mm}^5$$

$$F_{\bar{y}\bar{\omega}} = \int_F \bar{y}\bar{\omega} \, dF = \frac{30 \times 40(-1200)}{2} = -720\,000 \text{ mm}^5.$$

The location of the centroid S of the profile is found from eqns (2.4.8) and (2.4.9)

$$\bar{x}_s = \frac{F_{\bar{x}}}{F} = \frac{900}{100} = 9 \text{ mm}$$

$$\bar{y}_s = \frac{F_{\bar{y}}}{F} = \frac{2000}{100} = 20 \text{ mm}.$$

From eqn (2.4.26)

$$\bar{\omega}_0 = (\bar{\omega}_B)_{av} = \frac{F_{\bar{\omega}}}{F} = -\frac{18\,000}{100} = -180 \text{ mm}^2.$$

The origin is now moved to S (Fig. 2.4.3(e), (f), and (g)). The graph of $\bar{\omega}$ is simply that of $\bar{\omega} + 180$. At this point of the calculation all of the single integrals $F_{\bar{x}}$, $F_{\bar{y}}$, and $F_{\bar{\omega}}$ are zero. The product integrals could be calculated by using Fig. 2.4.3(e), (f), and (g) and Table 2.4.1 but this is not necessary since the transformation formulae in eqn (2.4.30) are available.

$$F_{\bar{x}\bar{y}} = F_{\bar{x}\bar{y}} - \frac{F_{\bar{x}}F_{\bar{y}}}{F} = 18\,000 - \frac{900 \times 2000}{100} = 0 \text{ mm}^4$$

$$F_{\bar{x}\bar{\omega}} = F_{\bar{x}\bar{\omega}} - \frac{F_{\bar{x}}F_{\bar{\omega}}}{F} = -360\,000 - \frac{900(-18\,000)}{100} = -198\,000 \text{ mm}^5$$

$$F_{\bar{y}\bar{\omega}} = F_{\bar{y}\bar{\omega}} - \frac{F_{\bar{y}}F_{\bar{\omega}}}{f} = -720\,000 - \frac{2000(-18\,000)}{100} = -360\,000 \text{ mm}^5$$

$$F_{\bar{x}\bar{x}} = F_{\bar{x}\bar{x}} - \frac{F_{\bar{x}}^2}{F} = 18\,000 - \frac{900^2}{100} = 9900 \text{ mm}^4$$

$$F_{\bar{y}\bar{y}} = F_{\bar{y}\bar{y}} - \frac{F_{\bar{y}}^2}{F} = 69\,333.3 - \frac{2000^2}{100} = 29\,333.3 \text{ mm}^4$$

$$F_{\bar{\omega}\bar{\omega}} = F_{\bar{\omega}\bar{\omega}} - \frac{F_{\bar{\omega}}^2}{F} = 14\,400\,000 - \frac{(-18\,000)^2}{100} = 11\,160\,000 \text{ mm}^6.$$

The reference axes, $\tilde{x}$ and $\tilde{y}$, may now be rotated into the direction of the principal axes, x and y, the angle of rotation ψ being given by eqn (2.4.13)

$$\tan 2\psi = \frac{2F_{\tilde{x}\tilde{y}}}{F_{\tilde{x}\tilde{x}} - F_{\tilde{y}\tilde{y}}} = 0$$

i.e. the $\tilde{x}$, $\tilde{y}$ and x, y axes coincide as can be expected from symmetry. The coordinates of the shear centre M are obtained from eqns (2.4.21) and (2.4.22)

$$\tilde{x}_M - \tilde{x}_B = \frac{(-360\,000)9900 - 0}{9900 \times 29\,333.3 - 0} = -12.272 \text{ mm}$$

$$\tilde{y}_M - \tilde{y}_B = \frac{0 - (-198\,000)(29\,333.3)}{9900 \times 29\,333.3 - 0} = 20 \text{ mm}.$$

The warping function with the pole at M is found from the transformation formulae, eqn (2.4.27)

$$\omega_M(s) = \tilde{\omega}_B(s) + (\tilde{y}_M - \tilde{y}_B)\tilde{x} - (\tilde{x}_M - \tilde{x}_B)\tilde{y}$$
$$= \tilde{\omega}_B(s) + 20\tilde{x} + 12.272\tilde{y}$$

where $\tilde{\omega}_B(s)$ is given in Fig. 2.4.3(g). As an example the value of $\omega_M(s)$ at the right-hand end of the upper flange is

$$180 + 20 \times 21 + 12.272(-20) = 354.6 \text{ mm}^2.$$

The product integrals in the principal coordinate system are found by using eqns (2.4.31) and (2.4.32)

$$\begin{matrix} F_{xx} \\ F_{yy} \end{matrix} = \tfrac{1}{2}(F_{\tilde{x}\tilde{x}} + F_{\tilde{y}\tilde{y}}) \mp \tfrac{1}{2}\sqrt{(F_{\tilde{x}\tilde{x}} - F_{\tilde{y}\tilde{y}})^2 + 4F_{\tilde{x}\tilde{y}}^2}]$$
$$= \tfrac{1}{2}(9900 + 29\,333.3) \mp \tfrac{1}{2}\sqrt{[(9900 - 29\,333.3)^2 + 0]}$$
$$= 9900 \text{ mm}^4 \text{ and } 29\,333.3 \text{ mm}^4, \text{ respectively.}$$

From eqn (2.4.33)

$$F_{\omega\omega} = F_{\tilde{\omega}\tilde{\omega}} + (\tilde{y}_M - \tilde{y}_B)F_{\tilde{\omega}\tilde{x}} - (\tilde{x}_M - \tilde{x}_B)F_{\tilde{\omega}\tilde{y}}$$
$$= 11\,160\,000 + 20(-198\,000) - (-12.272)(-360\,000)$$
$$= 2\,782\,080 \text{ mm}^6.$$

From eqn (1.4.20)

$$J_p = \tfrac{1}{3}(30 + 40 + 30)1^3 = 33.33 \text{ mm}^4.$$

Example 2.4.2

Find all of the section properties of the profile illustrated in Fig. 2.4.4 starting with the $\tilde{x}$, $\tilde{y}$ axes as shown.

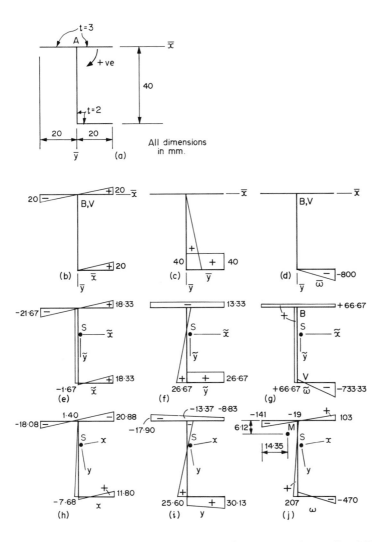

Fig. 2.4.4. Evaluation of coordinates in three coordinate systems for profile of Example 2.4.2.

The coordinates in the first system of axes are shown in Fig. 2.4.4(b), (c), and (d). The section properties are found as in the previous example by using Table 2.4.1.

F = area of section = 240 mm^2.

$$F_{\tilde{x}} = \frac{3 \times 20 \times 20 \times 1}{2} + \frac{3 \times 20(-20) \times 1}{2} + \frac{2 \times 20 \times 20 \times 1}{2} = 400 \text{ mm}^3$$

$$F_{\bar{y}} = \frac{2 \times 40 \times 40 \times 1}{2} + 2 \times 20 \times 40 \times 1 = 3200 \text{ mm}^3$$

$$F_{\bar{\omega}} = \frac{2 \times 20(-800) \times 1}{2} = -16\,000 \text{ mm}^4$$

$$F_{\bar{x}\bar{x}} = \frac{3 \times 20 \times 20 \times 20}{3} + \frac{3 \times 20(-20)(-20)}{3} + \frac{2 \times 20 \times 20 \times 20}{3}$$
$$= 21\,333 \text{ mm}^4$$

$$F_{\bar{y}\bar{y}} = \frac{2 \times 40 \times 40 \times 40}{3} + 2 \times 20 \times 40 \times 40 = 106\,667 \text{ mm}^4$$

$$F_{\bar{\omega}\bar{\omega}} = \frac{2 \times 20(-800)(-800)}{3} = 8\,533\,333 \text{ mm}^6$$

$$F_{\bar{x}\bar{y}} = \frac{2 \times 20 \times 20 \times 40}{2} = 16\,000 \text{ m}^4$$

$$F_{\bar{x}\bar{\omega}} = \frac{2 \times 20 \times 20 \times (-800)}{3} = -213\,333 \text{ mm}^5$$

$$F_{\bar{y}\bar{\omega}} = \frac{2 \times 20 \times 40 \times (-800)}{2} = -640\,000 \text{ mm}^5.$$

The coordinates of the centre of gravity are (eqns (2.4.8) and (2.4.9))

$$\bar{x}_s = \frac{400}{240} = 1.667 \text{ mm}$$

$$\bar{y}_s = \frac{3200}{240} = 13.33 \text{ mm}.$$

From eqn (2.4.26)

$$\bar{\omega}_0 = \frac{-16\,000}{240} = -66.667 \text{ mm}^2.$$

From the transformation formulae (2.4.30)

$$F_{\bar{x}\bar{y}} = 16\,000 - \frac{400 \times 3200}{240} = 10\,667 \text{ mm}^4$$

$$F_{\bar{x}\bar{\omega}} = -213\,333 - \frac{400(-16\,000)}{240} = -186\,667 \text{ mm}^5$$

$$F_{\bar{y}\bar{\omega}} = -640\,000 - \frac{3200(-16\,000)}{240} = -426\,667 \text{ mm}^5$$

$$F_{\bar{x}\bar{x}} = 21\,333 - \frac{400^2}{240} = 20\,667 \text{ mm}^4$$

$$F_{\bar{y}\bar{y}} = 106\,667 - \frac{3200^2}{240} = 64\,000\ \text{mm}^4$$

$$F_{\bar{\omega}\bar{\omega}} = 8\,533\,333 - \frac{(-16\,000)^2}{240} = 7\,466\,667\ \text{mm}^6.$$

The orientation of the principal axes is found from eqn (2.4.13)

$$\tan 2\psi = \frac{2 \times 10\,667}{20\,667 - 64\,000} = -0.492\,33$$

$$2\psi = -26.2°; \qquad \psi = -13.1°.$$

The transformation formulae (2.4.11) may now be employed to find the coordinates in the principal axis system.

$$x = 0.9740\bar{x} - 0.2268\bar{y}, \qquad y = 0.2268\bar{x} + 0.9740\bar{y}$$

and these functions are plotted in Fig. 2.4.4(h) and (i). The coordinates of the shear centre M are derived from eqns (2.4.21) and (2.4.22)

$$\bar{x}_M - \bar{x}_B = \frac{(-426\,667) \times 20\,667 - (-186\,667) \times 10\,667}{20\,667 \times 64\,000 - 10\,667^2} = -5.65\ \text{mm}$$

$$\bar{y}_M - \bar{y}_B = \frac{(-426\,667) \times 10\,667 - (-186\,667) \times 64\,000}{20\,667 \times 64\,000 - 10\,667^2} = 6.12\ \text{mm}$$

The warping function with the pole at M is (eqn 2.4.27)

$$\omega_M(s) = \tilde{\omega}_B(s) + 6.12\bar{x} + 5.65\bar{y}$$

and this is plotted in Fig. 2.4.4(j).

The product integrals in the principal coordinate system are obtained from eqns (2.4.31)–(2.4.33)

$$\frac{F_{xx}}{F_{yy}} = \tfrac{1}{2}(20\,667 + 64\,000) \mp \tfrac{1}{2}\sqrt{[(20\,667 - 64\,000)^2 + 4 \times 10\,667^2]}$$

$$= 18\,183\ \text{mm}^4 \text{ and } 66\,483\ \text{mm}^4 \text{ respectively}$$

$$F_{\omega\omega} = 7\,466\,667 + 6.12 \times (-186\,667) - (-5.65)(-426\,667)$$

$$= 3\,914\,000\ \text{mm}^6.$$

From eqn (1.4.20)

$$J_p = \tfrac{1}{3}(40 \times 3^3 + 60 \times 2^3) = 520\ \text{mm}^4.$$

Example 2.4.3

Find all of the section properties of the box girder whose profile is illustrated in Fig. 2.4.5(a) starting with the $\bar{x}$, $\bar{y}$ axes as shown.

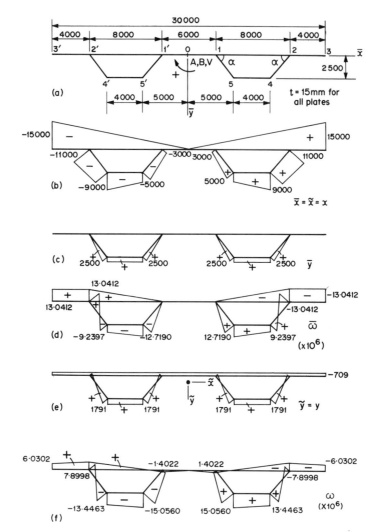

Fig. 2.4.5. Evaluation of coordinates in three coordinate systems for twin box girder of
Example 2.4.3.

The $\bar{x}$ and $\bar{y}$ coordinates of points in the profile are shown in Fig.
2.4.5(b) and (c).

$\bar{\omega}$ coordinate:
For member 1–5 sectorial radius $p_B = 3000 \sin \alpha = 2342.606$.
For member 2–4 sectorial radius $p_B = 11\,000 \sin \alpha = 8589.557$.

From eqn (1.4.29) for the hollow part of the profile

$$\Psi = 2A \bigg/ \oint \frac{ds}{t} = 2 \times 15 \times 10^6/(8000/15 + 4000/15 + 2 \times 3201.56/15)$$

$$= 24\,452.38.$$

Equations (2.3.37) and (2.3.46) are used for the open and closed parts of the profile, respectively.

At V $\bar{\omega} = 0$ and for plate 0 to 1 $p_B = 0$.

Integrating from V to 1, $\bar{\omega}_1 = \displaystyle\int_0^{3000} 0\,ds = 0.$

From 1 to 2, $\bar{\omega}_2 = 0 + \displaystyle\int_0^{8000} \left(0 - \frac{24\,452.38}{15}\right) ds = -13\,041\,242.$

From 2 to 4, $\bar{\omega}_4 = -13\,041\,242 + \displaystyle\int_0^{3201.56} \left(8589.557 - \frac{24\,452.38}{15}\right) ds$

$$= 9\,239\,689.$$

From 4 to 5, $\bar{\omega}_5 = 9\,239\,689 + \displaystyle\int_0^{4000} \left(2500 - \frac{24\,452.38}{15}\right) ds$

$$= 12\,719\,056.$$

From 5 to 1, $\bar{\omega}_1 = 12\,719\,056 + \displaystyle\int_0^{3201.56} \left(-2342.606 - \frac{24\,452.38}{15}\right) ds$

$$= 0. \quad \text{(which checks)}$$

In the last integration the sign of p_B only is changed as indicated because for plate 5 to 1 the sense is anticlockwise (i.e. negative) about B.

Also because $p_B = 0$ between 2 and 3

$$\bar{\omega}_3 = \bar{\omega}_2 = -13\,041\,242.$$

The $\bar{\omega}$ diagram Fig. 2.4.5(d) is antisymmetrical about the vertical centreline. The section properties are found as in the two previous examples by using Table 2.4.1.

$F = $ area of cross-section $= 762\,094$ mm^2

$F_{\bar{x}} = 0$ mm^3

$$F_{\bar{y}} = 2\left[15 \times 4000 \times 1 \times 2500 + \frac{2 \times 15 \times 3201.56 \times 1 \times 2500}{2}\right]$$

$$= 540\,117\,000 \text{ mm}^3$$

$F_{\bar{\omega}} = 0$ mm^4

$$F_{\bar{x}\bar{x}} = 2 \times 15 \left[\frac{15\,000^3}{3} + 3201.56\, \frac{3000(2 \times 3000 + 5000) + 5000(3000 \atop +2 \times 5000)}{6} \right.$$

$$+ 4000 \times \frac{5000(2 \times 5000 + 9000) + 9000(5000 + 2 \times 9000)}{6}$$

$$\left. + 3201.56\, \frac{9000(2 \times 9000 + 11\,000) + 11\,000(9000 + 2 \times 11\,000)}{6} \right]$$

$$= 50.995\,46 \times 10^{12}\ \text{mm}^4$$

$$F_{\bar{y}\bar{y}} = 2 \times 15 \left[\frac{2 \times 3201.56 \times 2500^2}{3} + 4000 \times 2500^2 \right]$$

$$= 1.150\,195 \times 10^{12}\ \text{mm}^4$$

$$F_{\bar{\omega}\bar{\omega}} = 2 \times 15 \left[\frac{8000 \times 13.0412^2}{3} + 4000 \times 13.0412^2 + \frac{3201.56 \times 12.7190^2}{3} \right.$$

$$+ 4000\, \frac{12.719(2 \times 12.719 + 9.2397) + 9.2397(12.719 + 2 \times 9.2397)}{6}$$

$$\left. + 3201.56\, \frac{9.2397(2 \times 9.2397 - 13.0412) - 13.0412(9.2397 \atop -2 \times 13.0412)}{6} \right]$$

$$\times 10^{12} = 58.100\,875 \times 10^{18}\ \text{mm}^6$$

$$F_{\bar{x}\bar{\omega}} = 2 \times 15 \left[\frac{3201.56 \times 12.7190(3000 + 2 \times 5000)}{6} \right.$$

$$4000 \times \frac{5000(2 \times 12.719 + 9.2397) + 9000(12.719 + 2 \times 9.2397)}{6}$$

$$+ 3201.56 \times \frac{9000(2 \times 9.2397 - 13.0412) + 11\,000(9.2397 \atop -2 \times 13.0412)}{6}$$

$$+ 8000 \times \frac{-13.0412(3000 + 2 \times 11\,000)}{6}$$

$$\left. + \frac{4000(-13.0412)(11\,000 + 15\,000)}{2} \right]$$

$$\times 10^6 = -23.837 \times 10^{15}\ \text{mm}^5$$

$$F_{\bar{x}\bar{y}} = F_{\bar{y}\bar{\omega}} = 0.$$

The coordinates of the centroid are (eqns (2.4.8) and (2.4.9))

$$\bar{x}_s = 0 \quad \text{and} \quad \bar{y}_s = \frac{540.117 \times 10^6}{0.762\,094 \times 10^6} = 708.73\ \text{mm}.$$

From eqn (2.4.26) $\bar{\omega}_0 = 0$. From the transformation formulae (2.4.30)

$$F_{\tilde{x}\tilde{y}} = 0 - 0 = 0$$

$$F_{\tilde{x}\tilde{\omega}} = -23.837 \times 10^{15} - 0 = -23.837 \times 10^{15} \text{ mm}^5$$

$$F_{\tilde{y}\tilde{\omega}} = 0 - 0 = 0$$

$$F_{\tilde{x}\tilde{x}} = 50.995\ 46 \times 10^{12} - 0 = 50.995\ 46 \times 10^{12} \text{ mm}^4$$

$$F_{\tilde{y}\tilde{y}} = 1.150\ 195 \times 10^{12} - \frac{(540.117 \times 10^6)^2}{0.762\ 09 \times 10^6} = 0.767\ 399 \times 10^{12} \text{ mm}^4$$

$$F_{\tilde{\omega}\tilde{\omega}} = 58.100\ 875 \times 10^{18} - 0 = 58.100\ 875 \times 10^{18} \text{ mm}^6.$$

The coordinates of the shear centre are found from eqns (2.4.21) and (2.4.22)

$$\tilde{x}_M - \tilde{x}_B = 0, \qquad \tilde{y}_M - \tilde{y}_B = \frac{23.837 \times 10^{15}}{50.995\ 46 \times 10^{12}} = 467.4 \text{ mm}.$$

$\tilde{\omega}_B\ (=\bar{\omega}_B)$ can be transformed to ω_M by using eqn (2.4.27)

$$\omega_M(s) = \tilde{\omega}_B(s) + 467.4\tilde{x}$$

and this is plotted in Fig. 2.4.5(f).

The product section properties are found from eqns (2.4.31)–(2.4.33), whence

$$F_{xx} = F_{\tilde{x}\tilde{x}} = 50.995\ 46 \times 10^{12} \text{ mm}^4$$

$$F_{yy} = F_{\tilde{y}\tilde{y}} = 0.767\ 399 \times 10^{12} \text{ mm}^4$$

$$F_{\omega\omega} = 58.100\ 875 \times 10^{18} - 467.4 \times 23.837 \times 10^{15} = 46.959 \times 10^{18} \text{ mm}^6.$$

The torsion constant J_p is obtained from eqns (1.4.15) and (1.4.20)

$$J_p = 2\left[\frac{4 \times (15 \times 10^6)^2}{8000/15 + 4000/15 + 2 \times 3201.56/15} + \tfrac{1}{3}(3000 + 4000)15^3\right]$$

$$= 2[733\ 571.26 + 7.88] \times 10^6$$

$$= 1.4671 \times 10^{12} \text{ mm}^4.$$

The last calculation again shows that it would be reasonable to neglect the open part of the profile when calculating J_p. The section properties of a three-cell box girder are evaluated in Section 2.6.1.

2.4.4. Expressions for the stresses and stress resultants using the principal axes and the shear centre for reference

The expressions for the longitudinal stress and the shear stress (eqns (2.3.56), (2.3.64)) have been derived in the original coordinate system.

While it is possible to transform these expressions in a formal manner by means of the coordinate transformations (eqns (2.4.7), (2.4.11), (2.4.26), and (2.4.27)) this is not necessary because of the uncoupling of the governing equations. This allows the axial deformation, the bending about the principal x-axis, the bending about the principal y-axis and the torsion about the shear centre M to be treated separately; these separate problems correspond to the four eqns (2.4.5). The stresses are derived from first principles. It is assumed that at the ends the axial load $N(z)$ is applied through the centroid S. If it is not applied through S it can be shifted there and two bending moments (one about the principal x-axis and one about the principal y-axis) added together with a bimoment M_{ω_M} (see Fig. 1.1.2). The formal definition of M_{ω_M} follows the same pattern as that for bending about the principal axes, e.g. about the principal x-axis,

$$M_x(z) = \int_F \sigma_z(z, s) y(s) \, dF. \tag{2.4.35}$$

The bimoment is defined as

$$M_{\omega_M}(z) = \int_F \sigma_{\omega_M}(z, s) \omega_M(s) \, dF. \tag{2.4.36}$$

But from eqn (2.3.56) the longitudinal warping stress due to twisting is

$$\sigma_{W_M}(z, s) = -E\omega_M(s)\phi''(z) \tag{2.4.37}$$

whence the bimoment is

$$M_{\omega_M}(z) = -E\phi''(z) \int_F \omega_M^2(s) \, dF = -E\phi''(z) F_{\omega_M \omega_M}. \tag{2.4.38}$$

In Chapter 1 it was shown in an intuitive manner that when the bimoment $M_{\omega_M}(z)$ varies along a beam a torque called the warping torsion moment, M_{DS}, occurs. M_{DS} is that part of the total torsion moment on the cross-section which arises from the restraint of warping, while the remainder of the torsion moment is M_{st}, i.e.

$$M_z = M_{DS} + M_{st}. \tag{2.4.39}$$

There is a simple relationship between $M_{\omega_M}(z)$ and $M_{DS}(z)$. From the equilibrium equation just before eqn (2.3.15) and eqn (2.4.36)

$$M'_{\omega_M}(z) = -\int_F \frac{\partial(\tau_{zs} t)}{\partial s} \omega_M(s) \, ds - \int_F p_z(z, s) \omega_M(s) \, ds.$$

In this expression the shear stress is the warping shear stress τ_w because the equation just before eqn (2.3.15) is one of equilibrium of an element in the longitudinal direction. Saint Venant shear stresses play no part in longitudinal equilibrium as can be seen from the derivation of eqn (1.4.11).

Another way of seeing this is to realise that τ_{zs} in eqn (2.3.15) can be either the total shear stress τ $(=\tau_{st}+\tau_{w})$ or the warping shear stress τ_{w} alone. The difference is then taken care of by a suitable adjustment to the constant $C(z)$ which represents a constant shear flow around the profile. In the following analysis τ_{zs} is interpreted as the warping shear stress alone.

It is shown in Chapter 3 that the second term on the right-hand side of the last equation is the first derivative of an externally applied bimoment. It arises in those beams which carry a distributed axial load p_z per unit area of surface. This is a fairly uncommon loading condition so for most practical cases $p_z = 0$. Integrating by parts and recalling that $d\omega_M = p_M(s)\,ds$, where $p_M(s)$ is the radius component

$$M'_{\omega_M}(z) = -[t\tau_w\omega_M]_{\text{edges}} + \int_F t\tau_w p_M(s)\,ds = M_{DS} \qquad (2.4.40)$$

since the shear stress on the edges is zero.

From eqns (2.4.39) and (2.4.40) an expression for the total torque at a point in a beam as a function of ϕ can be derived.

$$\begin{aligned} M_z(z) &= M_{st} + M'_{\omega_M} \\ &= GJ_p\phi' - EF_{\omega_M\omega_M}\phi'''. \end{aligned} \qquad (2.4.41)$$

This equation could have been derived by integrating from $z = 0$ to z on both sides of the last of eqns (2.4.5) because the total torque at any point along a beam is simply the sum of all of the externally applied torques up to that point. The right-hand side of the last of eqns (2.4.5) is the sum of the externally applied torques per unit length of beam.

For convenience some of the nomenclature and the relationships between the stress resultants and ϕ are summarized in Table 2.4.2.

Table 2.4.2. Section properties and stress resultants.

Symbol	Name	Function of ϕ	Units
$F_{\omega_M\omega_M}$	warping constant		mm^6
$EF_{\omega_M\omega_M}$	warping rigidity		$N\,mm^4$
J_p	torsion constant		mm^4
GJ_p	torsional rigidity		$N\,mm^2$
$M_{\omega_M}(z)$	bimoment	$= -E\phi''(z)F_{\omega_M\omega_M}$	$N\,mm^2$
$M_{st}(z)$	Saint Venant torsion moment	$= GJ_p\phi'(z)$	$N\,mm$
$M_{DS}(z)$	warping torsion moment	$= -EF_{\omega_M\omega_M}\phi'''(z)$	$N\,mm$
$M_z(z)$	total torsion moment	$= GJ_p\phi'(z) - EF_{\omega_M\omega_M}\phi'''(z)$	$N\,mm$

It is instructive at this stage to return to the I-beam illustrated in Fig. 1.1.6(e) and use eqn (2.4.36) to calculate the bimoment. The distribution of longitudinal stress is calculated as follows. The bending moment in the upper flange is $Pb/4$ and hence the bending stress is (tension stress is positive)

$$\sigma_W = \frac{Mx}{I_f} = +\frac{\left(\frac{Pb}{4}\right)x}{\frac{t_2 b^3}{12}} = +\frac{3Px}{t_2 b^2}.$$

This distribution is plotted in Fig. 2.4.6(a). The warping function $\omega_M(s)$ is calculated with the starting point V and the shear centre M at the centroid. For instance for point **1** the radius component from M is $a/2$ and twice the sectorial area (see eqn (2.3.37)) is $ab/4$. The warping

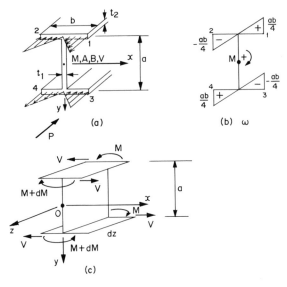

(a)

(b) ω

(c)

Fig. 2.4.6. (a) Stresses due to bimoment, $Pab/4$; (b) Warping function ω; (c) stress resultants in the flanges.

function is shown in Fig. 2.4.6(b). The bimoment M_{ω_M} at the end of the beam is now calculated from eqn (2.4.36) using a table of volume integrals, Table 2.4.1, as an aid.

Upper flange: $2 \cdot \dfrac{1}{3} \cdot \dfrac{b}{2} \cdot \dfrac{ab}{4} \cdot \dfrac{3Pb/2}{t_2 b^2} \cdot t_2 = \dfrac{Pab}{8}$

Web: $= 0$

Lower flange: $= \dfrac{Pab}{8}$

$$\overline{}$$

Total bimoment at end of beam $= \dfrac{Pab}{4}$.

$$\overline{}$$

Thus the stress distribution shown in Fig. 2.4.6(a) has an associated stress resultant M_ω which is that obtained in Chapter 1 by applying engineers' theory to the flanges separately.

In analysing the torsion of beams it is convenient to have relationships between the stresses and stress resultants. That between the longitudinal warping stress σ_w and the bimoment M_ω is obtained by eliminating ϕ'' from eqns (2.4.37) and (2.4.38), whence

$$\sigma_w(z, s) = \frac{M_{\omega_M}(z)\omega_M(s)}{F_{\omega_M \omega_M}}, \qquad (2.4.42)$$

Because of the uncoupling of the governing equations in Section 2.4.1 the four effects can be added to obtain the total longitudinal stress at any point in the profile. Thus the total longitudinal direct stress is

$$\sigma(s) = \frac{N}{F} + \frac{M_y x}{F_{xx}} + \frac{M_x y}{F_{yy}} + \frac{M_\omega \omega}{F_{\omega\omega}} \qquad (2.4.43)$$

The distribution of shear stress in an open profile is now derived. That for a closed profile is considered in Section 2.6. For an *open profile* the shear stress $\tau_{zs}(z, s)$ is obtained from eqns (2.4.43) and (2.3.15). [In the latter equation the variable $\bar{z} = z$ after a change of coordinate axes.] If there are no shear stresses at the edge $C = 0$, hence

$$\tau_{zs}(z, s_E) = -\frac{dM_y}{dz} \frac{1}{tF_{xx}} \int_0^{s_E} x(s)\, dF(s) - \frac{dM_x}{dz} \frac{1}{tF_{yy}} \int_0^{s_E} y(s)\, dF(s)$$

$$-\frac{dM_\omega}{dz} \frac{1}{tF_{\omega\omega}} \int_0^{s_E} \omega_M(s)\, dF(s) \qquad (2.4.44)$$

where these integrations are carried out from a free edge to a point s_E away from it.

Since the transverse shear forces V_x and V_y are the first derivatives of M_x and M_y and from eqn (2.4.40), eqn (2.4.44) can be written in the form

$$\tau_{zs}(z, s_E) = -\frac{V_x F_x(s_E)}{t F_{xx}} - \frac{V_y F_y(s_E)}{t F_{yy}} - \frac{M_{DS} F_\omega(s_E)}{t F_{\omega\omega}} \tag{2.4.45}$$

This is the shear stress due to transverse shear and due to warping, the total shear stress being obtained by adding the Saint Venant shear stress.

From eqns (2.4.43) and (2.4.45) it is seen that the last terms, which represent the stresses due to warping torsion, have a similar form to that of the earlier terms, which represent the usual stresses derived by engineers' theory.

So far no mention has been made of a sign convention for the bimoment M_ω and warping torsion moment M_{DS}. The sign convention can be deduced from eqns (2.4.36) and (2.4.40). From eqn (2.4.36) it is seen that a bimoment acting on a cross-section will be positive if the longitudinal stress $\sigma_w(z, s)$ and the warping function $\omega(s)$ have the same sign for every value of s. To see how this works in the case of an I-section the reader should again refer to Fig. 2.4.6. In Fig. 2.4.6(b) the positive sign for ω was fixed as soon as the indicated clockwise sense of rotation of the radius vector p_M was selected. To satisfy eqn (2.4.36) the positive bimoment on the right-hand end of the element in Fig. 2.4.6(c) must produce longitudinal stresses of the same sign as ω, tensile stresses being positive. Thus the stress resultants in the flanges must be as indicated in Fig. 2.4.6(c). It should be noted that in the case of the bimoment on the right-hand end, that is, Ma, the lower moment M is seen by the upper moment M as clockwise as does the lower M see the upper M. However, the reverse is the case at the left-hand end, i.e., they see one another as anticlockwise. A sign convention which can also be adopted is that the bimoment on the right-hand end of Fig. 2.4.6(c) is negative while that on the left-hand end is positive.

The transverse shear forces V are

$$V = \frac{dM}{dz} \tag{2.4.46}$$

and give rise to the warping torsion moment M_{DS} on each end

$$M_{DS} = Va = a\frac{dM}{dz} = \frac{dM_\omega}{dz}. \tag{2.4.47}$$

This twisting moment causes a clockwise (i.e. positive) rotation of the left-hand end of the element relative to the right-hand end and this is in agreement with the sign convention incorporated into equations such as eqn (2.4.39).

2.4.5. Solution of torsion eqn (2.4.5)

The first three of eqns (2.4.5) are solved by using engineers' theory in the normal way. The last equation relates to the twisting of the beam and its solution is one of the aims of Chapter 3 where several applications of this theory are presented. The last of eqns (2.4.5) is the governing torsion equation for open profiles. It is also applicable to closed profiles when the approximate theory (see Section 2.1) is being used. Its solution enables the internal stress resultants of torque, bimoment, etc. in a beam to be related to the externally applied twisting moments and bimoments. Once the internal torques and bimoments are evaluated the stress distribution at any cross-section may be found (see Section 2.6.1).

2.5. ENERGY METHOD FOR SETTING UP GOVERNING EQUATIONS

This section is not an essential part of the theory and may be omitted at a first reading. The energy methods (Matheson 1971) are not fundamentally different from the method based on equilibrium, compatibility, etc. but they are often more convenient and they can be used as a check on one's working. Furthermore, if it is desired to develop a second-order theory the energy theorems become very valuable tools. In this section the governing equations (2.4.5) which are, of course, the result of a *first-order analysis* only are derived. The total potential energy (Matheson 1971) U is the sum of the internal potential (strain) energy U_i and the potential energy of the external loads U_e, i.e.

$$U = U_i + U_e \qquad (2.5.1)$$

The internal potential energy of an element of the beam of volume dV is ($\varepsilon_x = \varepsilon_y = \gamma_{xy} = \gamma_{xy} = \gamma_{yz} = 0$ for open profiles—see Section 2.1)

$$dU_i = \frac{E}{2} \varepsilon_z^2 \, dV + \frac{G}{2} \gamma_{zs}^2 \, dV \qquad (2.5.2)$$

By integrating over the whole volume of the beam the internal strain energy is obtained.

$$U_i = \frac{E}{2} \int_V \varepsilon_z^2 \, dV + \frac{G}{2} \int_V \gamma_{zs}^2 \, dV. \qquad (2.5.3)$$

As discussed prior to the derivation of eqn (2.3.34), for open profiles the in-plane component of shear strain is zero and it is only necessary to consider the shear strain arising from the Saint Venant torsion. Hence the second part of the right-hand side of eqn (2.5.3) is (Fig. 1.4.7(c))

$$U_{i2} = \frac{G}{2} \int_V [r(\bar{z}, s)\phi'(\bar{z})]^2 \, d\bar{z} \, dF$$

$$= \frac{G}{2} \int_V [\phi'(\bar{z})]^2 \int_F r^2 \, dF \, d\bar{z}$$

$$= \frac{1}{2} \int_0^L GJ_p \phi'^2 \, d\bar{z}. \tag{2.5.4}$$

The first part of eqn (2.5.3) is developed by substituting from eqns (2.3.54) and (2.3.80), whence

$$U_{i1} = \frac{E}{2} \int_0^L [\zeta'^2 F - 2\zeta'\xi'' F_{\bar{x}} - 2\zeta'\eta'' F_{\bar{y}} - 2\zeta'\phi'' F_{\bar{\omega}}$$

$$+ \xi''^2 F_{\bar{x}\bar{x}} + \xi''\eta'' F_{\bar{x}\bar{y}} + 2\xi''\phi'' F_{\bar{x}\bar{\omega}}$$

$$+ \eta''^2 F_{\bar{y}\bar{y}} + 2\eta''\phi'' F_{\bar{y}\bar{\omega}} + \phi''^2 F_{\bar{\omega}\bar{\omega}}] \, d\bar{z} \tag{2.5.5}$$

If the principal axes are now chosen as the coordinate system for x and y and the shear centre M and an appropriate starting point, which makes $F_\omega = 0$, as the coordinate system for ω eqn (2.5.3) simplifies to

$$U_i = \frac{1}{2} \int_0^L [EF\zeta'^2 + EF_{xx}\xi_M''^2 + EF_{yy}\eta_M''^2 + EF_{\omega\omega}\phi''^2 + GJ_p\phi'^2] \, dz \tag{2.5.6}$$

As the beam deflects under the influence of the external loads they lose potential energy. Hence the quantity U_e will have a negative sign attached to it. The external loads acting on a beam in the original coordinate system are shown in Fig. 2.5.1(a). They consist of the line loads $q_{\bar{x}}(\bar{z})$, $q_{\bar{y}}(\bar{z})$, $q_{\bar{z}}(\bar{z})$, $m_{\bar{z}}(\bar{z})$, point loads at the point i $(i = 1, 2, 3, \ldots)$ $P_{\bar{x}i}$, $P_{\bar{y}i}$, $P_{\bar{z}i}$, $M_{\bar{z}i}$, and loads acting at the ends of the beam, that is, $M_{\bar{x}e}$, $M_{\bar{y}e}$, $m_{\bar{\omega}e}$, $M_{\bar{z}e}$. The line loads have been shown in Fig. 2.5.1(a) as acting through a common point on the profile for convenience. In fact they should be shown as having planes or lines of action. Figure 2.5.1(b) shows this at the cross-section $\bar{z}$. For example, $q_{\bar{x}}(\bar{z})$ acts along a plane parallel to the $\bar{x}$-axis and distance $\bar{y}_{qx}$ from it. The $\bar{x}$ component of the deflection of the point T through which $q_{\bar{x}}(\bar{z})$ acts is made up of two parts (Fig. 2.5.1(c)). The profile has a displacement of its shear centre M from M to M' and a rotation $\phi(\bar{z})$ about M'. Thus the $\bar{x}$ component of displacement is $\bar{\xi}_M - (\bar{y}_{qx} - \bar{y}_M)\phi(\bar{z})$ and the work done by the load $q_{\bar{x}}(\bar{z}) \, d\bar{z}$ which acts over a length $d\bar{z}$ is $q_{\bar{z}}(\bar{z})[\bar{\xi}_M - (\bar{y}_{qx} - \bar{y}_M)\phi(\bar{z})] \, dz$. This is a first-order analysis only; a second-order analysis, for example that published by Roik, Carl, and Lindner (1972), introduces additional terms. When this

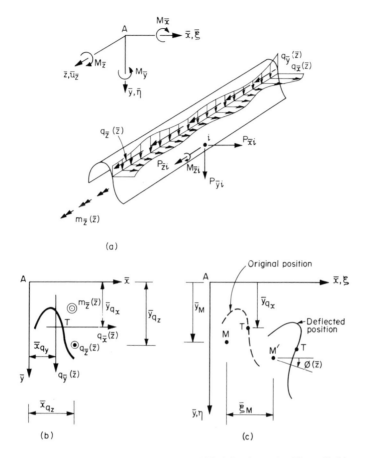

Fig. 2.5.1. (a) Loads, (b) load locations, and (c) deflections of a thin-walled beam.

analysis is extended to all components of loads the following expression for the change of potential energy of the loads is obtained.

$$U_e = -\left[\int_0^L \{q_{\bar{x}}[\bar{\xi}_M - (\bar{y}_{qx} - \bar{y}_M)\phi] + q_{\bar{y}}[\bar{\eta}_M + (\bar{x}_{qy} - \bar{x}_M)\phi] + q_{\bar{z}}\bar{\zeta}_{\bar{z}} + m_{\bar{z}}\phi\} \, d\bar{z}\right.$$

$$+ \sum_i \{P_{\bar{x}i}[\bar{\xi}_{Mi} - (\bar{y}_{Pi} - \bar{y}_{Mi})\phi_i] + P_{\bar{y}i}[\bar{\eta}_{Mi} + (\bar{x}_{Pi} - \bar{x}_M)\phi_i] + P_{\bar{z}i}\bar{\zeta}_{\bar{z}i} + M_{\bar{z}i}\phi_i\}$$

$$+ \left. \sum_e \{-M_{\bar{x}e}\bar{\eta}'_{Me} + M_{\bar{y}e}\bar{\xi}'_{Me} - M_{\bar{\omega}e}\phi'_e + M_{\bar{z}e}\phi_e\}\right]. \tag{2.5.7}$$

In arriving at this expression the only term which may cause difficulty to the reader is $-M_{\bar{\omega}e}\phi'_e$. Reference to eqn (2.4.43) shows that the component of the longitudinal stress which arises from warping is

$$\sigma_{\bar{z}e} = \frac{M_{\bar{\omega}e}\bar{\omega}_e}{F_{\bar{\omega}\bar{\omega}}} \qquad (2.5.8)$$

and reference to eqn (2.3.38) shows that the component of longitudinal deflection at e is

$$\bar{u}_{\bar{z}e} = -\phi'\bar{\omega}_B. \qquad (2.5.9)$$

Hence the work done by $M_{\bar{\omega}e}$ is

$$\delta U_{\bar{\omega}e} = \int_F \left(\frac{M_{\bar{\omega}e}\bar{\omega}_e}{F_{\bar{\omega}\bar{\omega}}}\right)(-\phi'_e\omega_B)\,\mathrm{d}F$$

$$= -M_{\bar{\omega}e}\phi'_e \qquad (2.5.10)$$

which justifies the appropriate term in eqn (2.5.7). Equation (2.5.7) can be transformed into the principal coordinates by simply dropping the bar over x, y, z, and ω and substituting it with eqn (2.5.6) into eqn (2.5.1). Thus, the total energy is

$$U = \frac{1}{2}\int_0^L (EF\zeta'^2 + EF_{xx}\xi_M''^2 + EF_{yy}\eta_M''^2 + EF_{\omega\omega}\phi''^2 + GJ_p\phi'^2)\,\mathrm{d}z$$

$$- \Bigg[\int_0^L \{q_x[\xi_M - (y_{qx} - y_M)\phi] + q_y[\eta_M + (x_{qy} - x_M)\phi] + q_z\zeta_z + m_z\phi\}\,\mathrm{d}z$$

$$+ \sum_i \{P_{xi}[\xi_{Mi} - (y_{Pi} - y_{Mi})\phi_i] + P_{yi}[\eta_{Mi} + (x_{Pi} - x_M)\phi_i] + P_{zi}\zeta_{zi} + M_{zi}\phi_i\}$$

$$+ \sum_e (-M_{xe}\eta'_{Me} + M_{ye}\xi'_{Me} - M_{\omega e}\phi'_e + M_{ze}\phi_e)\Bigg]. \qquad (2.5.11)$$

For equilibrium (see eqn (1.3.1))

$$\delta U = \frac{\partial U}{\partial \xi_M}\,\mathrm{d}\xi_M = \frac{\partial U}{\partial \eta_M}\,\mathrm{d}\eta_M = \frac{\partial U}{\partial \phi_M}\,\mathrm{d}\phi_M = 0 \qquad (2.5.12)$$

Because the terms in eqn (2.5.11) have been uncoupled, by virtue of the fact that we have referred the section properties, loads and deflections to the principal axes, differentiation of eqn (2.5.11) leads directly to eqns (2.4.5). The boundary conditions are to be found by integrating eqn (2.5.11) by parts and using the result to define the end conditions which conserve the total energy of the system.

$$[EF_{xx}\xi_M'' - M_{ye}]_0^L = 0 \quad \text{or} \quad [\delta\xi_M']_0^L = 0$$

$$[EF_{yy}\eta_M'' + M_{xe}]_0^L = 0 \quad \text{or} \quad [\delta\eta_M']_0^L = 0 \qquad (2.5.13)$$

$$[EF_{\omega\omega}\phi'' + M_{\omega e}]_0^L = 0 \quad \text{or} \quad [\delta\phi']_0^L = 0$$

These equations simply say that for no energy ($\delta U = 0$) to escape across the boundary of the system during a virtual displacement, either the total moments at the ends or the virtual slopes there must be zero. Similarly

$$[EF_{xx}\xi_M''' + P_{xe}]_0^L = 0 \quad \text{or} \quad [\delta\xi_M]_0^L = 0$$

$$[EF_{yy}\eta_M''' + P_{ye}]_0^L = 0 \quad \text{or} \quad [\delta\eta_M]_0^L = 0 \tag{2.5.14}$$

$$[GJ_p - EF_{\omega\omega}\phi''' = M_{ze} + P_{xe}y_{PM} - P_{ye}x_{PM}]_0^L = 0 \quad \text{or} \quad [\delta\phi]_0^L = 0$$

These equations make similar statements about the actual forces and the corresponding virtual displacements at the ends of the beam. Thus eqns (2.5.13) and (2.5.14) define the so-called natural boundary conditions of the problem.

2.6. TORSION ANALYSIS OF CLOSED PROFILES

An exact analysis of closed profiles which consist of one or more cells is usually very complicated and it is doubtful whether usable solutions can be obtained from it. In the earlier parts of this book an approximate theory for closed profiles due to von Karman and Christensen (1944) has been described. It follows very closely the theory of open profiles developed by Vlasov (1961). It has been seen that the crucial assumption for closed profiles is that the shear strain due to warping γ_w is zero and only the Saint Venant shear strain γ_s is significant. This means that the stress–strain relationship between τ_w and γ_w cannot be used to obtain τ_w which must be derived by considering an equilibrium condition in the longitudinal direction (Section 2.3.2). This is similar to that used in simple beam theory where the Bernoulli assumption implies zero shear strain and the shear stresses must be obtained from equilibrium of an element in the longitudinal direction (Section 1.4.1). In certain cases where the longitudinal warping stresses σ_w are large and change rapidly, this theory is not accurate. A 'more accurate analysis' was developed by Benscoter (1954) to overcome this problem. Both theories are detailed in the excellent volume published by Kollbrunner and Hajdin (1965). Benscoter's theory overcomes the difficulty in finding τ_w by starting with assumed forms for the components of deformation. In so doing the strains can be found by simple differentiations and then the stresses σ_w and τ_w are obtained by using the stress–strain relationship, thus yielding a more direct approach to the problem. The final solution bears a close resemblance to that of von Karman and Christensen; in fact, the governing torsion equations are identical except for the section properties, $F_{\omega\omega}$, etc.

In the following section the von Karman and Christensen theory is restated very briefly because much of it has already been covered. It is then applied to a few typical profiles. In Section 2.6.2 the theory due to

Benscoter is developed and a comparison made of the two theories when they are applied to some simple profiles.

2.6.1. Approximate torsion theory of closed profiles

It will be recalled from Section 1.4.2 that for a thin-walled closed profile undergoing *uniform* twist along its length the (Saint Venant) shear stresses are uniform through the wall thickness. The following is a summary of Saint Venant's analysis for hollow profiles. For a *single-cell* cross-section it was shown in Section 1.4.2 that the relationship between torque M and shear flow T is

$$M = 2AT \tag{2.6.1}$$

and that

$$\phi' = \frac{M}{G} \frac{\oint \frac{ds}{t}}{4A^2} = \frac{M}{GJ_p}. \tag{2.6.2}$$

Hence from eqns (2.6.1) and (2.6.2)

$$G\phi' = \frac{M}{2A} \frac{\oint \frac{ds}{t}}{2A} \tag{2.6.3}$$

$$= T/\Psi \tag{2.6.4}$$

where

$$\Psi = \frac{2A}{\oint \frac{ds}{t}}. \tag{2.6.5}$$

Ψ is a purely geometric quantity and is easily determined for a given single-cell profile (see Example 2.6.1).

For a multi-cell cross-section with n cells (Fig. 1.4.7) undergoing Saint Venant torsion the ϕ values for each cell are the same so a set of simultaneous equations in the shear flows $T_1, T_2, \ldots, T_i, \ldots, T_n$ can be obtained. Thus, for the ith cell,

$$\phi' = \frac{1}{2GA_i} \left[\oint \frac{T\,ds}{t} \right]_i$$

$$= \frac{1}{2GA_i} \left[T_i \oint_i \frac{ds}{t} - \sum_{k=1}^{m} T_k \int_k \frac{ds}{t} \right] \tag{2.6.6}$$

where the last terms refer to those parts of the boundary of the ith cell which it has in common with other cells. The n equations for $i = 1$ to n can be solved simultaneously for $T_i/G\phi'$. Comparison with eqn (2.6.4) shows that this quantity is Ψ_i, i.e. we solve for

$$\Psi_i = T_i/G\phi'. \tag{2.6.7}$$

For a profile with n cells joined together it is now assumed that the total torque M is the sum of the torques M_i ($i = 1$ to n).

From eqns (2.6.1) and (2.6.7)

$$M = G\phi'(2\Psi_1 A_1 + 2\Psi_2 A_2 + \dots 2\Psi_n A_n) \tag{2.6.8}$$

or

$$J_p = 2 \sum \Psi_i A_i. \tag{2.6.9}$$

It must be emphasized that for a single cell profile Ψ is found from equation (2.6.5) but for multi-cell profiles the Ψ_i values are found by solving a set of simultaneous equations. The following example shows how the aforegoing Saint Venant theories can be applied to three classes of cross-section.

Example 2.6.1

Three uniform beams whose cross-sections are illustrated in Fig. 2.6.1 are twisted with a uniform torque M. Derive expressions for the angle of rotation per unit length in terms of M/G.

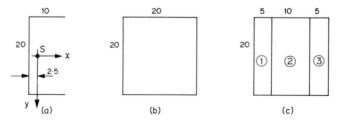

Fig. 2.6.1. Cross-sections analysed in Examples 2.6.1 and 2.6.2 ($t = 1$ for all plates).

For all three sections $\phi' = M/GJ_p$. Since the torque is constant the warping stresses are zero and only *Saint Venant torsion* is considered.

(a) Channel section

$$J_p = \sum \tfrac{1}{3} bt^3$$

$$= \tfrac{1}{3}(40) \times 1^3$$

$$= 13.33 \text{ mm}^4$$

$$\phi' = \frac{M}{13.33G} = 0.075 \frac{M}{G} \text{ radian per mm.}$$

(b) Square box section—single cell (refer eqn (2.6.2))

$$J_p = \frac{4A^2}{\oint \frac{ds}{t}} = \frac{4 \times (20 \times 20)^2}{80/1} = 8000 \text{ mm}^4$$

$$\phi' = \frac{M}{8000G} = 0.000 \ 125 \frac{M}{G} \text{ rad mm}^{-1}.$$

(c) 3-cell box section

For cell 1 eqn (2.6.6) is, after rearranging and using eqn (2.6.7),

$$200 = \Psi_1(50) - \Psi_2(20)$$

For cell 2

$$400 = -\Psi_1(20) + \Psi_2(60) - \Psi_3(20)$$

For cell 3

$$200 = \qquad -\Psi_2(20) + \Psi_3(50)$$

Solving these equations

$$\Psi_1 = \Psi_3 = 100/11 \quad \text{and} \quad \Psi_2 = 140/11$$

From eqn (2.6.9)

$$J_p = 2\left[\frac{100}{11} \times 100 + \frac{140}{11} \times 200 + \frac{100}{11} + 100\right]$$

$$= 96 \ 000/11$$

$$\phi' = \frac{11M}{96 \ 000G} = 0.000 \ 114 \ 583 \frac{M}{G} \text{ rad mm}^{-1}.$$

In the approximate solution it is now assumed that all of the out-of-plane distortion of the cross-section is due to Saint Venant shear strain γ_s. In other words, it is assumed that warping shear strain γ_w is negligibly small. The analysis follows as shown in Section 2.3.3 where it is seen in equation (2.3.45) that the warping function for a closed profile is

$$\omega(s) = \int_0^s \left[p_M(s) - \frac{\Psi_i}{t(s)}\right] ds \qquad (2.6.10)$$

For a single-cell profile Ψ is given by eqn (2.6.5) and for a multi-cell profile the Ψ_i values are found from eqn (2.6.7) and by solving a set of simultaneous equations of the form of eqn (2.6.6) (see also Example 2.6.1).

Example 2.6.2

Plot the warping functions $\omega_M(s)$ for each of the cross-sections shown in Fig. 2.6.1, where M is the shear centre. Calculate $F_{\omega\omega}$ and the ratio of $J_p/F_{\omega\omega}$ for each section (J_p from Example 2.6.1).

(a) Channel section

With the axes taken as shown in Fig. 2.6.1(a) at the centre of gravity S the $x(s)$, $y(s)$, and $\omega(s)$ values are plotted in Fig. 2.6.2. The coordinates of

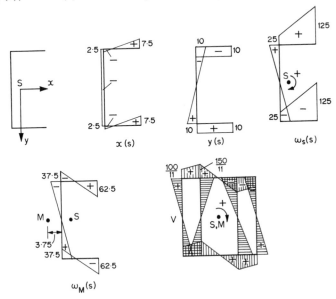

Fig. 2.6.2. Warping functions of channel section shown in Fig. 2.6.1(a) and 3-cell profile shown in Fig. 2.6.1(c).

the shear centre M are found by using eqn (2.4.21) when it is recognised that because of symmetry $F_{xy} = 0$. Hence with the use of eqns (2.4.29) or of the volume integral table (Table 2.4.1)

$$F_{\omega_s y} = \frac{10(-10)(25+125)}{2} + \frac{20(-10)(2\times25-25)+20(10)(25-2\times25)}{6}$$

$$+\frac{10(10)(-25-125)}{2}$$

$$= -16\,667 \text{ mm}^2$$

$$F_{yy} = 10(-10)(-10) + \frac{20(-10)(-2\times10+10)+20(10)(-10+2\times10)}{6}$$

$$+ 10(10)(10)$$

$$= 2667 \text{ mm}^4$$

Hence, $x_M = -\dfrac{16\,667}{2667} = -6.25$ mm and the warping function is easily calculated from eqn (2.3.37) (see Fig. 2.6.2). $F_{\omega\omega}$ $(=\int_F \omega^2 \, dF)$ is also found from volume integrals.

$$F_{\omega\omega} = 2\left[\frac{10(-37.5)[2(-37.5)+62.5]+10(62.5)[-37.5+2(62.5)]}{3} + 10(-37.5)(-37.5)\right]$$

$$= 29\,167 \text{ mm}^6$$

$$\frac{J_p}{F_{\omega\omega}} = \frac{13.33}{29\,167} = 0.000\,457\,1 \text{ mm}^{-2}.$$

(b) Square box section—single cell

Because of symmetry the centre of gravity S and the shear centre M lie at the centre of the profile.

$$\Psi = \frac{2A}{\displaystyle\oint \frac{ds}{t}} = \frac{2 \times 400}{(20+20+20+20)} = 10 \text{ mm}^2.$$

Substituting in eqn (2.6.10) this value and $p_M(s) = 10$ mm for all sides we see that for a square box the warping function $\omega_M(s) = 0$, i.e. there is no warping of the cross-section under uniform torsion. This result should be expected because of the assumptions made and the symmetry.

(c) 3-cell box section

For cells 1 and 3: $\qquad \displaystyle\oint \frac{ds}{t} = 5+20+5+20 = 50,$

and for cell 2: $\qquad \displaystyle\oint \frac{ds}{t} = 10+20+10+20 = 60.$

For the internal webs: $\displaystyle\int \frac{ds}{t} = 20.$

We can now set up three simultaneous equations from eqns (2.6.6) and (2.6.7) (as we did in Example 2.6.1).

For cell 1: $\quad 2 \times 5 \times 20 \;=\; 50\Psi_1 - 20\Psi_2.$

For cell 2: $\quad 2 \times 10 \times 20 = -20\Psi_1 + 60\Psi_2 - 20\Psi_3.$

For cell 3: $\quad 2 \times 5 \times 20 \;=\; \qquad -20\Psi_2 + 50\Psi_3.$

Solving these equations yields

$$\Psi_1 = \Psi_3 = \frac{100}{11} \quad \text{and} \quad \Psi_2 = \frac{140}{11}.$$

According to the argument which follows eqn (2.4.25) the starting point V should be located at the point where the horizontal or vertical axes of symmetry intersect the profile. Starting at V on the left-hand web (Fig. 2.6.2), the value of ω_M at the upper left-hand corner is:

$$\omega(10) = \int_0^{10} \left[p_M(s) - \frac{\Psi_1}{1} \right] dx = \int_0^{10} \left[10 - \frac{100}{11} \right] ds = \frac{100}{11}.$$

Integrating from this point to the point of intersection of the inner web and upper flange,

$$\omega = \frac{100}{11} + \int_0^5 \left[p_M(s) - \frac{\Psi_1}{1} \right] ds = \frac{100}{11} + \int_0^5 \left[10 - \frac{100}{11} \right] ds = \frac{150}{11}.$$

Along the middle section of the upper flange

$$\omega(s) = \frac{150}{11} + \int_0^s \left[10 - \frac{\Psi_2}{1} \right] ds = \frac{150}{11} + \int_0^s \left[10 - \frac{140}{11} \right] ds = \frac{150}{11} - \frac{30s}{11}$$

and where $s = 10$, $\omega(s) = -150/11$.

For the integration down the inner web $\Psi = \Psi_2 - \Psi_1 = 40/11$. Hence

$$\omega(s) = \frac{150}{11} - \int_0^s \left[5 - \frac{40}{11} \right] ds = \frac{150}{11} - \frac{15s}{11}$$

and when $s = 20$, $\omega = -150/11$. The minus sign in this last expression arises because the integration is carried out in the opposite sense to that indicated for ω in Fig. 2.6.2.

$$F_{\omega\omega} = 4\left\{ \left[10 \times \frac{100}{11} \times \frac{100}{11} \right] \Big/ 3 + \left[10 \times \frac{150}{11} \times \frac{150}{11} \right] \Big/ 3 \right.$$

$$+ \left[5 \times \frac{100}{11} \left(2 \times \frac{100}{11} + \frac{150}{11} \right) + 5 \times \frac{150}{11} \left(\frac{100}{11} + 2 \times \frac{150}{11} \right) \right] \Big/ 6$$

$$\left. + \left(5 \times \frac{150}{11} \times \frac{150}{11} \right) \Big/ 3 \right\}$$

$$= 7438 \text{ mm}^6.$$

$$\frac{J_p}{F_{\omega\omega}} = 1.1733 \text{ mm}^{-2}.$$

This ratio should be compared with 0.000 457 1 and infinity for the channel and square box sections, respectively.

Example 2.6.3

A crane runway has the cross-section illustrated in Fig. 2.6.3(a). Determine $F_{\bar{x}\bar{x}}$, $F_{\bar{y}\bar{y}}$, $F_{\bar{x}\bar{y}}$, $F_{\bar{x}\bar{\omega}}$, $F_{\bar{y}\bar{\omega}}$, $F_{\bar{\omega}\bar{\omega}}$, J_p, and the location of the shear centre. Evaluate $F_{\omega\omega}$.

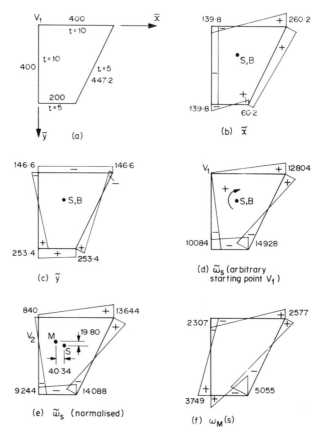

Fig. 2.6.3. Sectional properties of a crane girder.

Taking the original coordinates as shown in Fig. 2.6.3(a) the centroid S is at (139.8, 146.6). $F_{\bar{x}\bar{x}}$, $F_{\bar{y}\bar{y}}$, and $F_{\bar{x}\bar{y}}$ are found from the volume integral table (Table 2.4.1).

$$
\begin{aligned}
F_{\bar{x}\bar{x}} = {} & 400 \times 10 \times 139.8^2 + \{400 \times 10 \times (-139.8)[2(-139.8) + 260.2]\}/6 \\
& + \{400 \times 10 \times 260.2(-139.8 + 2 \times 260.2)\}/6 \\
& + \{200 \times 5 \times (-139.8)[2(-139.8) + 60.2]\}/6 \\
& + \{200 \times 5 \times 60.2(-139.8 + 2 \times 60.2)\}/6 \\
& + \{447.2 \times 5 \times 60.2(2 \times 60.2 + 260.2)\}/6 \\
& + \{447.2 \times 5 \times 260.2(60.2 + 2 \times 260.2)\}/6 \\
= {} & 215.76 \times 10^6 \text{ mm}^4.
\end{aligned}
$$

Similarly

$$F_{\bar{y}\bar{y}} = 251.107 \times 10^6 \text{ mm}^4$$

$$F_{\bar{x}\bar{y}} = -71.026 \times 10^6 \text{ mm}^4.$$

To plot $\bar{\omega}_S$ a starting point at V_1 is chosen and Ψ is calculated from eqn (2.6.5)

$$\Psi = (2 \times 120\,000) \Big/ \left(\frac{400}{10} + \frac{447.2}{5} + \frac{200}{5} + \frac{400}{10} \right) = 1145.9$$

On substituting in eqn (2.6.10) the values of $\bar{\omega}_S$ with starting point at V_1 are found and plotted in Fig. 2.6.3(d).

From eqn (2.4.26) the value of ω_0 can be calculated.

$$\omega_0 = \left[\frac{12\,804}{2} \times 400 \times 10 - \frac{(12\,804 - 14\,928)}{2} \times 447.2 \times 5 \right.$$

$$\left. - \frac{(10\,084 + 14\,928)}{2} \times 200 \times 5 - \frac{10\,084}{2} \times 400 \times 10 \right] \Big/ 11\,236$$

$$= -840.$$

Hence by adding 840 to all values of $\bar{\omega}_S$ shown in Fig. 2.6.3(d) the normalized values shown in Fig. 2.6.3(e) are obtained, i.e., $\int \omega_S \, dF = 0$.

From the table of volume integrals and Figs 2.6.3(b), (c), and (e)

$$F_{\bar{\omega}\bar{x}} = 7.138\,19 \times 10^9 \text{ mm}^5, \qquad F_{\bar{\omega}\bar{y}} = -11.537\,79 \times 10^9 \text{ mm}^5, \quad \text{and}$$

$$F_{\bar{\omega}\bar{\omega}} = 650.4200 \times 10^9 \text{ mm}^6.$$

The coordinates of the shear centre M are found from eqns (2.4.21) and (2.4.22) where $\tilde{x}_B$ and $\tilde{y}_B$ are zero because the pole B has been taken at the origin S.

$$\tilde{x}_M = \frac{-11.537\,79 \times 10^9 \times 215.76 \times 10^6 + 7.138\,19 \times 10^9 \times 71.026 \times 10^6}{215.76 \times 10^6 \times 251.107 \times 10^6 - (71.026 \times 10^6)^2}$$

$$= -40.34 \text{ mm}$$

$$\tilde{y}_M = \frac{11.537\,79 \times 10^9 \times 71.026 \times 10^6 - 7.138\,19 \times 10^9 \times 251.107 \times 10^6}{215.76 \times 10^6 \times 251.107 \times 10^6 - (71.026 \times 10^6)^2}$$

$$= -19.80 \text{ mm}.$$

To obtain the warping constant with pole M eqn (2.4.33) may be used.

$$F_{\omega\omega} = [650.4200 + (-19.80)(7.138\,19) - (-40.34)(-11.538\,89)] \times 10^9$$

$$= 43.65 \times 10^9 \text{ mm}^6.$$

From eqn (1.4.15)

$$J_p = \frac{4A^2}{\oint \frac{ds}{t}}$$

$$= \frac{4 \times (12 \times 10^4)^2}{\dfrac{400}{10} + \dfrac{447.2}{5} + \dfrac{200}{5} + \dfrac{400}{10}}$$

$$= 0.275\,02 \times 10^9 \text{ mm}^4.$$

Finally the values of $\omega_M(s)$ shown in Fig. 2.6.3(f) are obtained by using eqn (2.4.27).

The values of the section properties found in the last two examples could be used in the coefficients of the governing torsion equation to enable the internal stress resultants (torques M_{st} and M_{DS} and bimoment M_ω) to be expressed in terms of the externally applied loads. Such problems are solved in Section 3.2. Attention is now turned to the evaluation of the warping stresses (direct and shear) by using the approximate theory. So far only the distribution of Saint Venant torsion stresses has been considered. Since there are no longitudinal direct stresses in Saint Venant torsion these are only shear stresses τ_s.

To evaluate the warping shear stress τ_w it is necessary to turn to equation (2.3.15) where it is seen that it is the sum of three components. Usually p_z is zero or its contribution to longitudinal equilibrium can be ignored. For a *single-cell profile* an imaginary longitudinal cut can be introduced at an arbitrary point so the first part τ_0 of the shear stress is zero at the cut and the remainder τ_1 of the shear stress arises from a shear flow $C(z)$ $(=t\tau_1)$ around the profile (Fig. 2.6.4). Dealing firstly with τ_0 it is seen from the first term of eqn (2.3.15) that

$$\tau_0(z, s_E)t(s) = -\int_0^{s_E} \frac{\partial \sigma}{\partial z} \, dF$$

Fig. 2.6.4. The shear flow due to warping stress τ_w is $T = \tau_w t$, which is the sum of T_0 for an open profile and C which is constant around the profile.

Substituting from eqns (2.4.42) and (2.4.40)

$$\tau_0(z, s_E)t(s_E) = -\frac{M_{DS}(z)F_\omega(s_E)}{F_{\omega\omega}} \tag{2.6.11}$$

where

$$F_\omega(s_E) = \int_0^{s_E} \omega(s)\, dF$$

The magnitude of the shear flow $C(z)$ is obtained from the condition that there can be no longitudinal dislocations in the profile (see eqn (2.3.68))

$$C(z) = \frac{M_{DS}(z)}{F_{\omega\omega}} \frac{\oint F_\omega(s_E)/t\, ds_E}{\oint ds_E/t} \tag{2.6.12}$$

i.e. from eqn (2.3.66) the shear flow arising from the warping shear stress is

$$T(z, s_E) = \tau_0(z, s_E)t(s_E) + C(z)$$

$$= \frac{M_{DS}(z)}{F_{\omega\omega}} \left[\frac{\oint F_\omega(s_E)/t(s_E)\, ds}{\oint ds/t(s_E)} - F_\omega(s_E) \right] \tag{2.6.13}$$

The term in square brackets is a function only of the geometry of the cross-section so the warping shear stress is proportional to $M_{DS}(z)$.

In the above analysis it should be understood that $\tau_0(z, s_E)$ maintains the longitudinal equilibrium of an element which is under the action of the longitudinal warping stress $\sigma(z, s_E)$. The shear flow $C(z)$ which is constant at any profile z does not contribute to longitudinal equilibrium but it does contribute to M_{DS}. In other words M_{DS} consists of two parts, viz., $\int \tau_0 p_M\, dF$ and $\int \tau_1 t p_M\, ds = 2CA$. It is important to realise that the shear flow C is quite different from the Saint Venant shear flow described in Chapter 1. It is also important to understand that even though an imaginary cut is introduced into the profile the values of ω, F_ω, $F_{\omega\omega}$ and the location of the shear centre M all refer to the *uncut profile*. The following example shows how the analysis of a single-cell profile can be carried out. The analysis of multi-cell profiles is carried out in a similar manner but it is necessary to solve simultaneous equations for the shear flows C_n as suggested in eqn (2.3.68(a)). A multi-cell box girder is analysed in Chapter 3 (Example 3.3.2).

Example 2.6.4

Sketch the distribution of warping shear stress and warping longitudinal stress for the profile analysed in Example 2.6.3 when it is twisted about its shear centre M.

The distribution of warping shear stress is evaluated from eqn (2.6.13)

$$\tau(z, s_E) = \frac{M_{DS}(z)}{t(s_E)F_{\omega\omega}} \left[\frac{\oint \frac{F_\omega(s_E)}{t(s_E)} ds}{\oint \frac{ds}{t(s_E)}} - F_\omega(s_E) \right].$$

The value of $F_\omega(s_E) = \int_0^{s_E} \omega(s)\, dF$ is found by using the table of volume integrals (Table 2.4.1) and the warping function $\omega_M(s)$ shown on Fig. 2.6.3(f). The closed profile is first opened by a longitudinal cut at an arbitrary point: the top left hand corner is chosen and the integration is started there. The following are typical values of $F_\omega(s_E)$, (Fig. 2.6.5(a)).

$$F_{\omega b} = \frac{200 \times 10}{2}(-2307 + 135) = -2\,172\,000 \text{ mm}^4$$

$$F_{\omega c} = \frac{400 \times 10}{2}(-2307 + 2577) = 540\,000 \text{ mm}^4$$

$$F_{\omega d} = 540\,000 + \frac{223.6 \times 5}{2}(2577 - 1239) = 1\,287\,900 \text{ mm}^4.$$

The line integral of $F_\omega(s)/t(s)$ is also found by using Table 2.4.1 in the following manner. [This is merely Simpson's Rule.]

$$a\text{–}c: \quad \frac{10^6}{6 \times 10}[-4 \times 2.172 + 0.54] \times 400 \qquad = -54.32 \times 10^6$$

$$c\text{–}e: \quad \frac{10^6}{6 \times 5}[0.54 + 4 \times 1.2879 - 2.2304] \times 447.2 \quad = \quad 51.59 \times 10^6$$

$$e\text{–}g: \quad \frac{10^6}{6 \times 5}[-2.2304 - 4 \times 3.6574 - 2.8834] \times 200 \quad = -131.62 \times 10^6$$

$$g\text{–}i: \quad \frac{10^6}{6 \times 10}[-2.8834 + 4 \times 1.5866] \times 400 \qquad = \quad 23.09 \times 10^6$$

$$\oint \frac{F_\omega}{t} ds = -111.26 \times 10^6 \text{ mm}^4$$

$$\oint \frac{ds}{t} = \frac{400}{10} + \frac{447.2}{5} + \frac{200}{5} + \frac{400}{10} = 209.44.$$

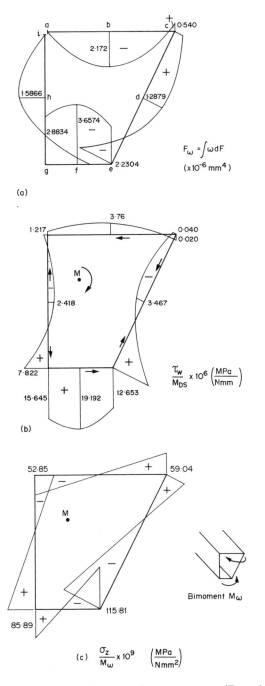

Fig. 2.6.5. Evaluation of warping stresses in a crane runway (Example 2.6.4).

Hence

$$\frac{\oint \frac{F_\omega}{t} ds}{\oint \frac{ds}{t}} = \frac{-111.26 \times 10^6}{209.44} = -0.5312 \times 10^6 \text{ mm}^4.$$

This value and the values of F_ω shown in Fig. 2.6.5(a) are substituted into the above expression for warping shear stress and plotted in Fig. 2.6.5(b). A typical value is that at b.

$$\tau(z, s) = \frac{M_{DS}(z)[-0.5312 - (-2.172)] \times 10^6}{10 \times 43.65 \times 10^9} = 3.76 M_{DS} \times 10^{-6} \text{ MPa}$$

when M_{DS} is measured in N mm. In practical cases a designer would have to add the shear stresses due to both bending and Saint Venant torsion to this warping shear stress.

The distribution of longitudinal warping stresses is obtained from the last term in eqn (2.4.43), i.e.

$$\sigma_z(s) = \frac{M_\omega(z)\omega(s)}{F_{\omega\omega}}.$$

This distribution is plotted in Fig. 2.6.5(c) from the ω values shown in Fig. 2.6.3(f).

The warping torsion moment M_{DS} and the bimoment M_ω in these expressions can be expressed in terms of the loads applied to a beam by solving the governing equations (eqn 2.4.5). This is done in Section 3.2.2 for a number of cases.

2.6.2. The torsion theory of closed profiles due to Benscoter (1954)

In this theory, which is an alternative theory to that developed in the previous section, the following assumptions are made.

(1) The cross-section preserves its geometry during twisting.
(2) The distribution of warping distortion is the same as in the case of Saint Venant torsion.
(3) The displacement $v(z, s)$, which is that tangential to the centre-line of the profile (Fig. 2.3.5), is, as a consequence of assumption (1) above, assumed to have the following form

$$v(z, s) = \phi(z)p_M(s) \qquad (2.6.14)$$

where $p_M(s)$ is the length of the perpendicular from the shear centre M to the tangent to the middle line of the profile. (Figure 2.3.8 shows a typical construction) and $\phi(z)$ is the angle of rotation about M.

(4) The displacement $u(z, s)$ which is the out-of-plane deflection of the cross-section is assumed to be proportional to the warping function $\omega(s)$, i.e.

$$u(z, s) = -\theta(z)\omega(s) \tag{2.6.15}$$

where $\theta(z)$ is a function of z, yet to be determined, and $\omega(s)$ is given by eqn (2.6.10), i.e.

$$\omega(s) = \int_0^s \left[p_M(s) - \frac{\Psi_i(z)}{t(s)} \right] ds. \tag{2.6.16}$$

It is of course understood that the shear centre M is the pole, that the starting point V has been chosen so that (see eqn (2.4.26))

$$\oint \omega(s)\, dF = 0, \tag{2.6.17}$$

and the reference axes x and y coincide with the principal axes of the cross-section.

The longitudinal and shear strains are derived from eqns (2.6.14) and (2.6.15).

$$\varepsilon_z = \frac{\partial u}{\partial z} = -\theta'\omega \tag{2.6.18}$$

$$\gamma_{zs} = \frac{\partial v}{\partial z} + \frac{\partial u}{\partial s} = \phi' p_M - \theta\dot{\omega} \tag{2.6.19}$$

where

$$\dot{\omega} = \frac{\partial \omega}{\partial s} = p_M - \frac{\Psi_i}{t}. \tag{2.6.20}$$

The stress resultants can now be determined in terms of the displacement, ϕ and θ from the stress–strain relationship, thus

$$\sigma_z = E\frac{\partial u}{\partial z} = -E\theta'\omega \tag{2.6.21}$$

$$\tau_{zs} = G(\phi' p_M - \theta\dot{\omega}). \tag{2.6.22}$$

The normal force N_z acting in the axial direction and bending moments M_x and M_y about the principal axes are therefore

$$N_z = \int_F \sigma_z\, dF = -E\theta'\int_F \omega\, dF = 0 \tag{2.6.23}$$

from eqn (2.6.17),

$$M_x = \int_F \sigma_z y \, dF = -E\theta' \int_F \omega y \, dF = 0 \qquad (2.6.24(a))$$

from eqn (2.4.16), and similarly

$$M_y = \int_F \sigma_z x \, dF = -E\theta' \int_F \omega x \, dF = 0. \qquad (2.6.24(b))$$

Since these bending moments are zero for pure torsion the transverse shear forces must also be zero to satisfy equilibrium, i.e.

$$V_x = V_y = 0. \qquad (2.6.25)$$

The torque M is found by integrating around the profile

$$M = \int_F \tau p_M \, dF = G\phi' \int_F p_M^2 \, dF - G\theta \int_F p_M \dot{\omega} \, dF \qquad (2.6.26)$$

and the bimoment M_ω is (see eqn (2.4.36))

$$M_\omega = \int_F \sigma_z \omega \, dF = -E\theta' \int_F \omega^2 \, dF. \qquad (2.6.27)$$

These equations can be simplified by introducing certain section properties. From eqn (2.4.38)

$$F_{\omega\omega} = \int_F \omega_M^2 \, dF. \qquad (2.6.28)$$

A further section property required is (see eqns (2.6.20) and (2.6.26))

$$\int_F \frac{\Psi_i}{t} p_M \, dF = \sum_{i=1}^n \Psi_i \int p_M \, ds = 2 \sum_{i=1}^n \Psi_i A_i \qquad (2.6.29)$$

where n is the number of cells in the profile. From eqn (2.6.9) it is seen that this expression is the torsion constant, J_p, i.e.

$$J_p = 2 \sum_{i=1}^n \Psi_i A_i = \sum_{i=1}^n \int_F \frac{\Psi_i}{t} p_M \, dF. \qquad (2.6.30)$$

A third section property is defined as

$$F_{pp} = \int_F p_M^2 \, dF. \qquad (2.6.31)$$

For convenience this section property is called the *polar constant*. It is not the same as the polar moment of inertia $(= \int_F r^2 \, dF$ where r is the radial distance from the pole to the element $dF)$.

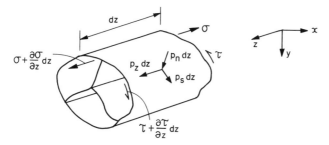

Fig. 2.6.6. Forces acting on an element of beam.

By inserting these expressions in eqns (2.6.26) and (2.6.27) the following expressions for the only non-zero stress resultants are obtained

$$M = GF_{pp}\phi' - G(F_{pp} - J_p)\theta \tag{2.6.32}$$

$$M_\omega = -EF_{\omega\omega}\theta' \tag{2.6.33}$$

The displacement functions ϕ and θ are now determined by setting up the governing equations. This is done by considering equilibrium of the element illustrated in Fig. 2.6.6. By first taking moments about an axis through the shear centre

$$\int_F \frac{\partial\tau}{\partial z} p_M \, dF + \int_s (p_s p_M + p_n p_N) \, ds = 0 \tag{2.6.34}$$

where

p_s = load per unit length acting in the tangential direction,
p_n = load per unit length acting in the normal direction,
p_M = distance from shear centre M to tangent to surface at point being considered,
p_N = distance from normal at point being considered to the shear centre M.

The last term is simply the external torque $m(z)$ applied to the surface of the element per unit length of beam.

The simplest way to obtain the second equilibrium equation, this time in the longitudinal direction, is to introduce a virtual displacement. With a virtual displacement $\theta(z) = -1$ the longitudinal displacement (eqn (2.6.15)) is

$$u = \omega. \tag{2.6.35}$$

The corresponding shear strain is

$$\gamma = \frac{\partial u}{\partial s} = \dot{\omega} = p_M - \frac{\Psi_i}{t}. \tag{2.6.36}$$

The sum of the internal energy and the external work done during this virtual displacement is zero.

$$\int_F \frac{\partial \sigma_z}{\partial z} \omega \, dF - \int_F \tau_{zs} \dot{\omega} \, dF + \int_s p_z \omega \, ds = 0. \tag{2.6.37}$$

As explained in Chapter 3 the last term is a bimoment $m_\omega(z)$ applied to the surface of the element per unit length of beam. On substituting from eqns (2.6.21) and (2.6.22) into eqns (2.6.34) and (2.6.37)

$$G\phi'' \int_F p_M^2 \, dF - G\theta' \int_F p_M \dot{\omega} \, dF + m(z) = 0 \tag{2.6.38}$$

$$-E\theta'' \int_F \omega^2 \, dF - G\phi' \int_F p_M \dot{\omega} \, dF + G\theta \int \dot{\omega}^2 \, dF + m_\omega(z) = 0. \tag{2.6.39}$$

On introducing the section properties defined in eqns (2.6.28), (2.6.30), and (2.6.31) and by using eqns (2.6.20) and (2.6.29) the following equations are obtained.†

$$GF_{pp}\phi'' - G(F_{pp} - J_p)\theta' + m(z) = 0 \tag{2.6.40}$$

$$-EF_{\omega\omega}\theta'' - G(F_{pp} - J_p)(\phi' - \theta) + m_\omega(z) = 0. \tag{2.6.41}$$

From eqn (2.6.40)

$$\theta' = \frac{F_{pp}}{F_{pp} - J_p}\phi'' + \frac{m(z)}{G(F_{pp} - J_p)}. \tag{2.6.42}$$

Differentiating eqn (2.6.41) once with respect to z gives

$$-EF_{\omega\omega}\theta''' + G(F_{pp} - J_p)(\phi'' - \theta') + m'_\omega(z) = 0. \tag{2.6.43}$$

Equations (2.6.40) and (2.6.43) are now added to obtain

$$-EF_{\omega\omega}\theta''' + GJ_p\phi'' + m'_\omega(z) + m(z) = 0. \tag{2.6.44}$$

Differentiating eqn (2.6.42) twice and substituting in (2.6.44) the following governing equation for the twisting of a beam is derived.

$$EF^0_{\omega\omega}\phi^{iv} - GJ_p\phi'' = m(z) + m'_\omega(z) - \frac{EF^0_{\omega\omega}}{GF_{pp}}m''(z) \tag{2.6.45}$$

† In deriving eqn (2.6.41) the expression

$$\int_F \dot{\omega}^2 \, dF = \int_F \left(p_M - \frac{\Psi_i}{t}\right)^2 dF = F_{pp} - 2J_p + \sum_{i=1}^n \Psi_i^2 \oint \frac{ds}{t} = F_{pp} - J_p$$

(from eqn (1.4.25)).

where

$$F^0_{\omega\omega} = F_{\omega\omega} \frac{F_{pp}}{F_{pp} - J_p}. \tag{2.6.46}$$

Comparison of eqns (2.6.45) and the last of eqns (2.4.5) show that they have the same form but the coefficients of ϕ^{iv} differ, that in eqn (2.6.45) having the greater value. The values of the external loading terms are identical except for those unusual cases when $m''(z) \neq 0$. Thus in practice the only difference between the approximate and more exact theories is that $F_{\omega\omega}$ is replaced by $F^0_{\omega\omega}$.

Example 2.6.5

Derive the section properties $F_{\omega\omega}$, F_{pp}, J_p, and $F^0_{\omega\omega}$ for a hollow box-girder of rectangular section $2a \times 2b$ $(a > b)$ and thickness t (Fig. 2.6.7(a)). Discuss the significance of these values as $\alpha = b/a$ varies and in the light of the two theories of torsion for hollow profiles.

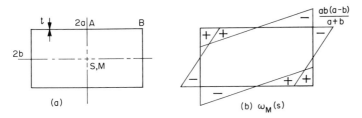

(a)   (b) $\omega_M(s)$

Fig. 2.6.7. Torsion analysis of a hollow rectangular box beam in Example 2.6.5.

From eqn (2.6.5)

$$\frac{\Psi}{t} = \frac{2A}{t \oint \dfrac{ds}{t}} = \frac{2 \times 4ab}{4(a+b)} = \frac{2ab}{a+b}.$$

From eqn (2.6.16)

$$\omega = \int_0^s \left[p_M - \frac{2ab}{a+b} \right] ds.$$

Obviously because of symmetry the starting points V must lie at the midpoint of each side. This ensures that $\oint \omega \, dF = 0$. The value of ω_M at point B is

$$\omega = \int_0^a \left(b - \frac{2ab}{a+b} \right) ds = -\frac{ab(a-b)}{a+b}.$$

The warping function is plotted in Fig. 2.6.7(b). Hence from Table 2.4.2 and eqn (2.6.28)

$$F_{\omega\omega} = 4\frac{a}{3}\left[\frac{ab(a-b)}{a+b}\right]^2 t + 4\frac{b}{3}\left[\frac{ab(a-b)}{a+b}\right]^2 t = \tfrac{4}{3}a^5 t\frac{\alpha^2(1-\alpha)^2}{1+\alpha}.$$

From eqn (1.4.15)

$$J_p = \frac{4(4ab)^2 t}{4(a+b)} = \frac{16a^3 t\alpha^2}{1+\alpha}.$$

From eqn (2.6.31)

$$F_{pp} = 2b^2\times 2at + 2a^2\times 2bt = 4a^3 t\alpha(1+\alpha).$$

From eqn (2.6.46)

$$F^0_{\omega\omega} = \tfrac{4}{3}a^5 t\frac{\alpha^2(1-\alpha)^2}{1+\alpha}\cdot\frac{1}{1-[(16a^3 t\alpha^2)/(1+\alpha)]/[4a^3 t\alpha(1+\alpha)]}$$

$$= \tfrac{4}{3}a^5 t\alpha^2(1+\alpha).$$

In the von Karman and Christensen theory the ratio of the coefficients is

$$\frac{F_{\omega\omega}}{J_p} = \frac{a^2}{12}(1-\alpha)^2$$

while in the Benscoter theory the ratio is

$$\frac{F^0_{\omega\omega}}{J_p} = \frac{a^2}{12}(1+\alpha)^2.$$

Thus it is seen that the greatest difference between the theories occurs when the cross-section is square ($\alpha = 1$). But this is just the value of α at which $\omega = 0$ at every point in the cross-section and hence there is no warping of the profile and the warping stresses are zero. A limited numerical study suggests that the differences between the two theories are not very great. In the following chapter the von Karman and Christensen theory is used as the basis for the analysis of beams and columns with closed profiles.

EXERCISES

2.1   Evaluate $\omega(s)$ at A and B and J_p and $F_{\omega\omega}$ for the channel illustrated. When the applied bimoment is $600\,\text{kN mm}^2$, what is the maximum longitudinal stress?

(Answer: $-337.5\,\text{mm}^2$, $562.5\,\text{mm}^2$, $320\,\text{mm}^4$, $14.175\times 10^6\,\text{mm}^6$, $23.8\,\text{MPa}$.)

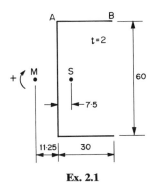

Ex. 2.1

2.2   Prove that for a single cell profile $\oint(p_B - \Psi/t)\,ds = 0$ i.e. that there cannot be any 'steps' in the value of ω. What is the physical interpretation of this equation? Extend the proof to a profile with more than one cell.

2.3   A hollow beam whose cross-section is an equilateral triangle of side 100 mm and wall thickness 10 mm is twisted by a uniform torque of 1000 kN mm. What is the shear stress? Evaluate Ψ and J_p and sketch $\omega(s)$. Can any general conclusions be drawn about the warping functions of such regular polygons?
(Answer: 11.547 MPa, 288.675 mm^2, 2 500 000, ω is zero everywhere.)

2.4   A thin cylinder of radius R and thickness t is slit longitudinally. Determining the following properties:
(a) Distance from longitudinal axis of cylinder to shear centre,
(b) J_p,
(c) $F_{\omega\omega}$.
(Answer: (a) $2R$; (b) $\frac{2}{3}\pi R t^3$; (c) $\pi R^5 t(\frac{2}{3}\pi^2 - 4)$.)

2.5   Locate the centroid and the shear centre of the cross-section illustrated and determine J_p and $F_{\omega\omega}$. What is the value of the longitudinal warping stress at the edges when $M_\omega = 100$ MN mm^2?
(Answer: $J_p = 4736$ mm^4, $F_{\omega\omega} = 20.417 \times 10^9$ mm^6, $\sigma_w = 47.03$ MPa.)

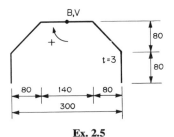

Ex. 2.5

2.6 Evaluate ω_1 and ω_2 and $F_{\omega\omega}$ for the cross-section illustrated (see also Exercise 1.6).

(Answer: -12.84, 66.23, $1.54 \times 10^6\,\mathrm{mm}^6$.)

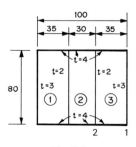

Ex. 2.6

2.7 Evaluate $F_{\omega\omega}$ and find the warping stresses at the points **1** to **4** in the crane runway illustrated when $M_\omega = 1\,\mathrm{kN\,m}^2$ and $M_{\mathrm{DS}} = 1\,\mathrm{kN\,mm}$.

(Answer: Coordinates of M (90, 154), ω_1 to $\omega_4 = 9300$, $10\,840$, $-14\,610$, $10\,910$.)

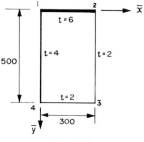

Ex. 2.7

3

APPLICATIONS OF THE THEORY OF THIN-WALLED STRUCTURES TO BENDING AND LATERAL BUCKLING OF BEAMS AND FLEXURAL TORSIONAL BUCKLING OF COLUMNS

3.1. INTRODUCTION

The purpose of this chapter is to show how the theory of thin-walled structures developed in the previous two chapters can be applied to practical problems. Here the emphasis is on analysis of given structures rather than on their design. The latter topic is treated in Chapter 7. Three types of problem are discussed in the present chapter. In Section 3.2 the twisting of thin-walled beams is analysed. It is seen in Chapter 2 that by choosing the reference axes correctly it is possible to uncouple the governing equations. Thus the bending of thin-walled beams can be treated by engineers' theory in the usual manner. Section 3.2 therefore deals only with the solution of the twisting equation which has the angle of rotation $\phi(z)$ as its unknown. It is shown that a closed-form solution, which depends upon the type of loading and the restraint conditions at the end of the beam, is obtainable. This solution is, in fact, the relationship between the given externally applied loads, torsions, and bimoments and the internal stress resultants (M_ω, M_{DS}, and M_{st}), and the angle of rotation ϕ. Several examples of beams with a single span showing how this closed-form solution may be used are presented in Section 3.2. After the internal stress resultants have been evaluated, an engineer is usually interested in obtaining the distribution of direct (warping) stresses and shear (both warping and Saint Venant) stresses. Section 3.3 shows how this may be done by using equations derived in Chapter 2.

When a thin-walled beam is continuous over two or more spans the relationship between given external loads, torsions and bimoments and the internal stress resultants at every point along the beam may be found by using a technique called *bimoment distribution*. It is shown in Section 3.4 that this technique is a parallel to moment distribution. Certain thin-walled columns are known to buckle by a combination of flexural and torsional buckling by the process of interaction of buckling modes (see Section 1.3.2). It is a small step to convert the governing equations for the bending and twisting of a beam into the equations which govern the buckling of a column. This is achieved in Section 3.5 by using a first-order stability analysis referred to in Section 1.3.1. It is found that the governing equations are coupled except in very special cases and this means that flexural and torsional buckling must occur simultaneously. An important result of this coupling is that the elastic critical load of such a column is below that predicted by the formulae commonly used by engineers. Some numerical examples illustrate the phenomenon of flexural–torsional buckling.

Another important application of the theory of thin-walled beams is to the study of lateral buckling of beams. Thin deep beams are often torsionally weak and when they are loaded they have a tendency to buckle sideways. This well-known phenomenon is studied in Section 3.6. The governing equations are set up by using an approach similar to that adopted in Section 3.5.

3.2. BENDING AND TWISTING OF PRISMATIC BEAMS

3.2.1. Bioments created by eccentric moments and loads

For the solution of practical beam problems there are two very useful theorems available in the book by Zbirohowski-Koscia (1967).

Theorem 3.2.1: 'A couple $Pd = M$ acting on a cross-section and in a plane parallel to the axis of the beam causes a bimoment M_ω whose value is equal to the product of M and the perpendicular distance from the shear centre to the plane of M.'

The meaning of this theorem is illustrated in Fig. 3.2.1 where it is seen that by adding two equal and opposing couples M_1 and M_2 at the shear

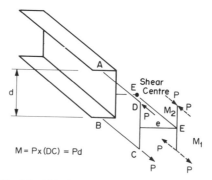

Fig. 3.2.1. A couple $Pd = M$ which does not act at the shear centre E can be replaced by one which does and by a bimoment Me.

centre the original couple M can be replaced by a couple $M_1\ (=M)$ at the shear centre and a bimoment Me. This theorem enables all of the applied couples to be related to the shear centre which becomes the origin of coordinates for torsion after the governing equations have been orthogonalized (see Section 2.4).

Theorem 3.2.2: 'An external force P acting parallel to the longitudinal axis of a beam causes a bimoment M_ω equal to the product of P and the sectorial coordinate ω_P of the point where P is applied.'

The meaning of this theorem is shown in Fig. 3.2.2. It can be proved by

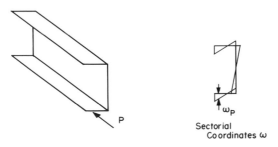

Fig. 3.2.2. A load P acting parallel to the axis of a beam causes a bimoment $P\omega_P$.

using the *principle of virtual displacements*. If two sets of loads are statically equivalent and an arbitrary displacement is imposed, the work done by one set of loads is equal to the work done on the other set. In this case the sets of 'loads' are the load P and the bimoment M_ω to which it is statically equivalent. In the discussion leading up to eqn (2.3.52) it is shown that the sectorial coordinate $\bar{\omega}_B$ is equal to the longitudinal displacement u when the angle of rotation ϕ' is -1 rad per unit length. When the centre of twist corresponds to the shear centre M, $\bar{\omega}_B$ is replaced by ω_M but the statement still holds true.

For convenience a virtual displacement is now introduced in the form of a twist of the beam of -1 rad per unit length and the principle of virtual displacements is applied. From eqn (2.5.10) the work done by the bimoment is $-M_\omega \phi'$, and the work done on the load P is Pu_P, i.e.

$$-M_\omega \phi' = Pu_P. \tag{3.2.1}$$

Hence

$$M_\omega = P\omega_{MP} \tag{3.2.2}$$

where ω_{MP} is the value of ω_M at the point of application of P. This proves the theorem.

In the next section the last of eqns (2.4.5), that is,

$$EF_{\omega\omega}\phi^{iv}(z) - GJ_p\phi''(z) = m_z(z) + \omega_M q_z'(z) - q_x(z)(y_q - y_M)$$
$$+ q_y(z)(x_q - x_M) \quad (3.2.3)$$

is solved. The first, third and fourth terms on the right are twisting moments per unit length. The second term is also a twisting moment per unit length as can be seen from eqns (3.2.2) and (2.4.40). This term is the warping torsion moment and it arises from the rate of change of bimoment which results from a non-uniform axial load q_z. Thus the right-hand side of eqn (3.2.3) can be calculated from the applied loads so it will be represented by a single symbol $m(z)$. Equation (3.2.3) is now written as

$$\phi^{iv}(z) - \lambda^2\phi''(z) = \frac{m(z)}{EF_{\omega\omega}} \quad (3.2.4)$$

where

$$\lambda^2 = \frac{GJ_p}{EF_{\omega\omega}}. \quad (3.2.5)$$

The right-hand side now physically means the torque per unit length and includes that which is externally applied and that arising from the rate of change of externally applied bimoment. $m(z)$ is also equal to the rate of change of the total torque, i.e.

$$m(z) = -\frac{dM_z(z)}{dz}. \quad (3.2.6)$$

3.2.2. General solution of eqn (3.2.4)

The general solution of eqn (3.2.4) is conveniently obtained by using Laplace transforms (Spiegel 1965). Let the transform of ϕ be Φ, i.e.

$$\Phi = L\phi \quad (3.2.7)$$

and eqn (3.2.4) becomes

$$p^4\Phi - p^3\phi_0 - p^2\phi_1 - p\phi_2 - \phi_3 - \lambda^2[p^2\Phi - p\phi_0 - \phi_1] = \frac{\bar{m}(p)}{EF_{\omega\omega}} \quad (3.2.8)$$

where ϕ_0, ϕ_1, ϕ_2 and ϕ_3 are the values of ϕ, $d\phi/dz$, $d^2\phi/dz^2$ and $d^3\phi/dz^3$, respectively, at $z = 0$ and $\bar{m}(p)$ is the Laplace transform of the torque $m(z)$ per unit length. Hence

$$\Phi = \phi_0\frac{1}{p} + \phi_1\frac{1}{p^2} + \phi_2\frac{1}{p(p^2-\lambda^2)} + \phi_3\frac{1}{p^2(p^2-\lambda^2)} + \frac{\bar{m}(p)}{p^2(p^2-\lambda^2)EF_{\omega\omega}}$$
$$(3.2.9)$$

The inverse transforms of the first two terms are obtained directly from tables in Spiegel's book (1965) and the next two terms are first written as partial fractions, for example

$$L^{-1}\left[\phi_2\frac{1}{p(p^2-\lambda^2)}\right]=L^{-1}\left\{\phi_2\left[-\frac{1}{\lambda^2p}+\frac{p}{\lambda^2(p^2-\lambda^2)}\right]\right\}$$

$$=-\frac{\phi_2}{\lambda^2}(1-\cosh\lambda z) \qquad (3.2.10)$$

and

$$L^{-1}\left[\phi_3\frac{1}{p^2(p^2-\lambda^2)}\right]=L^{-1}\left\{\phi_3\left[-\frac{1}{\lambda^2p^2}+\frac{1}{\lambda^2(p^2-\lambda^2)}\right]\right\}$$

$$=-\frac{\phi_3}{\lambda^3}(\lambda z-\sinh\lambda z). \qquad (3.2.11)$$

The inverse transform of the last term in eqn (3.2.9) is found by first expressing it as partial fractions

$$L^{-1}\left[\frac{\bar{m}(p)}{EF_{\omega\omega}}\frac{1}{p^2(p^2-\lambda^2)}\right]=L^{-1}\left\{\frac{\bar{m}(p)}{EF_{\omega\omega}}\left[-\frac{1}{\lambda^2p^2}+\frac{1}{\lambda^2(p^2-\lambda^2)}\right]\right\} \qquad (3.2.12)$$

and then using the Faltung theorem† (Spiegel 1965), thus,

$$L^{-1}\left[\frac{\bar{m}(p)}{EF_{\omega\omega}}\frac{1}{p^2(p^2-\lambda^2)}\right]=\frac{1}{\lambda^2EF_{\omega\omega}}\left[-\int_o^z(z-t)m(t)\,\mathrm{d}t\right.$$

$$\left.+\int_o^z\frac{m(t)}{\lambda}\sinh\lambda(z-t)\,\mathrm{d}t\right]$$

$$=-\frac{1}{\lambda GJ_p}\int_o^z[\lambda(z-t)$$

$$-\sinh\lambda(z-t)]m(t)\,\mathrm{d}t. \qquad (3.2.13)$$

The general solution of eqn (3.2.4) is now obtained by adding these results. Hence

$$\phi=\phi_0+\phi_1z-\frac{\phi_2}{\lambda^2}(1-\cosh\lambda z)-\frac{\phi_3}{\lambda^3}(\lambda z-\sinh\lambda z)$$

$$-\frac{1}{\lambda GJ_p}\int_o^z[\lambda(z-t)-\sinh\lambda(z-t)]m(t)\,\mathrm{d}t. \qquad (3.2.14)$$

This solution can be written in various alternative forms by using the

† This theorem is sometimes called the convolution theorem. It enables the inverse transformation of products of functions to be obtained.

results listed after eqn (2.4.38). Thus

$$\phi = \phi_0 + \frac{M_{st}(0)}{\lambda GJ_p}\lambda z + \frac{M_{DS}(0)}{\lambda GJ_p}(\lambda z - \sinh \lambda z) + \frac{M_{\omega_M}(0)}{GJ_p}(1 - \cosh \lambda z)$$

$$- \frac{1}{\lambda GJ_p}\int_0^z [\lambda(z-t) - \sinh \lambda(z-t)]m(t)\, dt \qquad (3.2.15)$$

or

$$\phi = \phi_0 + \phi_1 \frac{\sinh \lambda z}{\lambda} + \frac{M_z(0)}{\lambda GJ_p}(\lambda z - \sinh \lambda z)$$

$$+ \frac{M_{\omega_M}(0)}{GJ_p}(1 - \cosh \lambda z) - \frac{1}{\lambda GJ_p}\int_0^z [\lambda(z-t) - \sinh \lambda(z-t)]m(t)\, dt.$$

$$(3.2.16)$$

In these last two equations

$$M_z(0) = M_{st}(0) + M_{DS}(0) \qquad (3.2.17)$$

is the total torque applied at the end $z = 0$ and $M_{st}(0)$ and $M_{DS}(0)$ are the Saint Venant and warping torsion moments at that point.

The bimoment in a beam is obtained from eqns (2.4.38) and (3.2.16)

$$M_{\omega_M}(z) = -EF_{\omega\omega}\phi''(z) = -\frac{\phi_1 GJ_p}{\lambda}\sinh \lambda z$$

$$+ \frac{M_z(0)}{\lambda}\sinh \lambda z + M_{\omega_M}(0)\cosh \lambda z$$

$$- \frac{1}{\lambda}\int_0^z [\sinh \lambda(z-t)]\, m(t)\, dt \qquad (3.2.18)$$

where there have been two differentiations under the integral sign.†

The warping torsion moment is obtained from eqns (2.4.40) and

---

† The Leibnitz formula for differentiation under the integral sign is found in Sokolnikoff's book (1941). If

$$\Phi(\alpha) = \int_{u_0(\alpha)}^{u_1(\alpha)} f(x, \alpha)\, dx$$

then

$$\frac{d\Phi}{d\alpha} = f(u_1, \alpha)\frac{du_1}{d\alpha} - f(u_0, \alpha)\frac{du_0}{d\alpha} + \int_{u_0(\alpha)}^{u_1(\alpha)} \frac{\partial f(x, \alpha)}{\partial \alpha}\, dx.$$

It assumes continuity of all derivatives involved. It is a technique which is used when the indefinite integral is complicated or cannot be written down explicitly.

(3.2.18)

$$M_{DS}(z) = M'_{\omega_M}(z) = -\phi_1 GJ_p \cosh \lambda z$$
$$+ M_z(0) \cosh \lambda z + M_{\omega_M}(0)\lambda \sinh \lambda z$$
$$- \int_0^z [\cosh \lambda(z-t)]m(t)\,\mathrm{d}t \qquad\qquad (3.2.19)$$

and the Saint Venant torsion moment is

$$M_{st}(z) = GJ_p\phi'(z) = GJ_p\phi_1 \cosh \lambda z + M_z(0)(1 - \cosh \lambda z)$$
$$- M_\omega(0)\lambda \sinh \lambda z - \int_0^z [1 - \cosh \lambda(z-t)]m(t)\,\mathrm{d}t. \qquad (3.2.20)$$

As a check we see that by adding eqns (3.2.19) and (3.2.20) the total

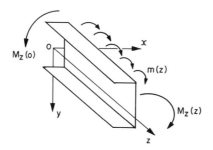

Fig. 3.2.3. For overall rotational equilibrium $M_z(z) + \int_0^z m(t)\,\mathrm{d}t = M_z(0)$.

moment acting at z is (see also Fig. 3.2.3)

$$M_z(z) = M_{st}(z) + M_{DS}(z) = M_z(0) - \int_0^z m(t)\,\mathrm{d}t \qquad (3.2.21)$$

3.2.3. Examples of warping torsion of beams of single span

The application of the solution (3.2.15) to (3.2.21) to practical problems is now straightforward. Boundary conditions can be applied in various ways as the following examples show.

3.2.3.1. Free–free beam with bimoment $M_{\omega 1}$ at one end

A beam with fork supports (Fig. 3.2.4) at each end is referred to as free–free because although the ends prevent rotation they do not impose longitudinal stresses and hence there are no bimoments at the supports.

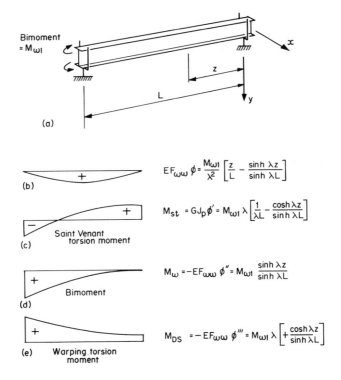

$$EF_{\omega\omega}\,\phi = \frac{M_{\omega 1}}{\lambda^2}\left[\frac{z}{L} - \frac{\sinh \lambda z}{\sinh \lambda L}\right]$$

$$M_{st} = GJ_p\phi' = M_{\omega 1}\,\lambda\left[\frac{1}{\lambda L} - \frac{\cosh \lambda z}{\sinh \lambda L}\right]$$

$$M_\omega = -EF_{\omega\omega}\,\phi'' = M_{\omega 1}\frac{\sinh \lambda z}{\sinh \lambda L}$$

$$M_{DS} = -EF_{\omega\omega}\,\phi''' = M_{\omega 1}\,\lambda\left[+\frac{\cosh \lambda z}{\sinh \lambda L}\right]$$

Fig. 3.2.4. Simply supported beam with bimoment applied at end.

In this problem therefore $m = 0$ and the boundary conditions are

$$\begin{aligned}
\text{at } z = 0 \qquad &\phi = M_\omega = 0 \\
\text{at } z = L \qquad &\phi = 0 \quad \text{and} \quad M_\omega = M_{\omega 1}
\end{aligned} \tag{3.2.22}$$

with $\phi_0 = M_\omega(0) = 0$ eqn (3.2.16) is satisfied. Substitution of conditions at $z = L$ in eqns (3.2.16) and (3.2.18) yield the following simultaneous equations

$$\begin{aligned}
\phi_1\frac{\sinh \lambda L}{\lambda} + \frac{M_z(0)}{\lambda GJ_p}(\lambda L - \sinh \lambda L) &= 0 \\
-\frac{\phi_1 GJ_p \sinh \lambda L}{\lambda} + \frac{M_z(0)}{\lambda}\sinh \lambda L &= M_{\omega 1}
\end{aligned} \tag{3.2.23}$$

whence

$$\phi_1 = -\frac{M_{\omega 1}(\lambda L - \sinh \lambda L)}{LGJ_p \sinh \lambda L} \quad \text{and} \quad M_z(0) = M_{\omega 1}/L. \tag{3.2.24}$$

Substitution in eqns (3.2.16), (3.2.18)–(3.2.21) gives

$$\phi = \frac{M_{\omega 1}}{GJ_p}\left[\frac{z}{L} - \frac{\sinh \lambda z}{\sinh \lambda L}\right]$$

$$M_{\omega_M}(z) = M_{\omega 1}\frac{\sinh \lambda z}{\sinh \lambda L}$$

$$M_{DS}(z) = M_{\omega 1}\lambda\frac{\cosh \lambda z}{\sinh \lambda L} \tag{3.2.25}$$

$$M_{st}(z) = M_{\omega 1}\lambda\left[\frac{1}{\lambda L} - \frac{\cosh \lambda z}{\sinh \lambda L}\right]$$

$$M_z(z) = M_{\omega 1}/L$$

These expressions are plotted in Fig. 3.2.4. It is left as an exercise for the reader to sketch the deformed shape and to verify that the signs are correct according to the convention established in Section 2.4.4.

3.2.3.2. Free–free beam with uniform twisting moment per unit length (Fig. 3.2.5)

The boundary conditions are

at $z = 0$ and L, $\phi = M_\omega = 0$. $\qquad$ (3.2.26)

It is obvious from symmetry that

$$M_z(0) = \frac{m_d L}{2}. \tag{3.2.27}$$

Substitution of these conditions in eqn (3.2.18) leads to

$$\phi_1 = \frac{m_d}{\lambda GJ_p}\left[\frac{\lambda L}{2} + \frac{1 - \cosh \lambda L}{\sinh \lambda L}\right] \tag{3.2.28}$$

and hence from eqns (3.2.16), (3.2.18)–(3.2.21)

$$\phi = \frac{m_d}{\lambda^2 GJ_p}\left[\frac{\lambda^2}{2}(Lz - z^2) - 1 + \frac{\sinh \lambda z + \sinh \lambda(L-z)}{\sinh \lambda L}\right]$$

$$M_{\omega_M}(z) = \frac{m_d}{\lambda^2}\left[1 - \frac{\sinh \lambda z + \sinh \lambda(L-z)}{\sinh \lambda L}\right]$$

$$M_{DS}(z) = -\frac{m_d}{\lambda}\left[\frac{\cosh \lambda z - \cosh \lambda(L-z)}{\sinh \lambda L}\right] \tag{3.2.29}$$

$$M_{st}(z) = \frac{m_d}{\lambda}\left[\lambda\left(\frac{L}{2} - z\right) + \frac{\cosh \lambda z - \cosh \lambda(L-z)}{\sinh \lambda L}\right]$$

$$M_z(z) = m_d\left(\frac{L}{2} - z\right).$$

These results are plotted in Fig. 3.2.5.

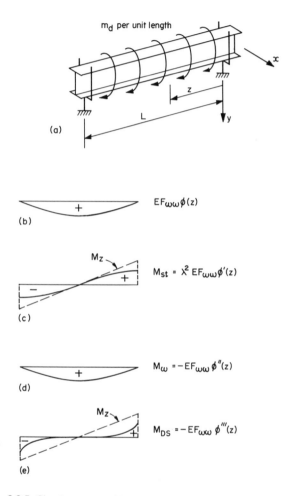

Fig. 3.2.5. Simply supported beam with uniform torque per unit length.

3.2.3.3. *Bimoment at end of free–free beam which is also allowed to rotate at $z = 0$ (Fig. 3.2.6)*

The boundary conditions are

at $z = 0$ $M_\omega(0) = M_z(0) = 0$

at $z = L$ $M_\omega = M_{\omega 1}, \phi = 0.$ (3.2.30)

Because there is no externally applied torque, i.e. $m(z) = 0$, it is obvious that at $z = L$, $M_z(z) = 0$. Also since the only external moments arise from the bimoment $M_{\omega 1}$ and they balance, no reactions of any kind occur at $z = 0$ and L.

Therefore from eqn (3.2.18)

$$\phi_1 = -\frac{\lambda M_{\omega 1}}{GJ_p \sinh \lambda L} \tag{3.2.31}$$

and from eqn (3.2.16)

$$\phi_0 = \frac{M_{\omega 1}}{GJ_p}. \tag{3.2.32}$$

Hence

$$\phi = \frac{M_{\omega 1}}{GJ_p}\left[1 - \frac{\sinh \lambda z}{\sinh \lambda L}\right] \qquad M_\omega(z) = M_{\omega 1}\frac{\sinh \lambda z}{\sinh \lambda L}$$

$$M_{DS}(z) = M_{\omega 1}\lambda\frac{\cosh \lambda z}{\sinh \lambda L} \qquad M_{st}(z) = -M_{\omega 1}\lambda\frac{\cosh \lambda z}{\sinh \lambda L} \tag{3.2.33}$$

$$M_z(z) = 0.$$

These results are plotted in Fig. 3.2.6. It is interesting to note that the angle of rotation at $z = 0$ is independent of the length L of the beam. As well as being an interesting violation of Saint Venant's principle it has important ramifications for practising engineers.

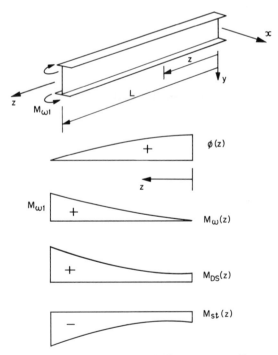

Fig. 3.2.6. Free beam with bimoment at $z = L$.

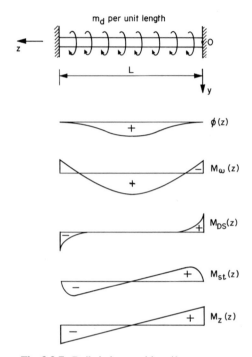

Fig. 3.2.7. Built-in beam with uniform moment.

3.2.3.4. *Built-in beam with uniform moment (Fig. 3.2.7)*

In this case $\phi_0 = \phi_1 = 0$ and $M_z(0) = m_d L/2$. Furthermore at $z = L$, $\phi = 0$, whence

$$M_\omega(0) = \frac{m_d}{\lambda^2}\left[\frac{\lambda L \sinh \lambda L}{2(1 - \cosh \lambda L)} + 1\right] \tag{3.2.34}$$

From eqns (3.2.16)–(3.2.21) we now obtain

$$\phi = \frac{m_d}{2\lambda^2 GJ_p}\left[\lambda^2(Lz - z^2) - \lambda L \sinh \lambda z + \frac{\lambda L \sinh \lambda L(1 - \cosh \lambda z)}{1 - \cosh \lambda L}\right]$$

$$M_\omega(z) = \frac{m_d}{\lambda^2}\left[\frac{\lambda L}{2}\sinh \lambda z + \frac{\lambda L \sinh \lambda L \cosh \lambda z}{2(1 - \cosh \lambda L)} + 1\right]$$

$$M_{DS}(z) = \frac{m_d L}{2}\left[\cosh \lambda z + \frac{\sinh \lambda L \sinh \lambda z}{1 - \cosh \lambda L}\right] \tag{3.2.35}$$

$$M_{st}(z) = m_d\left[\frac{L}{2} - z - \frac{L}{2}\cosh \lambda z - \frac{L \sinh \lambda L \sinh \lambda z}{2(1 - \cosh \lambda L)}\right]$$

$$M_z(z) = m_d\left(\frac{L}{2} - z\right).$$

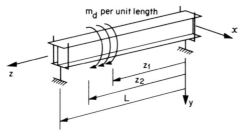

Fig. 3.2.8. Free–free beam with distributed torque over part of its span.

3.2.3.5. *Free–free beam with uniformly distributed moment m_d over part of its length (Fig. 3.2.8)*

In this case the ends of the beams are supported in the same manner as those illustrated in Fig. 3.2.4. The end conditions are

at $z = 0$ and L, $\phi = M_\omega = 0$.

Since the section properties of the beam are uniform along its length the end reactions $M_z(L)$ and $M_z(0)$ are found as if the moment $m_d(z_2 - z_1)$ were concentrated at the point $z = (z_2 + z_1)/2$.

$$M_z(0) = m_d(z_2 - z_1) \frac{(2L - z_2 - z_1)}{2L}. \tag{3.2.36}$$

From eqn (3.2.18) at $z = L$

$$(-\phi_1 GJ_p + M_z(0)) \frac{\sinh \lambda L}{\lambda} = -\frac{m_d}{\lambda^2} [\cosh \lambda(L - t)]_{z_1}^{z_2} \tag{3.2.37}$$

whence

$$-\phi_1 GJ_p + M_z(0) = -\frac{m_d}{\lambda \sinh \lambda L} [\cosh \lambda(L - z_2) - \cosh \lambda(L - z_1)] \tag{3.2.38}$$

and hence

$$M_\omega(z) = -\frac{m_d}{\lambda^2 \sinh \lambda L} [\cosh \lambda(L - z_2) - \cosh \lambda(L - z_1)] \sinh \lambda z$$

$$-\frac{1}{\lambda} \int_0^z m(t) \sinh \lambda(z - t) \, dt$$

$$= -\frac{m_d}{\lambda^2 \sinh \lambda L} [\cosh \lambda(L - z_2) - \cosh \lambda(L - z_1)] \sinh \lambda z$$

$$+ 0 \qquad\qquad\qquad\qquad \text{if} \quad 0 \leqslant z \leqslant z, \tag{3.2.39}$$

$$+ \frac{m_d}{\lambda^2} [1 - \cosh \lambda(z - z_1)] \qquad\qquad \text{if} \quad z_1 \leqslant z \leqslant z_2,$$

$$+ \frac{m_d}{\lambda^2} [\cosh \lambda(z - z_2) - \cosh \lambda(z - z_1)] \quad \text{if} \quad z_2 \leqslant z \leqslant L.$$

The warping torsion moment M_{DS} is obtained by differentiating eqn (3.2.39)

$$M_{DS}(z) = -\frac{m_d}{\lambda \sinh \lambda L} [\cosh \lambda (L - z_2) - \cosh \lambda (L - z_1)] \cosh \lambda z$$

$$+0 \qquad\qquad\qquad\qquad \text{if} \quad 0 \leq z \leq z_1,$$

$$-\frac{m_d}{\lambda} \sinh \lambda (z - z_1) \qquad\qquad \text{if} \quad z_1 \leq z \leq z_2, \qquad (3.2.40)$$

$$+\frac{m_d}{\lambda} [\sinh \lambda (z - z_2) - \sinh \lambda (z - z_1)] \quad \text{if} \quad z_2 \leq z \leq L.$$

From eqns (3.2.38) and (3.2.36)

$$\phi_1 = \frac{m_d}{GJ_p} \left[(z_2 - z_1)(2L - z_2 - z_1)/2L + \frac{\cosh \lambda (L - z_2) - \cosh \lambda (L - z_1)}{\lambda \sinh \lambda L} \right] \tag{3.2.41}$$

and from eqn (3.2.16)

$$\phi = \phi_1 \frac{\sinh \lambda z}{\lambda} + \frac{m_d(z_2 - z_1)(2L - z_2 - z_1)(\lambda z - \sinh \lambda z)}{2\lambda L G J_p}$$

$$+0 \qquad\qquad\qquad\qquad\qquad \text{if} \quad 0 \leq z \leq z_1,$$

$$-\frac{m_d}{\lambda^2 GJ_p} \left[\frac{\lambda^2}{2} (z - z_1)^2 + 1 - \cosh \lambda (z - z_1) \right] \quad \text{if} \quad z_1 \leq z \leq z_2,$$

$$-\frac{m_d}{\lambda^2 GJ_p} \left[\lambda^2 \left(z z_2 - z z_1 - \frac{z_2^2}{2} + \frac{z_1^2}{2} \right) + \cosh \lambda (z - z_2) - \cosh \lambda (z - z_1) \right]$$

$$\qquad\qquad\qquad\qquad\qquad \text{if} \quad z_2 \leq z \leq L. \tag{3.2.42}$$

The Saint Venant torsion moment is

$$M_{st}(z) = GJ_p \phi'$$

$$= \phi_1 GJ_p \cosh \lambda z + \frac{m_d(z_2 - z_1)(2L - z_2 - z_1)(1 - \cosh \lambda z)}{2L}$$

$$+0 \qquad\qquad\qquad\qquad\qquad \text{if} \quad 0 \leq z \leq z_1,$$

$$-\frac{m_d}{\lambda} [\lambda(z - z_1) - \sinh \lambda (z - z_1)] \qquad\qquad \text{if} \quad z_1 \leq z \leq z_2,$$

$$-\frac{m_d}{\lambda} [\lambda(z_2 - z_1) + \sinh \lambda (z - z_2) - \sinh \lambda (z - z_1)] \quad \text{if} \quad z_2 \leq z \leq L.$$

$$\tag{3.2.43}$$

Finally

$$M_z(z) = m_d(z_2 - z_1)\frac{(2L - z_2 - z_1)}{2L}$$

$$
\begin{aligned}
&+0 && \text{if} \quad 0 \leqslant z \leqslant z_1, \\
&-m_d(z - z_1) && \text{if} \quad z_1 \leqslant z \leqslant z_2, \\
&-m_d(z_2 - z_1) && \text{if} \quad z_2 \leqslant z < L.
\end{aligned}
\tag{3.2.44}
$$

The expressions given in eqn (3.2.29) can be checked by letting $z_1 = 0$ and $z_2 = L$ in eqns (3.2.40)–(3.2.44).

3.2.3.6. Free–free beam with concentrated torque M_1 at $z = z_1$ (Fig. 3.2.9)

A concentrated torque M_1 acting at $z = z_1$ can be treated as a uniformly distributed torque m_d acting between $z = z_1$ and z_2 as shown in Fig. 3.2.8 but $z_2 - z_1 = \Delta z$ is allowed to become very small and

$$m_d = M_1/\Delta z. \tag{3.2.45}$$

The integration terms in eqns (3.2.16)–(3.2.21) are treated by using the mean value theorem (Sokolnikoff 1941)

$$\int_{z_1}^{z_1 + \Delta z} f(z)\, dz = \Delta z \cdot f(z_1). \tag{3.2.46}$$

Since $f(z)$ is proportional to m_d, i.e.

$$f(z) = m_d \cdot f_1(z) \tag{3.2.47}$$

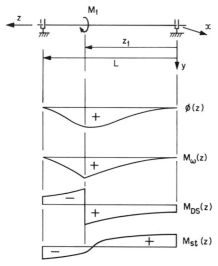

Fig. 3.2.9. Free–free beam with concentrated torque M_1 at $z = z_1$.

the integration terms can be written in the form

$$\int_{z_1}^{z_1+\Delta z} f(z)\,dz = M_1 f_1(z_1).$$ (3.2.48)

Because of the uniform section properties and the conditions of support

$$\text{at } z = 0, \qquad \phi = M_\omega(0) = 0 \quad \text{and} \quad M_z(0) = M_1 \frac{(L-z_1)}{L}$$

(3.2.49)

$$\text{at } z = L, \qquad \phi = M_\omega(L) = 0 \quad \text{and} \quad M_z(L) = -M_1 \frac{z_1}{L}.$$

From eqn (3.2.16)

$$0 = \phi_1 \frac{\sinh \lambda L}{\lambda} + \frac{M_1(L-z_1)}{\lambda L G J_p}(\lambda L - \sinh \lambda L)$$

$$- \frac{M_1}{\lambda G J_p}[\lambda(L-z_1) - \sinh \lambda(L-z_1)]$$

whence

$$\phi_1 = \frac{M_1}{G J_p}\left[\frac{L-z_1}{L} - \frac{\sinh \lambda(L-z_1)}{\sinh \lambda L}\right].$$ (3.2.50)

From eqns (3.2.50) and (3.2.16)

$$\phi = \frac{M_1}{\lambda G J_p}\left[z\frac{L-z_1}{L} - \frac{\sinh \lambda(L-z_1)\sinh \lambda z}{\sinh \lambda L}\right]$$

$$+0 \qquad\qquad\qquad 0 \leqslant z \leqslant z_1$$ (3.2.51)

$$- \frac{M_1}{\lambda G J_p}[\lambda(z-z_1) - \sinh \lambda(z-z_1)] \qquad z_1 \leqslant z \leqslant L.$$

From eqns (3.2.50), (3.2.49) and (3.2.18)

$$M_\omega(z) = \frac{M_1 \sinh \lambda(L-z_1)\sinh \lambda z}{\lambda \sinh \lambda L}$$

$$+0 \qquad\qquad \text{if} \quad 0 \leqslant z \leqslant z_1$$ (3.2.52)

$$- \frac{M_1}{\lambda}\sinh \lambda(z-z_1) \quad \text{if} \quad z_1 \leqslant z \leqslant L.$$

$$M_{DS} = M_\omega'(z)$$

$$= \frac{M_1 \sinh \lambda(L-z_1)\cosh \lambda z}{\sinh \lambda L}$$

$$+0 \qquad\qquad\qquad \text{if} \quad 0 \leqslant z \leqslant z_1$$ (3.2.53)

$$-M_1 \cosh \lambda(z-z_1) \quad \text{if} \quad z_1 \leqslant z \leqslant L.$$

From eqns (3.2.20), (3.2.50) and (3.2.49)

$$M_{st} = M_1 \left[\frac{L - z_1}{L} - \frac{\sinh \lambda (L - z_1) \cosh \lambda z}{\sinh \lambda L} \right]$$

$$+0 \qquad\qquad\qquad \text{if} \quad 0 \leq z \leq z_1 \qquad\qquad (3.2.54)$$

$$-M_1[1 - \cosh \lambda (z - z_1)] \quad \text{if} \quad z_1 \leq z \leq L$$

and from eqns (3.2.21), (3.2.53) and (3.2.54)

$$M_z(z) = M_1 \frac{L - z_1}{L} (= M_z(0)) \quad \text{if} \quad 0 \leq z \leq z_1$$

$$\text{or} \ -\frac{M_1 z_1}{L} (= M_z(L)) \quad \text{if} \quad z_1 \leq z \leq L \qquad\qquad (3.2.55)$$

(which checks).

3.2.3.7. Miscellaneous solutions for single-span beams

In this section a few solutions are quoted without the associated working. The solutions can be obtained in the same manner as those presented above. Further solutions for the bimoments at each end of a beam are given later in Table 3.4.1. The beam shown in Fig. 3.2.10(a) is fixed at

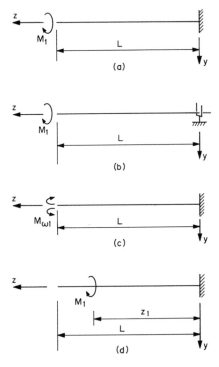

Fig. 3.2.10. Some single-span beams (see eqns 3.2.56–3.2.59 for results).

one end and free to warp and rotate at the other where a torque M_1 is applied.

$$\phi = \frac{M_1}{\lambda GJ_p}\left[\lambda z - \frac{\sinh \lambda L - \sinh \lambda(L-z)}{\cosh \lambda L}\right]$$

$$M_\omega(z) = -\frac{M_1 \sinh \lambda(L-z)}{\lambda \cosh \lambda L}$$

$$M_{DS}(z) = \frac{M_1 \cosh \lambda(L-z)}{\cosh \lambda L} \tag{3.2.56}$$

$$M_{st}(z) = M_1\left[1 - \frac{\cosh \lambda(L-z)}{\cosh \lambda L}\right]$$

$$M_z(z) = M_1.$$

The beam shown in Fig. 3.2.10(b) is allowed to warp at the right-hand end so it only experiences Saint Venant torsion.

$$\phi = \frac{Mz}{GJ_p}$$

$$M_\omega(z) = 0$$

$$M_{DS}(z) = 0 \tag{3.2.57}$$

$$M_{st}(z) = M_1$$

$$M_z(z) = M_1$$

The beam shown in Fig. 3.2.10(c) is built-in at $z = 0$ and is free to rotate and warp at $z = L$ where a bimoment $M_{\omega 1}$ is applied.

$$\phi = \frac{M_{\omega 1}(1 - \cos \lambda z)}{GJ_p \cosh \lambda L}$$

$$M_\omega(z) = M_{\omega 1}\frac{\cosh \lambda z}{\cosh \lambda L}$$

$$M_{DS}(z) = M_{\omega 1}\frac{\lambda \sinh \lambda z}{\cosh \lambda L} \tag{3.2.58}$$

$$M_{st}(z) = -M_{\omega 1}\frac{\lambda \sinh \lambda z}{\cosh \lambda L}$$

$$M_z(z) = 0$$

The beam shown in Fig. 3.2.10(d) is built-in at $z = 0$, carries a torque

M_1 at $z = z_1$ and is free at $z = L$.

$$\phi = \frac{M_1}{\lambda GJ_p}\left[\lambda z - \frac{\sinh \lambda z \cosh \lambda L - (1 - \cosh \lambda z)[\sinh \lambda (L - z_1) - \sinh \lambda L]}{\cosh \lambda L}\right]$$

$$+0 \qquad\qquad\qquad \text{if} \quad 0 \leqslant z \leqslant z_1,$$

$$-\frac{M_1}{\lambda GJ_p}[\lambda(z - z_1) - \sinh \lambda(z - z_1)] \quad \text{if} \quad z_1 \leqslant z \leqslant L.$$

$$M_\omega(z) = \frac{M_1}{\lambda}\left[\sinh \lambda z + \frac{(\sinh \lambda(L - z_1) - \sinh \lambda L)\cosh \lambda z}{\cosh \lambda L}\right]$$

$$+0 \qquad\qquad\qquad \text{if} \quad 0 \leqslant z \leqslant z_1,$$

$$-\frac{M_1}{\lambda}\sinh \lambda(z - z_1) \quad \text{if} \quad z_1 \leqslant z \leqslant L.$$

$$M_{DS}(z) = M_1\left[\cosh \lambda z + \frac{(\sinh \lambda(L - z_1) - \sinh \lambda L)\sinh \lambda z}{\cosh \lambda L}\right]$$

$$+0 \qquad\qquad\qquad \text{if} \quad 0 \leqslant z \leqslant z_1,$$

$$-M_1 \cosh \lambda(z - z_1) \quad \text{if} \quad z_1 \leqslant z \leqslant L.$$

$$M_{st}(z) = M_1\left[1 - \cosh \lambda z - \frac{(\sinh \lambda(L - z_1) - \sinh \lambda L)\sinh \lambda z}{\cosh \lambda L}\right]$$

$$+0 \qquad\qquad\qquad \text{if} \quad 0 \leqslant z \leqslant z_1,$$

$$-M_1[1 - \cosh \lambda(z - z_1)] \quad \text{if} \quad z_1 \leqslant z \leqslant L.$$

$$M_z(z) = \begin{cases} M_1 & \text{if} \quad 0 \leqslant z \leqslant z_1, \\ 0 & \text{if} \quad z_1 \leqslant z \leqslant L. \end{cases} \qquad (3.2.59)$$

In this case it is seen that if $z_1 \leqslant z \leqslant L$ the total torque is zero but that there are two sets of shear stresses, distributed through the profile in quite different ways, but they have equal and opposite stress resultants. For example, in an open profile the shear stresses arising from $M_{st}(z)$ are proportional to the distance from the middle plane of each of the plates in a profile whereas those arising from M_{DS} are uniform through the thickness of each plate.

Some of the problems treated in this section can also be solved conveniently by using Dirac δ-functions and Heaviside step-functions.

3.3. STRESSES IN PRISMATIC BEAMS DUE TO TWISTING AND WARPING

In the previous section the angle of rotation and the internal forces in beams were related to the externally applied loads and the section properties of the beam. In this section the distribution of normal and shear stresses is now derived from the internal forces and the geometry of the beam, i.e. its cross-section, its span and its end conditions. In Chapter 2 it was shown how the section properties of a given cross-section may be found and it is not proposed to repeat those calculations here.

Example 3.3.1

An I-beam (Fig. 3.3.1(a)) spans 1000 mm and is built-in at each end and carries a uniform torque of m N mm mm^{-1}. Determine the distribution of shear and direct stresses acting on the cross-section at each end and at the point $z = 250$ mm. $E = 206\,000$ MPa and $G = 82\,400$ MPa.

Section properties

$$F = 320 \text{ mm}^2$$

$$J_p = \sum \frac{at^3}{3} = \frac{160 \times 2^3}{3} = 426.7 \text{ mm}^4.$$

For ω see Fig. 3.3.1(b)

$$F_{\omega\omega} = \int \omega^2 \, dF = \frac{4 \times 20 \times 800 \times 800 \times 2}{3} = 34.133 \times 10^6 \text{ mm}^6.$$

$$\lambda = \sqrt{\left(\frac{GJ_p}{EF_{\omega\omega}}\right)} = \left(\frac{0.4 \times 426.7}{34.133 \times 10^6}\right)^{\frac{1}{2}} = 2.236\,16 \times 10^{-3} \text{ mm}^{-1}$$

$$\lambda L = 2.236\,16; \qquad \sinh \lambda L = 4.625\,209; \qquad \cosh \lambda L = 4.732\,077\,5.$$

Internal forces

From eqns (3.2.34) and (3.2.35)

(a) when $z = 0$

$$M_\omega(0) = \frac{mL^2}{(\lambda L)^2} \left[\frac{\lambda L \sinh \lambda L}{2(1 - \cosh \lambda L)} + 1\right]$$

$$= -\frac{m \times 10^6}{2.236\,16^2} \left[\frac{2.236\,16 \times 4.625\,209}{2 \times 3.732\,077\,5} - 1\right]$$

$$= -0.077\,124 m \times 10^6 \text{ N mm}^2.$$

$$M_{DS}(0) = \frac{mL}{2} = 500m \text{ N mm.}$$

$$M_{st}(0) = 0.$$

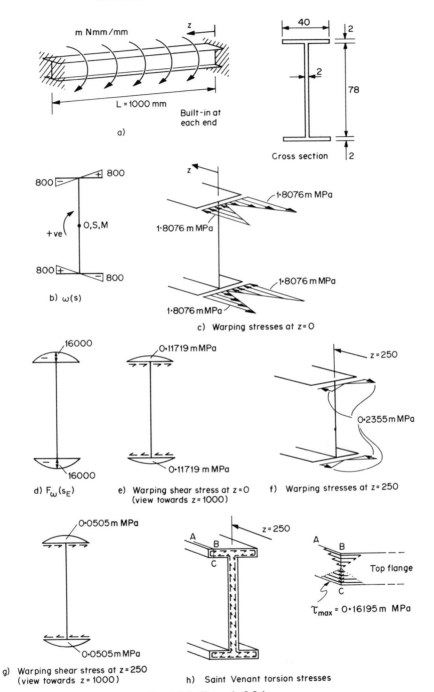

Fig. 3.3.1. Example 3.3.1.

(b) when $z = 250$

$\cosh 250\lambda = 1.160\,375;$ $\sinh 250\lambda = 0.588\,617$

$$M_\omega(250) = \frac{m}{2.236\,16^2 \times 10^{-6}} \left[\frac{2.236\,16}{2} \times 0.588\,617 \right.$$

$$\left. - \frac{2.236\,16 \times 4.625\,209 \times 1.160\,375}{2 \times 3.732\,077\,5} + 1 \right]$$

$$= 10\,049m \text{ N mm}^2.$$

$$M_{DS}(250) = m \times 500 \left(1.160\,375 - \frac{4.625\,209 \times 0.588\,617}{3.732\,077\,5} \right)$$

$$= 215.45m \text{ N mm}.$$

$$M_{st}(250) = m[500 - 250 - (500 \times 1.160\,375)$$

$$+ \frac{1000 \times 4.625\,209 \times 0.588\,617}{2 \times 3.732\,077\,5} \Big]$$

$$= 34.533m \text{ N mm}.$$

$$M_z(250) = M_{DS} + M_{st} = 250m \text{ N mm}.$$

(which checks).

Stresses

(a) when $z = 0$

At the tips of the flanges the longitudinal stress

$$= \sigma_w = \frac{M_\omega(0)\omega(s)}{F_{\omega\omega}} = \frac{0.077\,124\,m \times 10^6 \times 800}{34.133 \times 10^6} = 1.8076m \text{ MPa}.$$

The longitudinal stress away from the tips of the flanges follows the same pattern as $\omega(s)$ (see Fig. 3.3.1(c)).

The shear stress is found from the last term in eqn (2.4.45). The distribution of $F_\omega(s_E) = \int_0^{s_E} \omega(s)\,dF$ is shown in Fig. 3.3.1(d).

At the web-to-flange junction where $F_\omega(s) = 16\,000 \text{ mm}^4$ the shear stress is

$$\frac{M_{DS}F_\omega}{tF_{\omega\omega}} = \frac{500m \times 16\,000}{2 \times 34.133 \times 10^6} = 0.117\,19m \text{ MPa}$$

The Saint Venant shear stresses are zero at each end of the beam. Figure 3.3.1(e) shows the distribution of shear stresses arising from the warping torsion moment. They are uniformly distributed across the thickness of the flanges.

(b) when $z = 250$

At the tips of the flanges the longitudinal stress

$$= \sigma_w = \frac{10\,049m \times 800}{34.133 \times 10^6} = 0.2355m \text{ MPa}$$

The distribution of longitudinal stresses is shown in Fig. 3.3.1(f). The shear stress at the flange-to-web junction and due to M_{DS} is (see Fig. 3.3.1(g))

$$\frac{215.45m \times 16\,000}{2 \times 34.133 \times 10^6} = 0.0505m \text{ MPa}$$

The Saint Venant stresses are zero along the centreline of the plates forming the cross-section and have a maximum value on the outer skin. From eqns (1.4.17) and (1.4.19)

$$\tau_{max} = M_{st} \frac{t}{J_p} = \frac{34.553m \times 2}{426.7} = 0.161\,95m \text{ MPa}$$

The distribution of these stresses is shown in Fig. 3.3.1(h). It is linear through the plate thicknesses and there are, of course, complementary shear stresses acting in the longitudinal direction.

Example 3.3.2

The box-girder with three cells illustrated in Fig. 3.3.2(a) carries a uniformly distributed line load of 1 N mm^{-1} over a built-in span of 2000 mm. Determine the direct and shear stresses in the web at the point H, due to the transverse bending moment and shear force and the warping torsion. The following data are given: $F_{yy} = 605\,223 \text{ mm}^4$, $F_{\omega\omega} = 44.9653 \times 10^6 \text{ mm}^6$, $J_p = 1.2393 \times 10^6 \text{ mm}^4$, $\Psi_1 = \Psi_3 = 114.11 \text{ mm}^2$, $\Psi_2 = 138.65 \text{ mm}^2$, $\omega_H = -299.1 \text{ mm}^2$, $\omega_K = -193.2 \text{ mm}^2$, $G = 80\,000 \text{ MPa}$, $E = 206\,000 \text{ MPa}$.

The loading illustrated in Fig. 3.3.2(a) can be separated into the components shown in Fig. 3.3.2(b). In this example only the bending and torsional effects are considered and the combinations of loads causing distortion of the cross-section are ignored.

Transverse bending and shear

From elementary theory

$$\text{bending moment at support} = \frac{1 \times 2000^2}{12} = 0.3333 \times 10^6 \text{ N mm};$$

$$\text{compressive bending stress at H} = \frac{0.3333 \times 10^6 \times 25}{605\,233} = 13.769 \text{ MPa.}$$

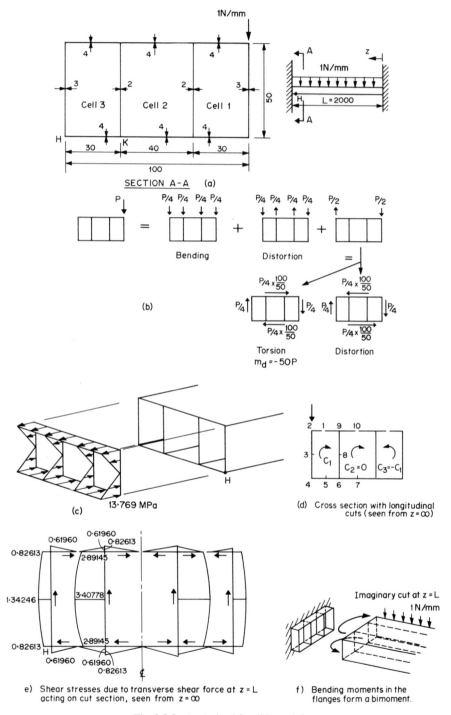

Fig. 3.3.2. Analysis of 3-cell box-girder.

Labels within figure:

1N/mm

A

1N/mm

L = 2000

z

4   4   4

3   2   2   3

Cell 3   Cell 2   Cell 1

50

4   4   4

H   30   K   40   30

100

SECTION A-A   (a)

P   P/4 P/4 P/4 P/4   P/4 P/4 P/4 P/4   P/2   P/2

=   +   +

Bending   Distortion

(b)

$P/4 \times \frac{100}{50}$ $P/4 \times \frac{100}{50}$

$P/4$ $P/4$ $P/4$ $P/4$

$P/4 \times \frac{100}{50}$ $P/4 \times \frac{100}{50}$

Torsion   Distortion

$m_d = -50P$

13·769 MPa

(c)

2   1   9   10

3   C_1 8 $C_2 = 0$ $C_3 = -C_1$

4   5   6   7

(d)   Cross section with longitudinal cuts (seen from $z = \infty$)

0·61960   0·61960   0·82613

0·82613   2·89145

1·34246   3·40778

0·82613   H   2·89145

0·61960   0·61960   0·82613   ₵

e)   Shear stresses due to transverse shear force at $z = L$ acting on cut section, seen from $z = \infty$

Imaginary cut at $z = L$

1N/mm

f)   Bending moments in the flanges form a bimoment.

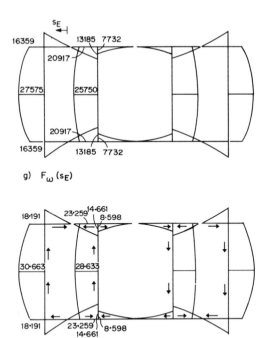

g) $F_\omega (s_E)$

(h) Shear flow in cut section due to warping torsion
     moment at z = L seen from z = ∞

Fig. 3.3.2. (continued).

The distribution of bending stresses over the whole cross-section at H is illustrated in Fig. 3.3.2(c).

The distribution of shear stresses due to the transverse shear force at H is found by the method used previously in Examples 1.4.3 and 1.4.4. The transverse shear force at $H = V = 1000$ N. Symmetrical cuts and shear flows are introduced as shown in Fig. 3.3.2(d). (Points 1, 3, 5, 7, 8 and 10 are at the mid-points of their respective plates.) The shear stress at each of the nodal points is found from eqn (1.4.6). For example, in the web at point **2**

$$\tau_2 = \frac{1000 \times 4 \times 15 \times 25}{605\,233 \times 3} = 0.826\,13 \text{ MPa}.$$

The distribution of this shear stress is illustrated in Fig. 3.3.2(e). Because the dislocation in cell 1 at **1** must be zero

$$\frac{1}{G} \oint \left(\tau + \frac{C_1}{t} \right) ds = 0,$$

whence

$$C_1 = -\oint \tau \, ds \Big/ \oint \frac{ds}{t}.$$

The numerator is found by using Simpson's Rule, thus

$$C_1 = 103.266/56.6667 = 1.822\ 34\ \text{N mm}^{-1}.$$

Hence the shear stress acting vertically up the web at H is

$$0.826\ 13 + 1.822\ 34/3 = 1.434\ \text{MPa}.$$

Saint Venant shear stresses

At a built-in support M_{st} $(= GJ_p\phi')$ is zero so the Saint Venant shear stresses are also zero.

Warping stresses

$$\lambda = [GJ_p/EF_{\omega\omega}]^{\frac{1}{2}} = 0.103\ 458\ \text{mm}^{-1}, \qquad \lambda L = 206.915$$

The terms $\sinh \lambda L$ and $\cosh \lambda L$ are both extremely large and almost equal, so from eqn (3.2.35)

$$\begin{aligned}
M_\omega(L) &= \frac{m_d}{\lambda^2}\left(1 - \frac{\lambda L}{2}\right) \\
&= \frac{-50}{(0.103\ 458)^2}(1 - 103.458) \\
&= 478\ 615\ \text{N mm}^2.
\end{aligned}$$

The positive sign of $M_\omega(L)$ indicates a clockwise bending moment in the upper flange when seen from above and an anticlockwise bending moment in the lower flange when seen from above; both moments are those acting on the cut end of the member at $z = L$ (Fig. 3.3.2(f)). The reader should check that these signs are correct by considering the physics of the problem (see Fig. 1.1.2(c)).

The direct warping stress at H is therefore

$$\begin{aligned}
\sigma_{\text{w,H}} &= \frac{M_\omega \omega_H}{F_{\omega\omega}} \\
&= \frac{478\ 615 \times (-299.1)}{44.9653 \times 10^6} \\
&= -3.184\ \text{MPa (i.e. compression)}.
\end{aligned}$$

The warping stresses are found by first twisting the closed profile about its shear centre and introducing the same longitudinal cuts as those shown in Fig. 3.3.2(d) and new constant shear flows C_1', C_2' and C_3' ($= C_1'$ from symmetry) which are assumed to act in an anticlockwise direction on the cells seen from $z = \infty$. These shear flows are also found from the requirements that the longitudinal dislocations must be zero for each cell.

From eqn (3.2.35) and with $m_d = -50$, $M_{DS}(L) = +50\,000$ N mm, i.e. clockwise on the end of the member. The shear flow in the cut profile before introducing C'_1, C'_2 and C'_3 is found from the last term of eqn (2.4.45). The values of $F_\omega(s_E)$ are obtained with the use of Table 2.4.1. For example the value of F_ω at node 3 is

$$F_{\omega,3} = \frac{4 \times 15}{2}(299.1 + 246.2) + \frac{3 \times 25}{2} \times 299.1 = 27\,575 \text{ mm}^4$$

where 246.2 is the value of ω at point 1 (Fig. 3.3.2(d)).

Figure 3.3.2(g) gives the calculated values of $F_\omega(s_E)$.

As seen from eqn (2.4.45) when these values are multiplied by $-M_{DS}/F_{\omega\omega}$ the shear flow in the cut profile is obtained, e.g. at point 3

$$T_{0,3} = -\frac{50\,000 \times 27\,575}{44.9653 \times 10^6} = -306.663 \text{ N mm}^{-1}.$$

The negative sign here indicates that the shear stress acts in an upwards direction, i.e. in an opposite sense to that of s_E (see Fig. 2.3.2). Figure 3.3.2(h) is a graph of the shear flow in the cut profile as seen from $z = \infty$.

The dislocations in the cells are reduced to zero by introducing C'_1, C'_2 and C'_3 as anticlockwise shear flows, e.g. in cell 1

$$\frac{1}{G}\left[\oint(\tau_0 + C'_1/t)\,ds - C'_2 \times \frac{50}{2}\right] = 0$$

where τ_0 is the shear stress in the cut profile (derived from Fig. 3.3.2(h)) and the last term is the effect of the shear flow C'_2 on the deformation of the web. The integration of τ_0 is again carried out by using Simpson's Rule and taking careful note of how the sign of τ_0 affects the sign of the dislocation. After integration the last equation becomes

$$\frac{1}{G}[-220.468 + 56.667C'_1 - 25C'_2] = 0$$

For cell 2 the corresponding equation is

$$\frac{1}{G}[1399.422 - 50C'_1 + 70C'_2] = 0$$

Solving these equations simultaneously gives

$$C'_1 = -7.197, \quad C'_2 = -25.133 \text{ N mm}^{-1} \text{ (i.e. both clockwise)}.$$

Thus the shear stress in the web at H is

$$\frac{18.191}{3} + \frac{7.197}{3} = 8.463 \text{ MPa (upwards)}.$$

Summary of results—stresses in web at H seen from $z = \infty$

Direct bending stress = 13.769 MPa   compressive
Direct warping stress = 3.184 MPa   compressive

  Total direct stress = 16.953 MPa   compressive

Shear stress due to
  bending   = 1.434 MPa   upwards
Saint Venant stress
  due to torsion   = 0
Shear stress due to
  warping torsion   = 8.463 MPa   upwards

  Total shear stress = 9.897 MPa   upwards

3.4. WARPING TORSION OF CONTINOUS BEAMS

In the two previous sections the warping torsion of beams with a single span has been considered. The same governing equation (eqn 3.2.4) could be used to study the behaviour of beams over many spans. All that is required is to satisfy the known conditions of continuity at each support. However, this is a rather tedious procedure because at each support there are four conditions to satisfy. Walker (1975) developed a technique of bimoment distribution which is applied as in moment distribution but it has to be corrected because the carry-over factors for bimoments are not the same as those for moments. Khan and Tottenham (1977) developed a more direct method but there is little to choose between the two methods. Here the method due to Khan and Tottenham will be used.

3.4.1. Bimoment distribution

Before describing the procedure for bimoment distribution it is necessary to establish a sign convention.

In the foregoing theory the origin has always been taken at the right-hand end, a clockwise rotation of the beam at any cross-section is positive and torques M_z, M_{DS}, and M_{st} are positive if they act on a section, which is viewed from $z = \infty$ by looking back towards the origin, in a clockwise manner.

The sign convention for a bimoment was established in Section 2.4.4. As stated there the sign convention can be looked at in two alternative ways which are consistent with one another. Firstly, the sign for stress is tension positive and the sign for the sectorial coordinate ω is established by swinging a radius vector in a clockwise sense about the pole (shear centre), so from eqn (2.4.36), that is,

$$M_\omega(z) = \int \sigma_\omega(z, s)\omega(s)\, \mathrm{d}F \qquad (3.4.1)$$

the positive sense of $M_\omega(z)$ is thereby defined. As an example, reference to Fig. 3.3.1(b) shows that according to this way of looking at the problem a positive bimoment would result in the pair of moments shown in Fig. 3.4.1. Thus when a solution to a problem is obtained by using the general warping equation (eqn (3.2.4)) and provided the positive sense of ω is clockwise about the pole and provided tensile stresses are positive, then the positive sense of M_ω can be obtained from the ω-diagram with tension stresses acting where ω is positive. The face over which these stresses act is that seen from $z = \infty$ by looking back towards the origin.

A second way of considering the sign convention for a bimoment is to think of it as a pair of moments as in Fig. 3.4.1. It is seen that on the right-hand side one of the moments appears as anticlockwise when viewed from the other. It has just been shown that according to the theory this is a positive bimoment and to be consistent it must be called positive according to this convention. However, it will be noted that the pair of moments to the left of the imaginary cut are negative according to this convention because one moment is clockwise when viewed from the other. Thus in bimoment distribution where this latter convention is used one must be careful to note that on one side of an imaginary cut the bimoment is positive and on the other side it is negative, i.e.

$$M_{\omega,\mathrm{L}} + M_{\omega,\mathrm{R}} = 0.$$

In obtaining solutions to eqn (3.2.4) this problem does not arise because then only the value of the bimoment acting on the face to the right of an imaginary cut is obtained. So far only the first way of looking at the sign convention has been used.

In bimoment distribution as in moment distribution of continuous beams imaginary cuts are made at the joints and attention is focussed on

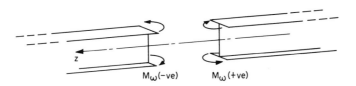

Fig. 3.4.1. At an imaginary cut in a beam the bimoments are equal and opposite.

the bimoments at each end of the now-isolated beams. To obtain the bimoment acting on the right hand end there will have to be a reversal of sign because then the bimoment is acting on the face seen from the origin. For example, the beam in Fig. 3.4.2 has a bimoment $M_{\omega,L}$ applied at the left-hand end. When eqn (3.2.4) is solved for the known boundary conditions the numerical value of the bimoment at the right-hand end can be evaluated. Since it is obtained from eqn (3.2.4) it is the bimoment acting on the face seen from infinity (see above). To evaluate the bimoment on the right-hand end of the beam, i.e. acting on the face seen from the origin the sign of this bimoment must be reversed.

This is the explanation for the signs in the following definitions (Fig. 3.4.2).

(a) Warping stiffness χ is defined as the bimoment required at the end of a beam to produce a unit warping at the same end. A *unit warping* is defined as one in which $\phi' = -1$ (the negative sign here ensures that the stiffness χ will be positive) which implies (see eqn (2.3.38)) that

$$u(s) = +\omega(s). \tag{3.4.2}$$

$\chi = M_{\omega,L}$ can be evaluated by solving the general warping equation of a beam (eqn 3.2.4) with $m = 0$.

(b) Carry-over bimoment is defined as the bimoment $-M_{\omega,R}$ developed at the fixed right-hand end of the beam when a unit warping is applied at the left-hand end L.

(c) Carry-over factor c is the ratio of the bimoment carried over to the right-hand end to the bimoment $M_{\omega,L}$ ($= \chi$) applied to the left-hand end.

$$c = -\frac{M_{\omega,R}}{M_{\omega,L}} = -\frac{M_{\omega,R}}{\chi}. \tag{3.4.3}$$

The minus sign is inserted to transfer $M_{\omega,R}$ to the left-hand side of the cut at the right-hand end of the member. In other words c is the ratio of the bimoments *on each end of the member* after it has been made into a free body by cuts at each end.

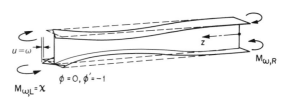

Fig. 3.4.2. Warping stiffness χ is defined as the bimoment $M_{\omega,L}$ required to produce a unit warping $[u(s) = \omega(s)]$.

(d) Distribution factor is the proportion of externally applied unit bimoment carried by one of the members at a joint. It is proportional to the warping stiffness of the members meeting at a joint. Thus, if two members ij and jk meet at j the distribution factors are

$$d_{ji} = \frac{\chi_{ji}}{\chi_{ji} + \chi_{jk}} \quad \text{and} \quad d_{jk} = \frac{\chi_{jk}}{\chi_{ij} + \chi_{jk}}. \tag{3.4.4}$$

The warping stiffnesses and carry-over bimoments of three beams (Fig. 3.4.3) with different restraint conditions at the right-hand end are easily derived from the general solution (eqns (3.2.16)–(3.2.21)) of eqn (3.2.4).

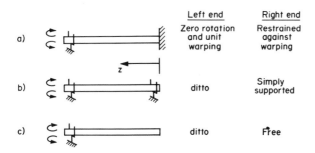

Fig. 3.4.3. Warping stiffness of beams with different end conditions.

Case (a)—right-hand end fully restrained against rotation and warping, left-hand end given unit warping.

The boundary conditions are

at $z = 0$, $\phi = \phi' = 0$ and at $z = L$, $\phi = 0$ and $\phi' = -1$.

Hence

$$M_{\omega,L} = \frac{\lambda L (\lambda L \cosh \lambda L - \sinh \lambda L)}{\lambda L \sinh \lambda L + 2(1 - \cosh \lambda L)} \frac{EF_{\omega\omega}}{L} \tag{3.4.5}$$

$$M_{\omega,R} = -\frac{\lambda L (\sinh \lambda L - \lambda L)}{\lambda L \sinh \lambda L + 2(1 - \cosh \lambda L)} \frac{EF_{\omega\omega}}{L}. \tag{3.4.6}$$

It is noted that here $M_{\omega,R}$ is the bimoment acting on the face at $z = 0$ viewed from $z = \infty$. Thus,

$$\chi = M_{\omega,L}$$

$$c = \frac{\sinh \lambda L - \lambda L}{\lambda L \cosh \lambda L - \sinh \lambda L}. \tag{3.4.7}$$

Case (b)—right-hand end restrained against rotation only

The boundary conditions are

at $z = 0$, $\phi = M_{\omega,R} = 0$ and at $z = L$, $\phi = 0$ and $\phi' = -1$.

Hence

$$M_{\omega,L} = \frac{(\lambda L)^2 \sinh \lambda L}{\lambda L \cosh \lambda L - \sinh \lambda L} \frac{EF_{\omega\omega}}{L} \qquad (3.4.8)$$

$$M_{\omega,R} = 0. \qquad (3.4.9)$$

Thus,

$$\chi = M_{\omega,L}$$
$$c = 0. \qquad (3.4.10)$$

Case (c)—right-hand end free (i.e. cantilever)

The boundary conditions are

at $z = 0$, $M_{\omega,R} = M_z (= M_{DS} + M_{st}) = 0$, at $z = L$, $\phi = 0$, $\phi' = -1$.

hence

$$M_{\omega,L} = \frac{\lambda L \sinh \lambda L}{\cosh \lambda L} \frac{EF_{\omega\omega}}{L}$$
$$M_{\omega,R} = 0 \qquad (3.4.11)$$

$$\chi = M_{\omega,L}$$
$$c = 0. \qquad (3.4.12)$$

The bimoments at the ends of some beams can be calculated from the formula in Section 3.2. Khan and Tottenham (1977) have conveniently summarized many practical cases which are presented here in Table 3.4.1. Only the notation and sign convention has been changed to bring them in line with those used in this text.

The steps in bimoment distribution follow the same pattern as that for moment distribution, that is,

(a) The beam is restrained against warping at each support point and the loads are applied. Each span is then a built-in beam whose end bimoments can be calculated.

(b) Each point is released in turn by imposing a balancing bimoment, distributing it and then carrying over bimoments to adjacent joints.

(c) After summing the bimoments at each joint the torques and stresses may be calculated. The signs of the bimoment on the right-hand ends of each span must be reversed before doing this last calculation.

Table 3.4.1. Bimoments at ends of single-span beam (Khan and Tottenham 1977).

Case	Bimoment at left end $M_{\omega,L}$	Bimoments at right end
	$\dfrac{m_d}{\lambda^2}\left[1-\dfrac{\lambda L\cosh(\lambda L/2)}{2\sinh(\lambda L/2)}\right]$	$-M_{\omega,L}$
	$\dfrac{ML}{2}\left[\dfrac{1-\cosh(\lambda L/2)}{\lambda L\sinh(\lambda L/2)}\right]$	$-M_{\omega,L}$
	$-\dfrac{M}{\lambda}\ \dfrac{\lambda a+\lambda b\cosh\lambda L-\lambda L\cosh\lambda b+\sinh\lambda b+\sinh\lambda a-\sinh\lambda L}{\lambda L\sinh\lambda L+2(1-\cosh\lambda L)}$	$-M_{\omega,L}$ but change a to b and b to a.
	$-\dfrac{m_d L}{2\lambda}\ \dfrac{\lambda L\sinh\lambda L-2\cosh\lambda L+2}{\lambda L\cosh\lambda L-\sinh\lambda L}$	0
	$-\dfrac{ML}{2}\ \dfrac{\sinh\lambda L-2\sinh(\lambda L/2)}{\lambda L\cosh\lambda L-\sinh\lambda L}$	0
	$\dfrac{M}{\lambda}\ \dfrac{\lambda L\sinh\lambda b-\lambda b\sinh\lambda L}{\lambda L\cosh\lambda L-\sinh\lambda L}$	0
	$\dfrac{m_d}{\lambda^2}\ \dfrac{\cosh\lambda L-\lambda L\sinh\lambda L-1}{\cosh\lambda L}$	0
	$\dfrac{M}{\lambda}\ \dfrac{\sinh(\lambda L/2)-\sinh\lambda L}{\cosh\lambda L}$	0
	$\dfrac{M}{\lambda}\ \dfrac{\sinh\lambda b-\sinh\lambda L}{\cosh\lambda L}$	0
	$\dfrac{-M_{\omega 1}}{\cosh\lambda L}$	$M_{\omega 1}$

Note: A positive bimoment is one in which one of the bending moments sees the other as anticlockwise.

Example 3.4.1

A beam whose cross-section is shown in Fig. 3.3.1(a) is continuous over three spans as shown in Fig. 3.4.4. Obtain the bimoments in the beam at

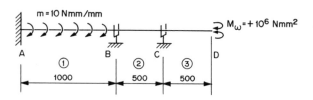

Fig. 3.4.4. Example 3.4.1.

each support and evaluate the maximum longitudinal stress.

From Example 3.3.1 $\lambda L_1 = 2.236\ 16$ whence $\lambda L_2 = \lambda L_3 = 1.118\ 08$

$\sinh \lambda L_1 = 4.625\ 209$ $\qquad$ $\sinh \lambda L_2 = 1.366\ 034\ 2$

$\cosh \lambda L_1 = 4.732\ 077\ 5$ $\qquad$ $\cosh \lambda L_2 = 1.692\ 941\ 1.$

From eqns (3.4.5) and (3.4.7)

$$\chi_1 = \frac{2.236\ 16(2.236\ 16 \times 4.732\ 08 - 4.625\ 21) \times 206\ 000 \times 34.133 \times 10^6}{[2.236\ 16 \times 4.625\ 209 + 2(1 - 4.732\ 08)] \times 1000}$$

$$= 3.253\ 57 \times 10^{10}$$

$$c_1 = \frac{4.625\ 21 - 2.236\ 16}{2.236\ 16 \times 4.732\ 08 - 4.625\ 21}$$

$$= 0.401\ 08$$

$$\chi_2 = \frac{1.118\ 08(1.118\ 08 \times 1.692\ 94 - 1.366\ 03)206\ 000 \times 34.133 \times 10^6}{[1.118\ 08 \times 1.366\ 03 + 2(1 - 1.692\ 24)] \times 500}$$

$$= 5.855\ 78 \times 10^{10}$$

$$c_2 = \frac{1.366\ 034 - 1.118\ 08}{0.526\ 809}$$

$$= 0.470\ 672.$$

From eqns (3.4.11) and (3.4.12)

$$\chi_3 = \frac{1.118\ 08 \times 1.366\ 03 \times 206\ 000 \times 34.133 \times 10^6}{1.692\ 941 \times 500}$$

$$= 1.339\ 899 \times 10^{10}$$

$$c_3 = 0.$$

The distribution factors are

$$d_{BA} = \frac{3.253\,57}{3.253\,57 + 5.855\,78} = 0.357\,17; \qquad d_{BC} = 0.642\,83$$

$$d_{CB} = \frac{5.855\,78}{5.885\,78 + 1.339\,899} = 0.813\,79; \qquad d_{CD} = 0.186\,21$$

The fixed-end bimoments in span 1 are found from eqn (3.2.34) or Table 3.4.1.

$$M_{\omega,L} = \frac{10 \times 1000^2}{2.236\,16^2} \left[\frac{2.236\,16 \times 4.625\,209}{2(1 - 4.732\,077\,5)} + 1 \right]$$

$$= -0.771\,237 \times 10^6 \text{ N mm}^2.$$

The fixed-end bimoments in span 2 are zero and the bimoments in span 3 are found from Table 3.4.1.

$$M_{\omega,L} = -\frac{10^6}{1.692\,941\,1} = -0.590\,69 \times 10^6 \text{ N mm}^2.$$

The bimoment distribution is carried out in Table 3.4.2.

Table 3.4.2. Bimoment distribution (Example 3.4.1)

	AB	BA	BC	CB	CD	DC
Distr. factors	1	0.357 17	0.642 83	0.813 79	0.186 21	
Fixed end M_ω	−0.7712	0.7712	0	0	−0.5907	1.000
Distr. & bal.		−0.2755	−0.4957	0.4807	0.1100	
Carry over	−0.1105		0.2263	−0.2333		
Distr. & bal.		−0.0808	−0.1454	0.1899	0.0434	
Carry over	−0.0324		0.0894	−0.0685		
Distr. & bal.		−0.0319	−0.0575	0.0557	0.0127	
Carry over	−0.0128		0.0262	−0.0271		
Dist. & bal.		−0.0094	−0.0169	0.0221	0.0050	
Total	−0.9269	0.3736	−0.3736	0.4195	−0.4196	1.000

The maximum bimoment is $0.9269 \times 10^6 \text{ N mm}^2$ and hence the maximum longitudinal stress due to warping is

$$\sigma_w = M_\omega \frac{\omega_{max}}{F_{\omega\omega}} = \frac{0.9269 \times 10^6 \times 800}{34.133 \times 10^6} = 21.72 \text{ MPa}.$$

3.5. FIRST-ORDER THEORY OF FLEXURAL TORSIONAL BUCKLING OF COLUMNS

Flexural torsional buckling is a phenomenon which occurs in certain columns made from thin-walled sections. Under some conditions they do not buckle by bending like simple Euler columns but they twist and bend simultaneously and hence the name flexural torsional buckling.

So far in this book structures have been analysed by using a first-order theory. It is possible to use first-order bending theory as the basis for a first-order theory of stability. The manner in which this can be achieved is first illustrated by showing how engineers' beam theory can be modified to account for stability effects.

The governing equation for the bending of a beam about the x-axis is well known, thus

$$EF_{yy}\frac{d^4\Delta(z)}{dz^4} = q(z) \tag{3.5.1}$$

where $\Delta(z)$ is the deflection

$q(z)$ is the applied load per unit length.

The buckling equation of a column can be obtained from this equation in a simple manner. Consider an element of the column of length dz shown in Fig. 3.5.1. Because of the curvature, the forces P acting on each end of the element and a downwards force, that is,

$$P\alpha = P\frac{dz}{R} = P\frac{d^2\Delta}{dz^2}\,dz,$$

are in equilibrium, i.e. the forces P at each end of the element can be

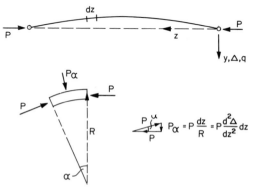

Fig. 3.5.1. A buckled column can be considered as a beam with a transverse load of $-P\dfrac{d^2\Delta}{dz^2}$ per unit length.

replaced by a transverse load of $P(\mathrm{d}^2\Delta/\mathrm{d}z^2)$ per unit length. Thus for the simple column eqn (3.5.1) becomes

$$EF_{yy}\frac{\mathrm{d}^4\Delta}{\mathrm{d}z^4}=-P\frac{\mathrm{d}^2\Delta}{\mathrm{d}z^2} \tag{3.5.2}$$

This is the general equation for column buckling. The Euler equation

$$EF_{yy}\frac{\mathrm{d}^2\Delta}{\mathrm{d}z^2}+P\Delta=0 \tag{3.5.3}$$

is a special case and is obtained by integrating eqn (3.5.2) twice while applying the boundary conditions for a pin-ended column.

This is the technique which is used here to obtain the governing equations for flexural torsional buckling. Figure 3.5.2 shows the cross-section of a column whose centroid is at the origin, whose principal axes

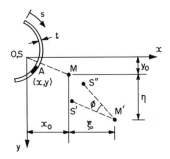

Fig. 3.5.2. Deflection and rotation of a cross-section of a column causes a torque.

coincide with the x- and y-axes and whose shear centre $M(x_0, y_0)$ moves to $M'(x_0+\xi,\ y_0+\eta)$ during buckling. The movement of S consists of a displacement (ξ, η) to S' and a rotation to S''. The movement of an element at $A(x, y)$ is $\xi+(y_0-y)\phi$ in the x direction and $\eta-(x_0-x)\phi$ in the y-direction. Hence the equivalent transverse loads acting on this element in the x- and y-directions, respectively, are

$$\mathrm{d}q_x=-(\sigma t\ \mathrm{d}s)\frac{\mathrm{d}^2}{\mathrm{d}z^2}[\xi+(y_0-y)\phi]$$

$$\mathrm{d}q_y=-(\sigma t\ \mathrm{d}s)\frac{\mathrm{d}^2}{\mathrm{d}z^2}[\eta-(x_0-x)\phi]. \tag{3.5.4}$$

These loads can be integrated over the cross-section to obtain the total equivalent transverse loads per unit length, that is,

$$q_x = -\sigma \frac{d^2\xi}{dz^2} \int t \, ds - \sigma y_0 \frac{d^2\phi}{dz^2} \int t \, ds + \sigma \frac{d^2\phi}{dz^2} \int yt \, ds$$

$$= -P\left(\frac{d^2\xi}{dz^2} + y_0 \frac{d^2\phi}{dz^2}\right) \tag{3.5.5}$$

where it has been recognised that the last term disappears because of the choice of origin. Similarly

$$q_y = -P\left(\frac{d^2\eta}{dz^2} - x_0 \frac{d^2\phi}{dz^2}\right) \tag{3.5.6}$$

On substituting into eqn (3.5.1), integrating twice and ignoring integration constants which represent rigid body movements two governing equations for flexural torsional buckling are obtained, that is,

$$EF_{xx} \frac{d^2\xi}{dz^2} + P\xi = -Py_0\phi \tag{3.5.7}$$

$$EF_{yy} \frac{d^2\eta}{dz^2} + P\eta = Px_0\phi \tag{3.5.8}$$

The elementary loads (3.5.4) also have a clockwise twisting effect about the axis through M', that is,

$$dm(z) = -(\sigma t \, ds)(y_0 - y)\frac{d^2}{dz^2}[\xi + (y_0 - y)\phi]$$

$$+ (\sigma t \, ds)(x_0 - x)\frac{d^2}{dz^2}[\eta - (x_0 - x)\phi] \tag{3.5.9}$$

On integrating over the whole cross-section the resulting clockwise torque per unit length is

$$m(z) = -y_0\sigma \frac{d^2\xi}{dz^2} \int t \, ds + x_0\sigma \frac{d^2\eta}{dz^2} \int t \, ds$$

$$+ \sigma \frac{d^2\xi}{dz^2} \int ty \, ds - \sigma \frac{d^2\eta}{dz^2} \int tx \, ds$$

$$- \sigma y_0^2 \frac{d^2\phi}{dz^2} \int t \, ds - \sigma x_0^2 \frac{d^2\eta}{dz^2} \int t \, ds$$

$$+ 2\sigma y_0 \frac{d^2\phi}{dz^2} \int ty \, ds + 2\sigma x_0 \frac{d^2\eta}{dz^2} \int tx \, ds$$

$$- \sigma \frac{d^2\phi}{dz^2} \int ty^2 \, ds - \sigma \frac{d^2\phi}{dz^2} \int tx^2 \, ds. \tag{3.5.10}$$

Recognising that many section properties (see eqn (2.3.80)) are contained in this expression and that S is the centroid this equation simplifies to

$$m(z) = P\left[x_0 \frac{d^2\eta}{dz^2} - y_0 \frac{d^2\xi}{dz^2}\right] - \frac{F_0}{F} P \frac{d^2\phi}{dz^2} \qquad (3.5.11)$$

where

F_0 = polar moment of inertia about the shear centre

$$= F_{yy} + F_{xx} + F(x_0^2 + y_0^2). \qquad (3.5.12)$$

On substituting eqn (3.5.11) into eqn (3.2.4) the third governing equation for torsional buckling is obtained

$$EF_{\omega\omega} \frac{d^4\phi}{dz^4} - \left(GJ_p - \frac{F_0 P}{F}\right) \frac{d^2\phi}{dz^2} - Px_0 \frac{d^2\eta}{dz^2} + Py_0 \frac{d^2\xi}{dz^2} = 0. \qquad (3.5.13)$$

The three simultaneous equations (3.5.7), (3.5.8) and (3.5.13) are homogeneous and represent an eigenvalue problem. It is noted that in general they are coupled but when the points S and M coincide, as they do for cross-sections with two axes of symmetry, $x_0 = y_0 = 0$ and the equations uncouple, thus

$$EF_{xx} \frac{d^2\xi}{dz^2} + P\xi = 0$$

$$EF_{yy} \frac{d^2\eta}{dz^2} + P_\eta = 0 \qquad (3.5.14)$$

$$EF_{\omega\omega} \frac{d^4\phi}{dz^4} - \left(GJ_p - \frac{F_0 P}{F}\right) \frac{d^2\phi}{dz^2} = 0$$

In this case there is no interaction between the critical modes for Euler buckling and for torsional buckling. Only the lowest of the three buckling loads is then of practical significance. For a pin-ended column in this class and which is free to warp at each end the boundary conditions are

at $z = 0$ and $z = L$, $\xi = \xi'' = \eta = \eta'' = \phi = \phi'' = 0$.

The solutions of eqns (3.5.14) are found by assuming buckling shapes in the form of sine functions which must satisfy these boundary conditions. Thus it is assumed that

$$\xi = C_1 \sin \frac{\pi z}{L}, \qquad \eta = C_2 \sin \frac{\pi z}{L}, \qquad \phi = C_3 \sin \frac{\pi z}{L} \qquad (3.5.15)$$

where C_1, C_2 and C_3 are constants. The critical load is therefore easily seen to be the lowest of

$$P_y = \frac{\pi^2 EF_{xx}}{L^2} \quad \text{or} \quad P_x = \frac{\pi^2 EF_{yy}}{L_2} \quad \text{or} \quad P_\phi = \frac{F}{F_0}\left(GJ_P + \frac{\pi^2 EF_{\omega\omega}}{L^2}\right)$$

(3.5.16)

where the notation reflects the axis about which buckling takes place.

For the more general case of a pin-ended column with ends which are free to warp when x_0 and y_0 are not zero a solution of the governing equations is obtained by again assuming a deformed shape given by eqn (3.5.15). After substitution in eqns (3.5.7), (3.5.8) and (3.5.13).

$$(P - P_y)C_1 + Py_0 C_3 = 0$$
$$(P - P_x)C_2 - Px_0 C_3 = 0$$

(3.5.17)

$$Py_0 C_1 - Px_0 C_2 - \left(\frac{\pi^2 EF_{\omega\omega}}{L^2} + GJ_P - \frac{F_0 P}{F}\right)C_3 = 0.$$

The critical value of the load is obtained when the determinant of these three equations is zero, i.e.

$$\begin{vmatrix} P - P_y & 0 & Py_0 \\ 0 & P - P_x & -Px_0 \\ Py_0 & -Px_0 & \dfrac{F_0}{F}(P - P_\phi) \end{vmatrix} = 0.$$

(3.5.18)

The determinant is expanded to yield the following cubic equation

$$\frac{F_{xx} + F_{yy}}{F_0} P^3 + \left[\frac{F}{F_0}(P_x y_0^2 + P_y x_0^2) - (P_x + P_y + P_\phi)\right]P^2$$

$$+ (P_x P_y + P_x P_\phi + P_y P_\phi)P - P_x P_y P_\phi = 0. \quad (3.5.19)$$

The left-hand side of this equation may be plotted for different values of P to obtain its three roots and Fig. 3.5.3 shows such a plot for the case when $P_x < P_y < P_\phi$. It is seen that the lowest critical load P_{cr1} is less than P_x. This is a simple example of the interaction of critical loads, discussed in Chapter 1. It arises because as the column starts to buckle as an Euler column some torsional buckling occurs simultaneously resulting in a lower critical load than if torsional effects were prevented in some way. This has important ramifications for designers because it might be supposed that the most efficient structure would be one in which all three buckling loads coincide. The following example illustrates the phenomenon of interaction of critical loads.

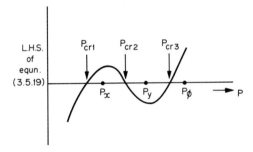

Fig. 3.5.3. A plot of the left-hand side of eqn (3.5.19) shows that the lowest critical load P_{cr1} is less than P_x.

Example 3.5.1

A pin-ended strut consists of a hollow tube (radius = 30 mm, thickness = 1 mm) which is axially loaded through its centre of area. The tube has a narrow longitudinal slit and the ends are free to warp.

(a) Find the length of the column at which P_ϕ (= the classical torsional buckling load) is equal to the Euler buckling load.

(b) Determine the percentage reduction in buckling load due to the interaction of the torsional and bending modes.

(c) What is the percentage reduction of buckling load if the length is decreased by 20%. $E = 206\,000$ MPa, $G = 80\,000$ MPa (see Exercise 2.4 for expressions of section properties).

Section properties

$$F_{xx} = F_{yy} = \pi R^3 t = 84\,823 \text{ mm}^4$$
$$F_{\omega\omega} = \pi R^5 t[\tfrac{2}{3}\pi^2 - 4] = 196.939 \times 10^6 \text{ mm}^6$$
$$J_p = \tfrac{2}{3}\pi R t^3 = 62.832 \text{ mm}^4$$
$$F = 2\pi R t = 188.496 \text{ mm}^2$$
$$x_0 = -2R = -60 \text{ mm}, \qquad y_0 = 0$$
$$F_0 = F_{xx} + F_{yy} + F(x_0^2 + y_0^2) = 848\,231.6 \text{ mm}^4.$$

Critical loads

From eqn (3.5.16)

$$P_x = P_y = \frac{\pi^2 E F_{yy}}{L^2} = 0.172\,457 \, L^{-2} \times 10^{12} \text{ N}$$

$$P_\phi = \frac{F}{F_0}\left(GJ_p + \frac{\pi^2 E F_{\omega\omega}}{L^2}\right) = 1117.01 + 88\,978.77 \, L^{-2} \times 10^6 \text{ N}.$$

(a) On equating these critical loads $L = 8645$ mm. For this value of L, $P_x = P_y = P_\phi = 2307.62$ N.

(b) Inspection of eqn (3.5.18) shows that when $y_0 = 0$ it is only necessary to equate a 2×2 determinant to zero, i.e.

$$\begin{vmatrix} P - P_x & -Px_0 \\ -Px_0 & \dfrac{F_0}{F}(P - P_\phi) \end{vmatrix} = 0$$

This forms a quadratic equation whose roots are

$$P = \dfrac{\dfrac{F_0}{F}(P_\phi + P_x) \pm \sqrt{\left[\dfrac{F_0^2}{F^2}(P_\phi + P_x)^2 - 4P_xP_\phi\dfrac{F_0}{F}\left(\dfrac{F_0}{F} - x_0^2\right)\right]}}{2\left(\dfrac{F_0}{F} - x_0^2\right)}$$

After inserting the values calculated above for the characteristic loads and section properties it is found that the lower root gives $P_{cr} = 0.5279P_\phi = 1218$ N, a decrease of 47% on P_ϕ.

(c) When $L = 6916$ mm $P_\phi = 2977.28$ N, $P_x = P_y = 3605.66$ N. By substituting in the above formula for P it is found that $P_{cr} = 1721$ N, a decrease of only 42% on P_ϕ.

If this member were part of a continuous beam or column over many spans it is seen that by reducing the spacing of the supports a considerable gain in strength can be achieved.

Example 3.5.1 and the discussion in Section 1.3.2 illustrate that there are two aspects to the interaction of critical loads. In some cases the critical modes of a perfect structure influence one another to reduce its critical load. This is not always the case as was seen for the case of a column whose cross-section has double symmetry. In these cases the governing equations (eqn 3.5.14) are uncoupled so there is no interaction. The other aspect is that the maximum load-capacity of an initially imperfect column may be reduced greatly as the imperfections are increased. Both of these aspects need to be considered by designers.

3.6. LATERAL BUCKLING OF BEAMS

Deep narrow beams have relatively high flexibility to lateral and twisting loads so that small disturbances of these kinds can cause lateral buckling of the beam. Beams in this class are designed to carry very large primary loads in the main plane of the beam but if a small lateral deflection or a small twisting deflection is introduced the primary loads can give rise to additional bending moments about the weak axis of the beam or additional twisting moments about its longitudinal axis. In Section 3.6.1 a simple case which does not involve twisting is analysed (Murray and Grundy 1970) and in Section 3.6.2 the more general case is treated (Timoshenko and Gere 1961). In both sections first-order stability analyses are again used, i.e. the solution to an eigenvalue problem is obtained.

3.6.1. Lateral buckling of a deep narrow beam or truss without twisting

When a deep narrow beam (Fig. 3.6.1) is supported by a pair of knife edges, which are aligned and lie in the plane of the beam, a vertical load P together with the self-weight of the beam can cause it to buckle laterally in a manner which involves only bending about its weaker axis, i.e. without twisting. It is found that when P reaches its critical value the beam simultaneously buckles sideways and rotates about the axis through the knife edges. In so doing the load P and the self-weight of the beam lose potential energy and the beam gains some strain energy by virtue of the bending about its minor axis. The analysis is conveniently carried out by using the Rayleigh–Ritz method.

A deep narrow beam (Fig. 3.6.2) is centrally loaded with P at a distance d below the knife edges. When P reaches its critical value it can be assumed that the beam first buckles laterally with the approximate form $a \sin \pi x/L$ (Fig. 3.6.2(b)) and then rotates about the knife edges until P is vertically below them. Thus the load is lowered an amount $\Delta = [d^2 + a^2]^{\frac{1}{2}} - d$, which for small a is close to $\Delta = a^2/2d$. The loss of potential energy of P is therefore

$$U_{e1} = Pa^2/2d. \tag{3.6.1}$$

If the weight of the beam is significant P will not lie directly below the knife edges but the error should be small. However, it is then necessary to consider the loss of potential energy of the self-weight of the beam.

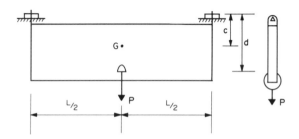

Fig. 3.6.1. A deep narrow beam supported by two knife edges.

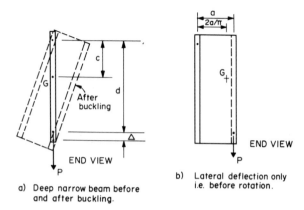

Fig. 3.6.2. Analysis of a deep narrow beam supported in such a way that there is no twisting.

The distance of the centre of gravity G from the original plane of the beam when it has deflected laterally is $2a/\pi$ and it is a distance c below the knife edges. Thus G is lowered through $[c^2+(2a/\pi)^2]^{\frac{1}{2}}-c$ which is approximately $2a^2/\pi^2c$ for small a.

Hence the loss of potential energy of the self-weight W of the beam is

$$U_{e2} = 2Wa^2/\pi^2c. \tag{3.6.2}$$

The strain energy absorbed by the beam is

$$U_i = \frac{EF_{xx}}{2} \int_0^L \xi'' \, dz = \frac{EF_{xx}\pi^4 a^2}{4L^3} \tag{3.6.3}$$

where F_{xx} is the second moment of area of the beam about its vertical axis.

Two extreme cases are now considered, firstly when $P \gg W$. In passing from the unbuckled to the buckled form $U_i = U_{e1}$, when

$$P_{cr} = \frac{\pi^4 EF_{xx} d}{2L^3}. \tag{3.6.4}$$

The other extreme case occurs when $P=0$ and the length of the beam is at its critical value. Equating U_i and U_{e2}

$$L_{cr} = \frac{\pi^2}{2} \sqrt[3]{\left(\frac{EF_{xx}c}{W}\right)}. \tag{3.6.5}$$

It should be noted that in both cases the stability of the beam can be improved by arranging for the loads P or W to be as low as possible below the knife edge supports or alternatively by increasing F_{xx} in some way. For the cases where P and W are roughly equal a more exact analysis is required (Murray and Grundy 1970). This problem has considerable interest to engineers engaged on the erection of structures. The same problem occurs when a plane roof truss or similar framed structure is lifted. In this case the main difficulty is to evaluate F_{xx} accurately. Possibly the simplest way is to obtain it experimentally prior to lifting. The truss can be laid on its side and loaded as a simple beam.

3.6.2. Lateral torsional buckling of a deep narrow beam or truss

When the ends of a perfectly straight beam are prevented from rotating about a longitudinal axis lateral buckling can still occur but there will be simultaneous bending about the weak axis and twisting about the longitudinal axis passing through the shear centre M.

The governing equations for this case are derived by considering a beam (Fig. 3.6.3) with two axes of symmetry in its buckled position. The applied loads will, of course, move along their lines of action as the beam buckles but this will not affect the internal stress resultants at an arbitrary cross-section z from the origin. In other words the bending moment and shear force diagrams for the beam in its undeformed and deformed positions are identical. However, the deflection does mean that the position of the three principal axes at z have changed by displacements ξ, η and ζ in the x, y and z directions, respectively, and by rotations $-d\eta/dz$, $d\xi/dz$ and ϕ about the x, y and z axes, respectively, where the signs conform to the left-hand screw rule. Because the governing equations for torsion (eqns 2.4.5) refer to the principal axes it is necessary to resolve the stress resultants which are of interest, that is, the bending and twisting moment, at z into the direction of the principal axes x_1, y_1 and z_1

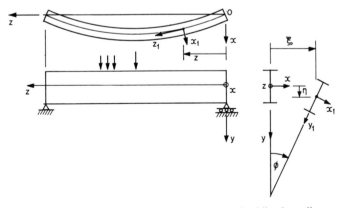

Fig. 3.6.3. Beam in its deformed position after buckling laterally.

(Fig. 3.6.3). This is easily achieved by resolving the bending and twisting moments as vectors and then by introducing approximations which are valid when ξ, η, ζ and ϕ are small.

Thus the curvatures in the $x-z$ and $y-z$ planes are $d^2\xi/dz^2$ and $d^2\eta/dz^2$, respectively, and the curvatures in the x_1-z_1 and y_1-z_1 planes are therefore the products of these values and $\cos\phi$, respectively. For small ϕ, $\cos\phi \doteq 1$ and hence the bending equations about the two principal axes in the element at z in its new position are

$$EF_{xx}\frac{d^2\xi}{dz^2} = M_{y1} \tag{3.6.6}$$

$$EF_{yy}\frac{d^2\eta}{dz^2} = M_{x1}. \tag{3.6.7}$$

The moments M_{y1} and M_{x1} are the resolved components of the bending moment at z and act about the y_1 and x_1 axes respectively. The positive senses of these moments are indicated in Fig. 3.6.4.

The governing equations for the twisting of the beam are eqns (3.2.4) and (3.2.5), that is,

$$\phi^{iv}(z) - \lambda^2\phi''(z) = \frac{m_z(z)}{EF_{\omega\omega}} \tag{3.6.8}$$

$$\lambda^2 = \frac{GJ_p}{EF_{\omega\omega}}. \tag{3.6.9}$$

In these equations $m_z(z)$ is the resolved part of the bending and twisting moments in the direction of the principal axis z_1. The sign convention for positive twisting moment is the same as that used previously (Fig. 3.6.4). The last four equations are now applied to some examples.

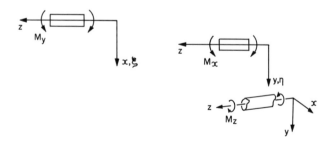

Fig. 3.6.4. Positive senses of M_y, M_x, and M_z.

Example 3.6.1

The I-beam shown in Fig. 3.6.5 carries a uniform bending moment M_0 at every cross-section before buckling. It is desired to find an expression for the critical value of M_0.

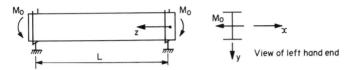

Fig. 3.6.5. I-beam carrying equal moments M_0 (Example 3.6.2.1).

In this case the bending moment at every cross-section is M_0 and after buckling M_0 must be resolved into its components in the x_1, y_1, and z_1 (principal) directions (Fig. 3.6.3), thus

$$M_{x1} = M_0 \cos \left(\frac{\mathrm{d}\xi}{\mathrm{d}z} \right) \doteqdot M_0; \qquad M_{y1} = M_0 \sin \phi \doteqdot M_0 \phi;$$

$$M_{z1} = -\frac{\mathrm{d}\xi}{\mathrm{d}z} M_0. \tag{3.6.10}$$

Hence from eqns (3.6.6)–(3.6.8)

$$EF_{xx} \frac{\mathrm{d}^2 \xi}{\mathrm{d}z^2} - \phi M_0 = 0 \tag{3.6.11}$$

$$EF_{yy} \frac{\mathrm{d}^2 \eta}{\mathrm{d}z^2} - M_0 = 0 \tag{3.6.12}$$

$$\frac{\mathrm{d}^4 \phi}{\mathrm{d}z^4} - \lambda^2 \frac{\mathrm{d}^2 \phi}{\mathrm{d}z^2} - \frac{\mathrm{d}^2 \xi}{\mathrm{d}z^2} \frac{M_0}{EF_{\omega\omega}} = 0 \tag{3.6.13}$$

where it has been recognised in deriving eqn (3.6.13) that (see eqn (3.2.6))

$$m_z(z) = -\frac{\mathrm{d}M}{\mathrm{d}z}. \tag{3.6.14}$$

Equations (3.6.11)–(3.6.13) are partially uncoupled as follows

$$EF_{xx} \frac{\mathrm{d}^2 \xi}{\mathrm{d}z^2} - \phi M_0 = 0 \tag{3.6.15}$$

$$EF_{yy} \frac{\mathrm{d}^2 \eta}{\mathrm{d}z^2} - M_0 = 0 \tag{3.6.16}$$

$$\frac{\mathrm{d}^4 \phi}{\mathrm{d}z^4} - \lambda^2 \frac{\mathrm{d}^2 \phi}{\mathrm{d}z^2} - \beta \phi = 0 \tag{3.6.17}$$

where

$$\beta = \frac{M_0^2}{E^2 F_{\omega\omega} F_{xx}}. \tag{3.6.18}$$

The general solution of eqn (3.6.17) is obtained by assuming it has the form $A e^{kz}$. After substitution

$$\phi = A_1 \sin mz + A_2 \cos mz + A_3 \sinh nz + A_4 \cosh nz \tag{3.6.19}$$

where

$$m = \sqrt{\left[\frac{-\lambda^2 + \sqrt{(\lambda^4 + 4\beta)}}{2}\right]} \quad \text{and} \quad n = \sqrt{\left[\frac{\lambda^2 + \sqrt{(\lambda^4 + 4\beta)}}{2}\right]}. \tag{3.6.20}$$

The constants of integration can be found from the boundary conditions. Since each end is free to warp, the boundary conditions are $\phi = \phi'' = 0$ at $z = 0$ and L. From those at $z = 0$ it is seen that A_2 and $A_4 = 0$. Conditions at $z = L$ result in the following simultaneous equations

$$A_1 \sin mL + A_3 \sinh nL = 0$$
$$-A_1 m^2 \sin mL + A_3 n^2 \sinh nL = 0 \tag{3.6.21}$$

These equations are satisfied if the determinant is zero, which occurs when $m = \pi/L$, i.e. A_3 must also be zero. Thus, the buckled mode is $\phi = A_1 \sin mz$ and the lowest buckling moment occurs when

$$\left[\frac{-\lambda^2 + \sqrt{(\lambda^4 + 4\beta)}}{2}\right] = \frac{\pi^2}{L^2}. \tag{3.6.22}$$

After using eqns (3.6.9) and (3.6.18) it is found that

$$(M_0)_{cr} = \frac{\pi}{L} \sqrt{\left[EGF_{xx}J_p\left(\frac{EF_{\omega\omega}}{GJ_p}\frac{\pi^2}{L^2} + 1\right)\right]}. \tag{3.6.23}$$

This critical moment does not depend upon the second moment of area about the x-axis, that is, F_{yy}, but more elaborate investigation shows that when the width-to-depth ratio is not small the constant factor outside of the root sign in eqn (3.6.23) should be replaced by a factor which depends upon b/h. This effect is important only for the case of long beams.

Example 3.6.2

Derive an expression for the critical moment of the deep narrow beam illustrated in Fig. 3.6.6(a). When $t = 1$, $b = 10$, $h = 40$ and $L = 500$ evaluate the critical moment and compare it with that obtained from the following approximate formulae for deep narrow beams.

$$(M_0)_{cr} = \frac{\pi}{L} \sqrt{(EGF_{xx}J_p)}.$$

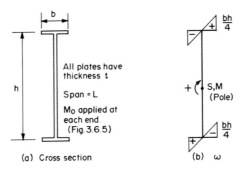

(a) Cross section   (b) ω

Fig. 3.6.6. Beam analysed in Example 3.6.2.

Section properties

$J_p = \frac{1}{3}(2b + h)t^3$ (eqn 1.4.20). With the pole at S and M (Fig. 3.6.6(b)) the warping function is plotted in the usual way (see for example Fig. 2.4.7(b)).

$$F_{\omega\omega} = \int_F \omega^2 \, dF$$

$$= 4\left[\frac{1}{3} \times \frac{b}{2} \times \left(\frac{bh}{4}\right)^2 t\right] \qquad \text{(Table 2.4.1)}$$

$$= \frac{b^3 h^2 t}{24}$$

$$F_{xx} = \frac{b^3 t}{6}.$$

Critical moment (eqn 3.6.23)

$$(M_0)_{cr} = \frac{\pi}{L} \sqrt{\left\{206\,000 \times 80\,000\right.}$$

$$\times \frac{b^3 t}{6} \cdot \frac{(2b + h)t^3}{3}\left[2.6 \times \frac{b^3 h^2 t}{24} \times \frac{3}{(2b + h)t^3} \cdot \frac{\pi^2}{L^2} + 1\right]\left.\right\}$$

$$= 95\,059 \sqrt{\left\{\frac{b^3 t^4 (2b + h)}{L^2}\right\}} \sqrt{\left[3.2076 \frac{b^3 h}{t^2 L^2 \left(\frac{2b}{h} + 1\right)} + 1\right]}.$$

When $t = 1$, $b = 10$, $h = 40$ and $L = 500$

$$(M_0)_{cr} = 95\,059 \sqrt{\left\{\frac{1000 \times 1 \times 60}{500^2}\right\}} \times 1.16 = 54.02 \times 10^3 \text{ N mm.}$$

The approximate formula is the same as this when the factor 1.16 is omitted.

$$(M_0)_{cr}(\text{approximate}) = 46.57 \times 10^3 \text{ N mm.}$$

Thus it is seen that the approximate formula should be used only if *both* b/h and h/L are small.

Example 3.6.3

Determine the critical value of the load P which is centrally applied to the beam illustrated in Fig. 3.6.7(a).

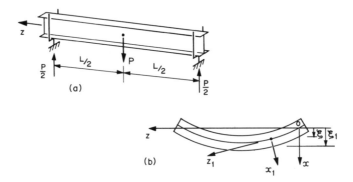

Fig. 3.6.7. Lateral buckling of I-beam with central load P.

After selecting the origin of coordinates at the right-hand end of the beam it is only necessary to consider the right-hand half (Fig. 3.6.7(b)). In the global coordinate system

$$M_x = -\frac{Pz}{2}, \qquad M_y = 0, \qquad M_z = \frac{P}{2}(\xi_1 - \xi) \tag{3.6.24}$$

where the sign convention is that shown in Fig. 3.6.4. These moments can be resolved into the local coordinate directions x_1, y_1 and z_1 as follows

$$M_{x1} = -\frac{Pz}{2}, \qquad M_{y1} = -\frac{Pz\phi}{2}, \qquad M_{z1} = \frac{Pz}{2}\frac{d\xi}{dz} + \frac{P}{2}(\xi_1 - \xi) \tag{3.6.25}$$

where the usual approximations which apply when the deflections are small have been used. These expressions are substituted into eqns (3.6.6)–(3.6.8) and eqn (3.6.14) is employed. Thus,

$$EF_{xx}\frac{d^2\xi}{dz^2} + \frac{Pz}{2}\phi = 0 \tag{3.6.26}$$

$$EF_{yy}\frac{d^2\eta}{dz^2} + \frac{Pz}{2} = 0 \tag{3.6.27}$$

$$\frac{d^4\phi}{dz^4} - \lambda^2 \frac{d^2\phi}{dz^2} + \frac{Pz}{2EF_{\omega\omega}} \frac{d^2\xi}{dz^2} = 0. \tag{3.6.28}$$

After eliminating ξ between eqns (3.6.26) and (3.6.28)

$$\phi^{iv} - \lambda^2 \phi'' - Kz^2\phi = 0 \tag{3.6.29}$$

where

$$K = \frac{P^2}{4E^2 F_{\omega\omega} F_{xx}}. \tag{3.6.30}$$

A solution of this equation was first obtained by Timoshenko (1907). There is a theorem in the theory of differential equations which states that the solution of an equation such as eqn (3.6.29) can be represented by a series of the form

$$\phi = a_0 + a_1 z + a_2 z^2 + a_3 z^3 + a_4 z^4 + \ldots \tag{3.6.31}$$

However it is known that at $z = 0$, ϕ and ϕ'' are zero because the rotation and bimoment are zero there.

Thus an appropriate form of solution is

$$\phi = a_1 z + a_3 z^3 + a_4 z^4 + a_5 z^5 + \ldots \tag{3.6.32}$$

Since eqn (3.6.29) must be satisfied for all values of z the coefficients of the various powers of z, obtained by substitution, must be zero. It is found that the coefficients a_4, a_6, a_8 etc. are zero and the remaining equations are as follows

Coefficient of z: $-6\lambda^2 a_3 + 120 a_5$ $= 0$

Coefficient of z^3: $-\lambda^2 K a_1$ $-20\lambda^2 a_5 + 420 a_7$ $= 0$

Coefficient of z^5: $-\lambda^2 K a_3$ $-42\lambda^2 a_7 + 3024 a_9 = 0$

... and so on.   (3.6.33)

To these equations must be added two further boundary conditions, that is, at $z = L/2$, $\phi' = \phi''' = 0$. These are conditions of symmetry; the latter occurs because the total torque $(= -EF_{\omega\omega}\phi''' + GJ_p\phi')$ there must be zero. Thus

$$a_1 + 3a_3(L/2)^2 + 5a_5(L/2)^4 + \ldots \qquad = 0$$
$$6a_3 \qquad + 60a_5(L/2)^2 + 210a_7(L/2)^4 = 0. \tag{3.6.34}$$

Equations (3.6.33) and (3.6.34) can be truncated after a convenient number of terms and written in matrix form. The determinant is equated to zero and its lowest root gives the critical value of K. Hence P_{cr} is found from eqn (3.6.30) and the result is conveniently expressed in the form

$$P_{cr} = \gamma(EGF_{xx}J_p)^{\frac{1}{2}}/L^2 \tag{3.6.35}$$

where γ depends upon $L^2 GJ_p/EF_{\omega\omega}$. Table 3.6.1 gives some values of γ (Timoshenko and Gere 1961).

Table 3.6.1. Values of γ for a central load at mid height.

$L^2 GJ_p/EF_{\omega\omega}$	0.4	4	8	16	24	48	96	400
γ	86.4	31.9	25.6	21.8	20.3	18.8	17.9	17.2

Timoshenko and Gere (1961) present further results which show the very important influence of the height of the load point, especially when the span is relatively small. When the load is attached to the bottom flange the value of γ in the first column of Table 3.6.1 changes from 86.4 to 147 and when it is applied to the upper flange γ reduces to 51.5. The corresponding figures in the last column are, however, 18.7 and 15.8, indicating a relatively small influence of the load position on P_{cr}. This case occurs when the beam is very long or if $F_{\omega\omega}$ is very small. In the limit when $F_{\omega\omega} = 0$ eqn (3.6.29) assumes the form

$$\phi'' + K_1 z^2 \phi = 0 \tag{3.6.36}$$

where

$$K_1 = \frac{P^2}{4EGF_{xx}J_p}. \tag{3.6.37}$$

The solution of eqn (3.6.36) is also found by assuming a series for ϕ of the form given in eqn (3.6.32). The result of this analysis is

$$P_{cr} = \frac{16.936}{L^2}(EGF_{xx}J_p)^{\frac{1}{2}} \tag{3.6.38}$$

The result of the analysis of other loading cases and boundary conditions will be found in the monograph by Timoshenko and Gere (1961). The way in which these studies are incorporated into codes of practice is described in Chapter 7.

3.7. GENERAL COMMENTS ON ANALYTICAL SOLUTIONS OF BENDING AND BUCKLING PROBLEMS

While there are many analytical solutions of bending and buckling problems in the literature they are seldom exactly what the designer requires for his particular structure. In practice beams and columns are never simply supported or built-in, the shape of their cross-section is usually more complicated than is assumed for the analytical solutions and initial imperfections may play an important role. However, perhaps the most important influence which is ignored is that of local buckling. This is yet another buckling mode and it involves changes in the shape of the cross-section. It is usually the case that it interacts with Euler and torsional buckling and this causes a reduction in the elastic critical load. For these problems it is necessary to resort to numerical techniques except for the simplest of cases. Finite elements enable such analyses to be carried out. A special class of element, called the finite strip, is more efficient for that class of thin-walled structure which is long and has a constant geometry of cross-section throughout its length. Such structures are called folded plate structures. It is the purpose of the next chapter to show how these structures can be analysed for bending and buckling. This does not mean that analytical solutions are obsolete; on the contrary, when they are available they are the most convenient way of analysing a structure.

EXERCISES

3.1 (a) A concentrated force P of 1 kN is applied to the column at A. Calculate the bimoment acting on the cross-section at $z = 0$.

(b) The load P is now replaced by a distributed load q at A which increases uniformly down the length of the column. If its maximum value is 500 N mm^{-1} what is the twisting moment per unit length arising from it?

(Answer: (a) 200 000 N mm^2

(b) 100 N.)

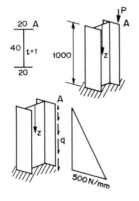

Ex. 3.1

3.2 A beam is supported by an axle which allows free rotation at $z = L$ and it is built-in at $z = 0$. It carries a uniformly distributed twisting moment m_d per unit length.

(a) Define the boundary conditions.

(b) Obtain an expression for the bimoment at $z = 0$.

(Answer: (a) at $z = 0$, $\phi = \phi' = 0$;   at $z = L$, $\phi'' = M_z = 0$

$$(b) \quad \frac{m_d}{\lambda^2 \cosh \lambda L} (\cosh \lambda L - \lambda L \sinh \lambda L - 1).\Big)$$

3.3 (a) Derive eqns (3.5.56) from the general solution given in Section 3.2.2.

(b) Calculate the maximum values of each of τ_{st}, σ_w and τ_w and state where each may be found for the beam illustrated.

(c) Evaluate τ_{st}, σ_w and τ_w at the point A for $z = L/2$.

(Note: The section properties are evaluated in Example 2.6.1)

(Answer: (a) 75 MPa at $z = L$ on outer surface, 160.85 at MPa at $z = 0$ at tip of flange.

(b) 6.67 MPa at $z = 0$ near midpoint of flange.

(c) 74.9 MPa, 0.0617 MPa, 0.0007 MPa.)

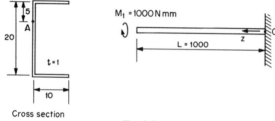

Cross section

Ex. 3.3

3.4 The lighting pole illustrated consists of a hollow tube of 120 mm

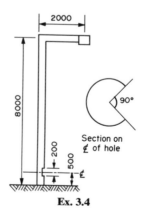

Section on ₵ of hole

Ex. 3.4

mean diameter and 3 mm thickness. An unreinforced inspection hole whose dimensions and location are shown is cut near the base of the vertical leg. Determine the bending and longitudinal warping stresses at the edge of the hole and the deflection at the end of the arm when the wind load is 0.2 N per mm along the vertical pole and horizontal arm.
(Answer: Longitudinal bending stress = 198 MPa.
Longitudinal warping stress =   87 MPa
Deflection   = 298 mm.)

3.5   A beam whose cross-section is shown in Fig. 3.3.1(a) is supported and loaded as shown in the diagram. Calculate the bimoments at each support and evaluate the maximum longitudinal stress.
(Answer: $M_{\omega,\mathrm{A}} = 0.8817 \times 10^6$ N mm^2, $M_{\omega,\mathrm{B}}(\text{left}) = 0.5979 \times 10^6$ N mm^2
$M_{\omega,\mathrm{C}} = 1.3821 \times 10^6$ N mm^2, $\sigma_\mathrm{w} = 32.393$ MPa.)

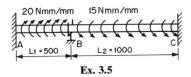

20 Nmm/mm   15 Nmm/mm

A   L₁ = 500   B   L₂ = 1000   C

Ex. 3.5

3.6   Analyse the beam treated in Exercise 3.5 when the built-in support at A is replaced by a fork (i.e. $\phi = \phi'' = 0$).
(Answer: $M_{\omega,\mathrm{A}} = 0$, $M_{\omega,\mathrm{B}}(\text{left}) = 0.2278 \times 10^6$ N mm^2,
$M_{\omega,\mathrm{C}} = 1.5295 \times 10^6$ N mm^2, $\sigma_\mathrm{w,max} = 35.848$ MPa.)

3.7   An angle section whose dimensions are 90 mm × 90 mm × 2 mm thick carries an axial load as a pin-ended column through its centroid. Find the length of the column at which P_ϕ, the torsional buckling load, is equal to the Euler load for buckling about the major principal axis. If buckling about the minor principal axis is prevented what is the percentage reduction in buckling load due to the interaction of the torsional and Euler modes?
(Answer: 8.336 m, $P_\phi = P_x = 14\,220$, $P_\mathrm{cr} = 8819$, 38% reduction.)

3.8   A deep narrow cantilever of length L carries a concentrated load P at mid-height at its free end. Assume that $F_{\omega\omega} = 0$ and show that the governing equation for torsional buckling is

$$\phi'' + \frac{P^2 z^2}{EGF_{xx}J_\mathrm{p}} \phi = 0$$

where z is the distance along the centreline of the beam from the loaded end. Assume a power series (eqn 3.6.31) as the solution and show that

$$P_\mathrm{cr} = \frac{4.013}{L^2} (EGF_{xx}J_\mathrm{p})^{\frac{1}{2}}.$$

3.9   Derive eqns (3.2.38)–(3.2.44) by solving eqn (3.2.4) with the use of Laplace transforms and Heaviside step-functions.

3.10   Derive eqns (3.2.51)–(3.2.55) by solving eqn (3.2.4) with the use of Laplace transforms and a Dirac delta-function.

4

ELASTIC ANALYSIS OF BOX-GIRDERS, STIFFENED PLATES AND OTHER FOLDED PLATE STRUCTURES BY THE FINITE STRIP METHOD

4.1. INTRODUCTION

In the two previous chapters it is shown how the behaviour of thin-walled beams and columns can be studied by using analytical methods. Although these methods are generally applicable to simple problems it is difficult to extend them to certain classes of thin-walled structures. For these structures powerful numerical methods are often available and this chapter is concerned with one of these.

A folded plate structure is, as the name suggests, a prismatic structure which can be formed by folding a flat rectangular plate along lines parallel to its length. For the purposes of this book the definition also includes similar structures which could be manufactured by folding and joining with other plates along lines parallel to its length. For example, a three-cell box girder (Fig. 1.4.13) requires both folding and joining. Another example is the very important thin-walled structure called the stiffened plate which consists of a flat plate and longitudinal stiffening ribs. Figure 4.1.1 shows a few examples of stiffened plates used in practice while Fig. 1.1.1(a) shows one case where a stiffened plate is used for the walls of a large box girder and a tower.

The reason why it is usually more convenient to resort to numerical methods for these structures is that they are used in practical situations where there is a need to introduce irregularities. For example, transverse stiffeners, framing or diaphragms are required to preserve the shape of a box girder. Also local wheel loads on the deck of a box girder can cause high stresses which the designer would like to evaluate. Furthermore these structures may experience local or global buckling and may have a reduced buckling load because of the interaction of these buckling modes. Local buckling is a type of instability which involves distortion of the cross-section. Figure 4.1.2 shows typical local buckles in a stiffened plate observed during an axial compression test.

A numerical method which can deal with all of these problems must be very powerful. While the finite element method is in theory able to solve these problems, in practice, the writing of the program and the data preparation for folded plate problems is often formidable and expensive.

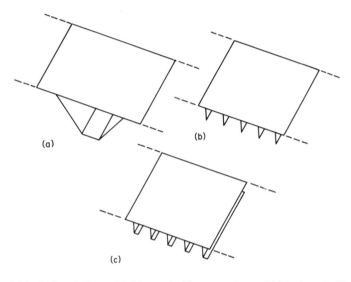

Fig. 4.1.1. Stiffened plates with (a) trough, (b) rectangular, and (c) L-shaped stiffeners.

Fig. 4.1.2. A flat plate with bulb-flat stiffeners is loaded axially through Macalloy bars and exhibits local buckling. The plate buckles in a chequerboard pattern and the stiffeners twist in sympathy with it (compare waviness of bulb-flat tip with stringline).

It is found that a specialized finite element method called the *finite strip method* is very powerful, easier to set up, and gives accurate results. In this method the elements are longitudinal strips of plate which are joined to one another along nodal lines running the full length of the structure. The finite strip method is easily adapted to bending problems of, say, box girders and stiffened plates as well as to the buckling analysis of these structures.

There are two approaches to the finite-element method; one uses the stiffness (or displacement) method while the other uses the flexibility (or force) method. In Section 4.2 the stiffness method of matrix analysis of structures is very briefly described. This is done simply for the sake of completeness and readers who are familiar with this technique may omit this section. In Section 4.3 the linear analysis of a folded plate structure by the finite strip (stiffness) method is described. This method was first developed by Cheung (1976) who applied it to the analysis of box girders and other bending structures. An interesting application of the finite strip method is to determine the stress distribution in a box girder with diaphragms and Loo's (1976) analysis of this problem is described in Section 4.4.

The main aim of these analyses is to enable a designer to evaluate stress levels at any point in a thin-walled structure and to check them according to the requirements of a code of practice. He may also need these analyses to determine the stress levels on the boundaries of small plate panels of the structure prior to carrying out a check on their buckling stresses. Methods for evaluating the buckling stresses of small plate panels with arbitrarily distributed boundary stresses are described in Chapters 5 and 7. However, the finite strip method can also be used to evaluate the buckling stresses of folded plate structures. Stanley and Sved (1975) have used this method to find the buckling stresses of stiffened plates and, more recently, the same method has been used by Murray and Thierauf (1981) to produce design tables of the buckling stresses of stiffened plate panels. The method automatically accounts for the effects of interaction of critical loads and the results present a designer with all of the basic information he requires to design a stiffened panel which is loaded axially as a pin-ended column. This work is described in Section 4.5. It is based upon a first-order stability analysis, i.e. only the critical stresses are determined. One of the features of this analysis is that it defines the conditions under which a column of given cross-section will buckle locally. Whereas the torsional and lateral buckling analyses presented in Chapter 3 were based upon the assumption that the cross-section did not distort during buckling, the finite strip method relaxes that restriction. The critical stress is evaluated even when local, Euler and torsional buckling modes are interacting. Results indicate that when this happens the critical stress can be greatly reduced.

4.2. INTRODUCTION TO FINITE ELEMENTS USING THE STIFFNESS METHOD

The linear relationship between the force F and displacement δ of a simple spring can be expressed by the equation

$$F = S\delta \qquad (4.2.1)$$

where S is the stiffness of the spring. In this case it is noted that F, S and δ are scalar quantities, one number only being required to define each of them. However, when a cantilever (Fig. 4.2.1(a)) is loaded the forces applied at the end (a node) can be, say, a transverse load F_1 and a moment, to which for convenience the symbol F_2 is attached. The

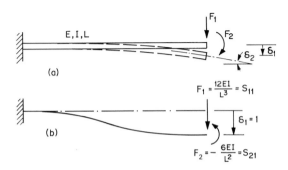

Fig. 4.2.1. Elements of the stiffness matrix of a cantilever can be found by imposing unit displacements.

corresponding displacements are, respectively, the vertical displacement δ_1 and the end rotation which is given the symbol δ_2. The force and displacement are now vector quantities

$$\begin{bmatrix} F_1 \\ F_2 \end{bmatrix} \quad \text{and} \quad \begin{bmatrix} \delta_1 \\ \delta_2 \end{bmatrix}$$

and the constant of proportionality between them is a 2×2 stiffness matrix S.

Thus

$$\begin{bmatrix} F_1 \\ F_2 \end{bmatrix} = \begin{bmatrix} S_{11} & S_{12} \\ S_{21} & S_{22} \end{bmatrix} \begin{bmatrix} \delta_1 \\ \delta_2 \end{bmatrix}. \qquad (4.2.2)$$

To evaluate the elements S_{11} and S_{21} of S a unit vertical deflection $\delta_1 = 1$ is introduced, but rotation δ_2 is prevented (i.e. $\delta_2 = 0$). The forces required at the right-hand end of the cantilever to sustain this deflection pattern (Fig. 4.2.1(b)) are easily evaluated and it is seen that eqn (4.2.2) degenerates into

$$\begin{bmatrix} F_1 \\ F_2 \end{bmatrix} = \begin{bmatrix} S_{11} \\ S_{21} \end{bmatrix}.$$ (4.2.3)

In this case, therefore,

$$S_{11} = \frac{12EI}{L^3}, \qquad S_{21} = -\frac{6EI}{L^2}.$$ (4.2.4)

Similarly if $\delta_1 = 0$ and $\delta_2 = 1$ from elementary beam theory

$$S_{12} = -\frac{6EI}{L^2}, \qquad S_{22} = \frac{4EI}{L}.$$ (4.2.5)

It is seen that $S_{12} = S_{21}$. This is a general result which is a consequence of Maxwell's reciprocal theorem.

A large structure can be divided into elements each of which has nodal points at which the analyst wishes to evaluate the deflections. By writing down two column vectors $\{F\}$ and $\{\delta\}$ for the known forces and the unknown displacements, respectively, at the nodal points of the structure and by introducing unit displacements in turn it is possible to establish the stiffness matrix of the structure. Such a technique is described in many elementary texts on the matrix analysis of structures.

In practice it is usually more convenient to separate a typical element from the structure and write down the forces and the associated displacements which can occur at each node in that element. The linear relationship between these quantities is written as a small matrix equation in the local coordinates of the element. For example the end moments and rotations of a beam element may be chosen as the forces and displacements and the element stiffness equation is

$$\begin{bmatrix} F_1^\alpha \\ F_2^\alpha \end{bmatrix} = \begin{bmatrix} S_{11}^\alpha & S_{12}^\alpha \\ S_{21}^\alpha & S_{22}^\alpha \end{bmatrix} \begin{bmatrix} \delta_1^\alpha \\ \delta_2^\alpha \end{bmatrix} \quad \text{or} \quad F^\alpha = S^\alpha \delta^\alpha$$ (4.2.6)

where the α indicates that the element and not the structure is being treated.

However, for many structures the local and global (structure) coordinates do not coincide and it is necessary to transform both the forces and displacements to global coordinates so that the structure (or global) stiffness matrix can be assembled at a later stage. It is a matter of simple geometrical analysis to establish the following linear relationship between the displacements $\delta^{\alpha'}$ in local coordinates and δ^α the same displacements expressed in global coordinates,

$$\delta^{\alpha'} = C \delta^\alpha$$ (4.2.7)

where C is a matrix. A simple equilibrium analysis or the application of the principle of virtual displacements shows that the forces expressed in local coordinates are transformed to the global coordinate system, i.e. to F^α by the following relationship

$$F^\alpha = C^\mathrm{T} F^{\alpha'} \tag{4.2.8}$$

where C^T is the transpose of C. This result is at first surprising but it should be realised that it is simply a requirement of equilibrium. A set of forces expressed in one coordinate system is being replaced by the same set but they are expressed in a different coordinate system. These relationships expressed in eqns (4.2.7) and (4.2.8) are known as the *contragradient law*. The element stiffness equation in global coordinates is now

$$F^\alpha = C^\mathrm{T} F^{\alpha'} = C^\mathrm{T} S^{\alpha'} \delta^{\alpha'} = C^\mathrm{T} S^{\alpha'} C \delta^\alpha \tag{4.2.9}$$

where eqns (4.2.8), (4.2.6) and (4.2.7) have been used. The product $C^\mathrm{T} S^{\alpha'} C$ is the constant of proportionality between F^α and δ^α, i.e. it is the element stiffness matrix S^α expressed in global coordinates.

The structure stiffness matrix is now assembled by writing down equations of equilibrium at each node. The element stiffness matrices are slotted into the global stiffness matrix at the appropriate locations. Boundary conditions are also inserted to complete the establishment of the global stiffness matrix, S. The equilibrium equations become

$$F = S\delta \tag{4.2.10}$$

where F are the loads applied at the nodes of the structure and δ are the corresponding unknown displacements. Equation (4.2.10) is solved by means of a computer using matrix inversion and the element forces and displacements are found by back-substitution into eqns (4.2.8) and (4.2.7).

The same technique can be applied to continua problems (e.g. plates, shells, etc.) The only really new concept to be considered then is the use of *generalized coordinates*. The displacements at points within the elements are expressed as polynomials. For example a sheet or lamina shown in Fig. 4.2.2(a) has been divided into triangular elements, one of which is shown in Fig. 4.2.2(b). The displacements in the x- and z-directions of an arbitrary point inside of the element are expressed as linear functions of x and z.

$$
\begin{aligned}
u(x, z) &= C_1 x + C_2 z + C_3 \\
v(x, z) &= C_4 x + C_5 z + C_6
\end{aligned}
\tag{4.2.11}
$$

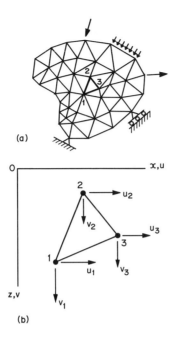

Fig. 4.2.2. (a) A lamina is divided into finite elements; (b) nodal displacements of a triangular element.

or in matrix form

$$\begin{bmatrix} u \\ v \end{bmatrix} = \begin{bmatrix} x & z & 1 & 0 & 0 & 0 \\ 0 & 0 & 0 & x & z & 1 \end{bmatrix} \begin{bmatrix} C_1 \\ C_2 \\ \cdot \\ \cdot \\ \cdot \\ C_6 \end{bmatrix} \qquad (4.2.12)$$

The constants C_1–C_6 can be related to the six nodal displacements u_1, v_1, etc. (Fig. 4.2.2) by substituting the coordinates $(a_1, b_1), (a_2, b_2)$ and (a_3, b_3) into eqn (4.2.12). Thus

$$\begin{bmatrix} a_1 & b_1 & 1 & & & \\ a_2 & b_2 & 1 & & 0 & \\ a_3 & b_3 & 1 & & & \\ & & & a_1 & b_1 & 1 \\ & 0 & & a_2 & b_2 & 1 \\ & & & a_3 & b_3 & 1 \end{bmatrix} \begin{bmatrix} C_1 \\ C_2 \\ C_3 \\ C_4 \\ C_5 \\ C_6 \end{bmatrix} = \begin{bmatrix} u_1 \\ u_2 \\ u_3 \\ v_1 \\ v_2 \\ v_3 \end{bmatrix} \qquad (4.2.13)$$

This equation can be solved for C_1–C_6 by inversion and these values are then substituted into eqn (4.2.12). From this point onwards all quantities are expressed in terms of the joint displacement vector which stands on the right-hand side of eqn (4.2.13). Expressions for the strain components ε_x, ε_z and γ_{xz} are first obtained from the usual first-order expressions, that is,

$$
\begin{bmatrix} \varepsilon_x \\ \varepsilon_z \\ \gamma_{xz} \end{bmatrix} = \begin{bmatrix} \dfrac{\partial u}{\partial x} \\ \dfrac{\partial v}{\partial z} \\ \dfrac{\partial u}{\partial z} + \dfrac{\partial v}{\partial x} \end{bmatrix}
\tag{4.2.14}
$$

and the stresses from the stress–strain relationship. For example for a plane stress problem

$$
\begin{bmatrix} \sigma_x \\ \sigma_z \\ \tau_{xz} \end{bmatrix} = \frac{E}{(1-v^2)} \begin{bmatrix} 1 & v & 0 \\ v & 1 & 0 \\ 0 & 0 & \dfrac{1-v}{2} \end{bmatrix} \begin{bmatrix} \varepsilon_x \\ \varepsilon_z \\ \gamma_{xz} \end{bmatrix}
\tag{4.2.15}
$$

This stress distribution enables statically equivalent forces to be established at each node in each of the x- and z-directions so that an element stiffness matrix has been established. The procedure from this point onwards is identical to that described previously for a framework.

This description is of a finite element technique called the direct method which is probably the simplest available. The vast literature available describes how the approach may be refined. In the next section the finite element technique known as 'the finite strip method' is described.

4.3. LINEAR ANALYSIS OF FOLDED PLATE STRUCTURES BY FINITE STRIP METHOD

A simply-supported folded plate structure is illustrated in Fig. 4.3.1. It has end diaphragms which are assumed to be rigid within their own plane and perfectly flexible normal to it. The structure is divided into longitudinal strips one of which is shown in Fig. 4.3.1. Each strip has constant properties, but they can vary from strip to strip. On the boundaries of each strip there will be, in general, both in-plane stresses (σ_x, σ_z, and τ_{xz}) and out-of-plane stresses arising from the bending and twisting moments (M_x, M_z, M_{xz}). It is fortunate that in first-order theory the in-plane and out-of-plane effects are not coupled. In other words, these two effects can be treated separately by two sets of equations. In the next two sections

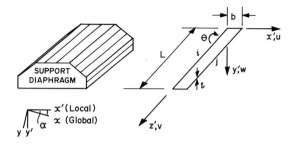

Fig. 4.3.1. Folded plate problem and strip ij.

the in-plane and out-of-plane stiffness equations of the element are derived and in Section 4.3.3 the element stiffness matrices are assembled to form the stiffness matrix of the entire structure.

4.3.1. In-plane stiffness of a strip

The boundary conditions for the in-plane forces and displacements are summarized as follows:

at $z = 0$ and L where the structure is simply supported

$$u = 0; \qquad \sigma_z = 0 \quad \text{or} \quad \frac{\partial v}{\partial z} + v \frac{\partial u}{\partial x} = 0 \tag{4.3.1}$$

at $x = 0$

$$u = u_i; \qquad v = v_i \tag{4.3.2}$$

at $x = b$

$$u = u_j; \qquad v = v_j. \tag{4.3.3}$$

A set of displacement functions which satisfy these conditions are the following truncated Fourier Series.

$$u = \sum_{m=1,2}^{r} \left[\left(1 - \frac{x}{b}\right) u_{im} + \left(\frac{x}{b}\right) u_{jm} \right] \sin k_m z$$

$$v = \sum_{m=1,2}^{r} \left[\left(1 - \frac{x}{b}\right) v_{im} + \left(\frac{x}{b}\right) v_{jm} \right] \cos k_m z \tag{4.3.4}$$

where

$$k_m = \frac{m\pi}{L}. \tag{4.3.5}$$

and u_{im}, v_{im}, etc. are the generalized displacement parameters of the mth term along the nodal lines. Equations (4.3.4) can be written in matrix form as follows

$$\begin{bmatrix} u \\ v \end{bmatrix} = \{C_{p1} C_{p2} \dots C_{pr}\} \begin{bmatrix} \delta_{p1} \\ \delta_{p2} \\ \vdots \\ \delta_{pr} \end{bmatrix} \tag{4.3.6}$$

where

$$C_{pm} = \begin{bmatrix} \left(1 - \dfrac{x}{b}\right)\sin k_m z & 0 & \dfrac{x}{b}\sin k_m z & 0 \\[2mm] 0 & \left(1 - \dfrac{x}{b}\right)\cos k_m z & 0 & \dfrac{x}{b}\cos k_m z \end{bmatrix} \tag{4.3.7}$$

$$\delta_{pm} = \{u_{im} \quad v_{im} \quad u_{jm} \quad v_{jm}\}^{\mathrm{T}}. \tag{4.3.8}$$

The strain matrix is now defined and then eqn (4.3.6) is used.

$$\varepsilon = \begin{bmatrix} \varepsilon_x \\ \varepsilon_z \\ \gamma_{xz} \end{bmatrix} = \begin{bmatrix} \dfrac{\partial u}{\partial x} \\[2mm] \dfrac{\partial v}{\partial z} \\[2mm] \dfrac{\partial u}{\partial z} + \dfrac{\partial v}{\partial x} \end{bmatrix} = \{B_{p1} B_{p2} \dots B_{pr}\} \begin{bmatrix} \delta_{p1} \\ \delta_{p2} \\ \vdots \\ \delta_{pr} \end{bmatrix} \tag{4.3.9}$$

where

$$B_{pm} = \begin{bmatrix} -\dfrac{1}{b}\sin k_m z & 0 & \dfrac{1}{b}\sin k_m z & 0 \\[2mm] 0 & -\left(1 - \dfrac{x}{b}\right)k_m \sin k_m z & 0 & -\dfrac{x}{b}k_m \sin k_m z \\[2mm] \left(1 - \dfrac{x}{b}\right)k_m \cos k_m z & -\dfrac{1}{b}\cos k_m z & \dfrac{x}{b}k_m \cos k_m z & \dfrac{1}{b}\cos k_m z \end{bmatrix}. \tag{4.3.10}$$

For plane stress problems the stress matrix is related to ε as follows

$$\sigma = \begin{bmatrix} \sigma_x \\ \sigma_y \\ \tau_{xy} \end{bmatrix} = \begin{bmatrix} \dfrac{E}{1-v^2} & \dfrac{vE}{1-v^2} & 0 \\[2mm] \dfrac{vE}{1-v^2} & \dfrac{E}{1-v^2} & 0 \\[2mm] 0 & 0 & G \end{bmatrix} \begin{bmatrix} \varepsilon_x \\ \varepsilon_z \\ \gamma_{xz} \end{bmatrix} \tag{4.3.11}$$

i.e.

$$\sigma = D_p \varepsilon = D_p B_p \delta_p. \tag{4.3.12}$$

This equation represents the sum of all the harmonics m.
    The total potential energy of the strip is the sum of three quantities.

$$U_p = \frac{t}{2} \int_0^L \int_0^b (\sigma_x \varepsilon_x + \sigma_z \varepsilon_z + \tau_{xz} \gamma_{xz})\, dx\, dz$$

$$- \int_0^L \int_0^b (Xu + Zv)\, dx\, dz - \int_0^L (X_i u_i + Z_i v_i + X_j u_j + Z_j v_j)\, dz. \tag{4.3.13}$$

The first term is the strain energy gained by the strip during deformation, the second term is the potential energy lost by the surface loads X and Z (forces per unit area) and the third is the potential energy lost by the loads X_i, Z_i, X_j, and Z_j which act along the nodal lines i and j. In matrix notation eqn (4.3.13) becomes

$$U_p = \frac{t}{2} \int_0^L \int_0^b \varepsilon^T \sigma\, dx\, dz - \int_0^L \int_0^b \{u\, v\} \begin{bmatrix} X \\ Z \end{bmatrix} dx\, dz$$

$$- \int_0^L \{u_i\, v_i\, u_j\, v_j\} \begin{bmatrix} X_i \\ Z_i \\ X_j \\ Z_j \end{bmatrix} dz. \tag{4.3.14}$$

By now employing eqns (4.3.9), (4.3.11), (4.3.6) and (4.3.8) eqn (4.3.14) can be written as follows

$$U_p = \frac{t}{2} \int_0^L \int_0^b \delta_p^T B_p^T D_p B_p \delta_p\, dx\, dz - \int_0^L \int_0^b \{C_p \delta_p\}^T \begin{bmatrix} X \\ Z \end{bmatrix} dx\, dz$$

$$- \sum_m^r \delta_{pm}^T \int_0^L R_m^T \begin{bmatrix} X_i \\ Z_i \\ X_j \\ Z_j \end{bmatrix} dz \tag{4.3.15}$$

where

$$R_m^T = \begin{bmatrix} \sin k_m z & 0 & 0 & 0 \\ 0 & \cos k_m z & 0 & 0 \\ 0 & 0 & \sin k_m z & 0 \\ 0 & 0 & 0 & \cos k_m z \end{bmatrix}. \tag{4.3.16}$$

The loads X and Z can also be expressed as Fourier series, thus

$$X = \sum_{n=1,2}^{r} X_n \sin k_n z; \qquad Z = \sum_{n=1,2}^{r} Z_n \cos k_n z. \qquad (4.3.17)$$

This form for Z has been chosen so that the loads are self-equilibrating. The coefficients X_n and Z_n are obtained by Fourier analysis. For distributed loads from $z = c$ to $z = d$

$$X_n = \frac{\displaystyle\int_c^d X \sin k_n z \, dz}{\displaystyle\int_0^L \sin^2 k_n z \, dz}. \qquad (4.3.18)$$

For a concentrated load X_c at $z = c$

$$X_n = \frac{X_c \sin k_n c}{\displaystyle\int_0^L \sin^2 k_n z \, dz}. \qquad (4.3.19)$$

The coefficients Z_n can also be found by Fourier analysis.
Equation (4.3.15) can now be written as

$$U_p = \sum_m^r \sum_n^r \left[\frac{t}{2} \delta_{pm}^T \left(\int_0^L \int_0^b B_{pm}^T D_p B_{pn} \, dx \, dz \right) \delta_{pn} \right.$$

$$- \delta_{pm}^T \int_0^L \int_0^b C_{pm}^T \begin{bmatrix} X_n \sin k_n z \\ Z_n \cos k_n z \end{bmatrix} dx \, dz$$

$$\left. - \delta_{pm}^T \int_0^L R_m^T R_n \begin{bmatrix} Z_{in} \\ X_{in} \\ X_{jn} \\ Z_{jn} \end{bmatrix} dz \right]. \qquad (4.3.20)$$

All of the integrals turn out to be orthogonal, i.e.

$$\int_0^L \sin k_m z \sin k_n z \, dz = \int_0^L \cos k_m z \cos k_n z \, dz = 0 \ \text{(for } m \neq n) \qquad (4.3.21)$$

$$= \frac{L}{2} \text{(for } m = n)$$

and, hence, the suffix n can be dropped. From the principle of minimum total potential energy, that is,

$$\frac{\partial U_p}{\partial \delta_{pm}} = 0 \qquad\qquad (4.3.22)$$

and from eqn (4.3.20)

$$\frac{\partial U_p}{\partial \delta_{pm}} = 0 = \sum_m^r \left[t\delta_{pm}^T \int_0^L \int_o^b B_{pm}^T D_p B_{pm} \, dx \, dz \right.$$

$$\left. - \int_0^L \int_o^b C_{pm}^T \begin{bmatrix} X_m \sin k_m z \\ Z_m \cos k_m z \end{bmatrix} dx \, dz - \int_0^L R_m^T R_m \begin{bmatrix} X_{im} \\ Z_{im} \\ X_{jm} \\ Z_{jm} \end{bmatrix} dz \right]. \qquad (4.3.23)$$

This is in reality a set of four simultaneous equations for each value of m. They can be written in the form

$$S_{pm}\delta_{pm} - F_{pm} - F_{pm}^a = 0 \qquad\qquad (4.3.24)$$

where S_{pm} is a 4×4 square symmetrical stiffness matrix, F_{pm} is the matrix for the distributed loads and F_{pm}^a is that for the nodal loads. After some algebraic manipulation and tedious but simple integrations the matrix for S_{pm} is found to be

$$S_{pm} = \begin{bmatrix} S_1 + S_2 & & \text{symmetrical} & \\ S_3 - S_4 & S_5 + S_6 & & \\ -S_1 + S_2/2 & -S_3 - S_4 & S_1 + S_2 & \\ S_3 + S_4 & S_5/2 - S_6 & -S_3 + S_4 & S_5 + S_6 \end{bmatrix}$$

where

$$S_1 = \frac{LEt}{2(1 - \nu^2)b} \qquad\qquad S_4 = \frac{Lk_m Gt}{4}$$

$$S_2 = \frac{Lbk_m^2 Gt}{6} \qquad\qquad S_5 = \frac{Lbk_m^2 Et}{6(1 - \nu^2)}$$

$$S_3 = \frac{Lk_m \nu Et}{4(1 - \nu^2)} \qquad\qquad S_6 = \frac{LGt}{2b} \qquad\qquad (4.3.25)$$

$$G = \frac{E}{2(1 + \nu)} = \text{shear modulus.}$$

The remaining terms in eqn (4.3.24) must be obtained by first finding the Fourier coefficients from eqn (4.3.18) and (4.3.19) and then substituting them in eqn (4.3.23).

4.3.2. Bending stiffness of a strip

The boundary conditions for the bending forces and displacements at $z = 0$ and L are

$$w = 0, \quad M_z = 0 \quad \text{or} \quad \left(\frac{\partial^2 w}{\partial z^2} + v \frac{\partial^2 w}{\partial x^2} \right) = 0 \tag{4.3.26}$$

and at $x = 0$

$$w = w_i, \quad \frac{\partial w}{\partial x} = \theta_i \tag{4.3.27}$$

and at $x = b$

$$w = w_j, \quad \frac{\partial w}{\partial x} = \theta_j. \tag{4.3.28}$$

These boundary conditions can be satisfied by the following displacement function

$$
w = \sum_{m=1,2}^{r} \left[\left(1 - \frac{3x^2}{b^2} + \frac{2x^3}{b^3} \right) w_{im} + \left(x - \frac{2x^2}{b} + \frac{x^3}{b^2} \right) \theta_{im} \right.
$$
$$
\left. + \left(\frac{3x^2}{b^2} - \frac{2x^3}{b^3} \right) w_{jm} + \left(\frac{x^3}{b^2} - \frac{x^2}{b} \right) \theta_{jm} \right] \sin k_m z \tag{4.3.29}
$$

$$= \sum_{m=1,2}^{r} C_{bm} \delta_{bm} \tag{4.3.30}$$

where

$$C_{bm} = \left\{ 1 - \frac{3x^2}{b^2} + \frac{2x^3}{b^3}, \; x - \frac{2x^2}{b} + \frac{x^3}{b^2}, \; \frac{3x^2}{b^2} - \frac{2x^3}{b^3}, \; \frac{x^3}{b^2} - \frac{x^2}{b} \right\} \sin k_m z \tag{4.3.31}$$

$$\delta_{bm} = \{ w_{im} \theta_{im} w_{jm} \theta_{jm} \}^{\mathrm{T}}. \tag{4.3.32}$$

The curvatures are written as

$$
\chi = \begin{bmatrix} -\dfrac{\partial^2 w}{\partial x^2} \\[2mm] -\dfrac{\partial^2 w}{\partial z^2} \\[2mm] 2\dfrac{\partial^2 w}{\partial x \, \partial z} \end{bmatrix} = B_b \delta_b = \sum_{m=1,2}^{r} B_{bm} \delta_{bm}. \tag{4.3.33}
$$

From these relationships

$$
B_{bm} = \begin{bmatrix} \left(\dfrac{6}{b^2}-\dfrac{12x}{b^3}\right)\sin k_m z & \left(\dfrac{4}{b}-\dfrac{6x}{b^2}\right)\sin k_m z \\[2mm] \left(1-\dfrac{3x^2}{b^2}+\dfrac{2x^3}{b^3}\right)k_m^2\sin k_m z & \left(x-\dfrac{2x^2}{b}+\dfrac{x^3}{b^2}\right)k_m^2\sin k_m z \\[2mm] 2\left(-\dfrac{6x}{b^2}+\dfrac{6x^2}{b^3}\right)k_m\cos k_m z & 2\left(1-\dfrac{4x}{b}+\dfrac{3x^2}{b^2}\right)k_m\cos k_m z \end{bmatrix}
$$

$$
\begin{bmatrix} \left(-\dfrac{6}{b^2}+\dfrac{12x}{b^3}\right)\sin k_m z & \left(-\dfrac{6x}{b^2}+\dfrac{2}{b}\right)\sin k_m z \\[2mm] \left(\dfrac{3x^2}{b^2}-\dfrac{2x^3}{b^3}\right)k_m^2\sin k_m z & \left(\dfrac{x^3}{b^2}-\dfrac{x^2}{b}\right)k_m^2\sin k_m z \\[2mm] 2\left(\dfrac{6x}{b^2}-\dfrac{6x^2}{b^3}\right)k_m\cos k_m z & 2\left(\dfrac{3x^2}{b^2}-\dfrac{2x}{b}\right)k_m\cos k_m z \end{bmatrix}. \quad (4.3.34)
$$

The relationship between moment and curvature (Timoshenko and Woinowsky-Kreiger 1959) is

$$
M = \begin{bmatrix} M_x \\ M_z \\ M_{xz} \end{bmatrix} = D \begin{bmatrix} 1 & \nu & 0 \\ \nu & 1 & 0 \\ 0 & 0 & \dfrac{1-\nu}{2} \end{bmatrix} \begin{bmatrix} -\dfrac{\partial^2 w}{\partial x^2} \\ -\dfrac{\partial^2 w}{\partial z^2} \\ 2\dfrac{\partial^2 w}{\partial x\,\partial z} \end{bmatrix} = D_b B_b \delta_d \quad (4.3.35)
$$

where

$$
D = \frac{Et^3}{12(1-\nu^2)}.
$$

The total potential energy for bending is

$$
U_b = \frac{1}{2}\int_0^L \int_0^b \left(-M_x\frac{\partial^2 w}{\partial x^2}-M_z\frac{\partial^2 w}{\partial z^2}+2M_{xz}\frac{\partial^2 w}{\partial x\,\partial z}\right)dx\,dz
$$

$$
-\int_0^L \int_0^b Yw\,dx\,dz - \int_0^L \{w_i\ \theta_i\ w_j\ \theta_j\}\begin{bmatrix} Y_i \\ M_i \\ Y_j \\ M_j \end{bmatrix}dz. \quad (4.3.36)
$$

Thus, the equation which corresponds to eqn (4.3.23) is

$$
S_{bm}\delta_{bm} - F_{bm} - F_{bm}^a = 0 \quad (4.3.37)
$$

where the stiffness matrix S_{bm} is given by eqn (4.3.38) and again F_{bm} and F_{bm}^a must be derived from a Fourier analysis.

$$S_{bm} = \begin{bmatrix} S_7 & & \text{symmetrical} \\ S_8 & S_{11} & & \\ S_9 & -S_{10} & S_7 & \\ S_{10} & S_{12} & -S_8 & S_{11} \end{bmatrix}$$

where

$$S_7 = DL \left(\frac{13}{70} bk_m^4 + \frac{6}{5b} k_m^2 + \frac{6}{b^3} \right) \qquad S_{10} = DL \left(-\frac{13}{840} b^2 k_m^4 + \frac{1}{10} k_m^2 + \frac{3}{b^2} \right)$$

$$S_8 = DL \left(\frac{11}{420} b^2 k_m^4 + \left(\frac{\nu}{2} + \frac{1}{10}\right) k_m^2 + \frac{3}{b^2} \right) \quad S_{11} = DL \left(\frac{b^3 k_m^4}{210} + \frac{2}{15} bk_m^2 + \frac{2}{b} \right)$$

$$S_9 = DL \left(\frac{9}{140} bk_m^4 - \frac{6}{5b} k_m^2 - \frac{6}{b^3} \right) \qquad S_{12} = DL \left(-\frac{b^3 k_m^4}{280} - \frac{1}{30} bk_m^2 + \frac{1}{b} \right).$$

$$(4.3.38)$$

4.3.3 Assembly and solution

So far only the element (a strip) has been considered and its stiffness has been analysed in two parts separately because of the uncoupling of in-plane and out-of-plane effects. When both sets of loads act the element stiffness equation is obtained by combining eqns (4.3.25) and (4.3.38). The element deflection matrix becomes

$$\delta_m = \{u_{im} v_{im} w_{im} \theta_{im} u_{jm} v_{jm} w_{jm} \theta_{jm}\}^T \qquad (4.3.39)$$

and the element load matrix to which it corresponds is obtained by combining F_{pm}, F_{pm}^a, F_{bm}, and F_{bm}^a from eqns (4.3.23) and (4.3.36).

The load matrix then assumes the form

$$F_m = \{U_{im} V_{im} W_{im} M_{im} U_{jm} V_{jm} W_{jm} M_{jm}\}^T \qquad (4.3.40)$$

and the stiffness equation then becomes

$$S_m \delta_m = F_m \qquad (4.3.41)$$

where S_m is the comprehensive element stiffness matrix. The element $(s_m)_{ij}$ of S_m is made up of appropriate elements from S_{pm} and S_{bm}, thus

$$\begin{array}{c} (s_m)_{ij} \\ (4 \times 4) \end{array} = \begin{bmatrix} \begin{array}{c} (s_{pm})_{ij} \\ (2 \times 2) \end{array} & 0 \\ 0 & \begin{array}{c} (s_{bm})_{ij} \\ (2 \times 2) \end{array} \end{bmatrix} \qquad (4.3.42)$$

Finally, it is necessary to assemble the element stiffness matrices to form the structure stiffness matrix. This involves a transformation from local coordinates to global coordinates. The contragradient transformation (eqns (4.2.7) and (4.2.8)) of forces and displacements (Fig. 4.3.1) between the local element coordinates x', y', z' and the global (structure coordinates x, y, z is

$$F_m = C^T F'_m \tag{4.3.43}$$

$$\delta'_m = C \delta_m \tag{4.3.44}$$

The transformation matrix is

$$C = \begin{bmatrix} c & 0 \\ 0 & c \end{bmatrix} \tag{4.3.45}$$

and

$$c = \begin{bmatrix} \cos \alpha & 0 & -\sin \alpha & 0 \\ 0 & 1 & 0 & 0 \\ \sin \alpha & 0 & \cos \alpha & 0 \\ 0 & 0 & 0 & 1 \end{bmatrix}. \tag{4.3.46}$$

From eqns (4.3.43), (4.3.44) and (4.3.45)

$$F_m = (C^T S'_m C) \delta_m \tag{4.3.47}$$

whence

$$S_m = C^T S'_m C. \tag{4.3.48}$$

It is now possible to assemble the stiffness matrix of the whole structure in the normal way.

Example 4.3.1

A plate of length L, thickness t and width $2b$ (Fig. 4.3.2) is divided into two strips each of width b. Set up the assembled stiffness matrix S_m for the mth term in the Fourier series. Describe how the stiffness equation is affected if

(a) nodes 1 and 3 are pinned;
(b) nodes 1 and 3 are built-in.

For the first element the deflection and load vectors for the mth term in the Fourier series are

$$\delta_m = \{u_{1m} v_{1m} w_{1m} \theta_{1m} u_{2m} v_{2m} w_{2m} \theta_{2m}\}^T$$

$$F_m = \{U_{1m} V_{1m} W_{1m} M_{1m} U_{2m} V_{2m} W_{2m} M_{2m}\}^T$$

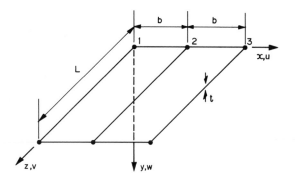

Fig. 4.3.2. Plate with two elements developed in Example 4.3.1.

The stiffness matrix for the first element is

$$
S_m =
\begin{bmatrix}
S_{p11} & S_{p12} & & & S_{p13} & S_{p14} & & \\
 & & & 0 & & & & 0 \\
S_{p21} & S_{p22} & & & S_{p23} & S_{p24} & & \\
 & & S_{b11} & S_{b12} & & & S_{b13} & S_{b14} \\
 & 0 & & & & 0 & & \\
 & & S_{b21} & S_{b22} & & & S_{b23} & S_{b24} \\
S_{p31} & S_{p32} & & & S_{p33} & S_{p34} & & \\
 & & & 0 & & & & 0 \\
S_{p41} & S_{p42} & & & S_{p43} & S_{p44} & & \\
 & & S_{b31} & S_{b32} & & & S_{b33} & S_{b34} \\
 & 0 & & & & 0 & & \\
 & & S_{b41} & S_{b42} & & & S_{b43} & S_{b44}
\end{bmatrix}
$$

where S_{p11}, S_{p12}, etc., are the elements of the in-plane stiffness matrix S_{pm} (eqn (4.3.25)) and S_{b11}, S_{b12}, etc., are those of the bending stiffness matrix S_{bm} (eqn (4.3.38)). There is an identical stiffness matrix for the second strip 2-3.

The deflection and load vectors for the entire plate are

$$
\delta_m = \{u_{1m} v_{1m} w_{1m} \theta_{1m} u_{2m} v_{2m} w_{2m} \theta_{2m} u_{3m} v_{3m} w_{3m} \theta_{3m}\}^{\mathrm{T}}
$$
$$
F_m = \{U_{1m} V_{1m} W_{1m} M_{1m} U_{2m} V_{2m} W_{2m} M_{2m} U_{3m} V_{3m} W_{3m} M_{3m}\}^{\mathrm{T}}
$$

and the corresponding (structural) stiffness matrix is obtained by overlaying the two element stiffness matrices as follows

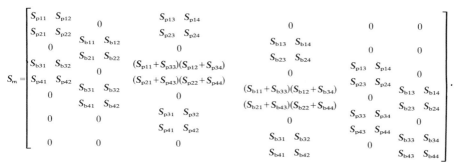

(a) When nodes 1 and 3 are pinned the following elements of δ_m and F_m are zero, w_{1m}, w_{3m}, M_{1m}, and M_{3m}. The stiffness equation is solved after deleting the third row and column and eleventh row and column of S_m and inserting $M_{1m} = M_{3m} = 0$ in F_m.

(b) When nodes 1 and 3 are built-in the following elements of δ_m are zero, w_{1m}, θ_{1m}, w_{3m}, and θ_{3m}. The stiffness equation is solved after eliminating the third, fourth, eleventh, and twelfth rows and columns of S_m.

4.4. ANALYSIS OF BOX-GIRDERS WITH DIAPHRAGMS OR FRAMEWORKS BY THE FINITE STRIP METHOD

In this section it is shown how the finite strip method may be used to analyse the linear elastic behaviour of box-girders. The results of this analysis are the deflections and distribution of stresses in the cross-section for the given distribution of applied loads and geometry of the girder. When the girder has no diaphragms the finite strip method described in Section 4.3 may be used directly, or alternatively, the analytical method of Section 3.2.2 may be applied. These two methods will not give quite the same answers because the latter method assumes that the cross-section does not change its shape whereas the former allows the cross-section to distort. The additional stresses which arise from the distortion of a box girder can be calculated by using the theory outlined in Section 7.2.1.

The method used here to analyse box girders with diaphragms is that developed by Loo (1976). The analysis of the assembled structure is essentially an application of the flexibility (force) method, but some of the load-deflection relationships are first evaluated by the finite strip method which is a stiffness (displacement) method.

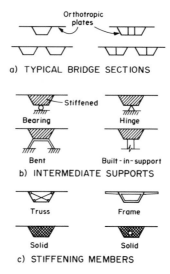

a) TYPICAL BRIDGE SECTIONS

b) INTERMEDIATE SUPPORTS

c) STIFFENING MEMBERS

Fig. 4.4.1. Some typical features of multi-span continuous box-girder bridges.

Figure 4.4.1 shows some typical cross-sections and other features of multi-span continuous box-girder bridges. Intermediate supports and stiffening members must be considered in the analysis. It is assumed that in the entire girder, i.e. over several spans, there are M stiffening diaphragms or frameworks which are connected to the box at discrete points (Fig. 4.4.2) called restrained points. It is also assumed that the diaphragms have perfect flexibility out of their plane so they offer no resistance to displacements in the spanwise direction. Thus there are only three redundant forces at each restrained point, that is, a vertical force, a horizontal force in the transverse direction, and a moment in the plane of the girder cross-section. Sliding and hinged connections at the supports [Fig. 4.4.2(b) and (c)] contribute a further one and two redundancies, respectively, at each support. In the flexibility method, releases are introduced at the redundancies so that the structure is reduced to one which is statically determinate. On this structure the unknown redundant forces are applied and at a restrained section I the redundant forces are

$$[r]_\mathrm{I} = \{r_1, r_2, \ldots, r_N\}_\mathrm{I}^\mathrm{T}. \tag{4.4.1}$$

For a fully built-in diaphragm or framework the number of redundant forces N is equal to $3n$ where n is the number of restrained points. The deflections at the corresponding points induced by the applied loading on

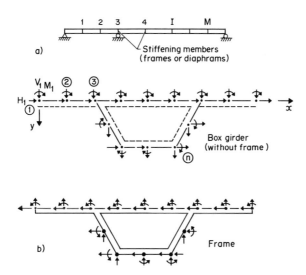

Fig. 4.4.2. (a) Elevation of box-girder with transverse stiffeners; (b) location of redundant forces between box-girder and frame. (Other redundant forces occur at the supports.)

the released structures are

$$[\delta]_{\mathrm{I}} = \{\delta_1, \delta_2, \ldots, \delta_N\}_{\mathrm{I}}^{\mathrm{T}} \tag{4.4.2}$$

These displacements are in fact the relative deflections at the nodal points between the box-girder with the frames (or diaphragms) removed and the frames (or diaphragms) themselves. The displacements of the box-girder are obtained by using the basic finite strip method described in Section 4.3 and those of the frames by ordinary structural or stress analysis. In the case of diaphragms these latter deflections will undoubtedly be small and can probably by neglected.

The compatibility condition at the M restrained sections is

$$[f][r] = -[\delta] \tag{4.4.3}$$

where

$$\{r\} = [\{r\}_1 \{r\}_2 \ldots \{r\}_M]^{\mathrm{T}} \tag{4.4.4}$$

$$\{\delta\} = [\{\delta\}_1 \{\delta\}_2 \ldots \{\delta\}_M]^{\mathrm{T}} \tag{4.4.5}$$

and $[f]$ is the flexibility matrix of the released structure. The elements of $[f]$ are evaluated in the usual way, i.e. by introducing unit values of $\{r\}$ (e.g. $\{1 \quad 0 \quad 0 \ldots 0\}^{\mathrm{T}}$, to obtain the first column of $[f]$, $\{0 \quad 1 \quad 0 \quad 0 \ldots 0\}^{\mathrm{T}}$ to obtain the second and so on) and by calculating the resulting relative deflections between the box-girder and the frames. Thus the flexibility matrix is the sum of that of the released box-girder and that of the framework, i.e.

$$
[f] = \begin{bmatrix} [f_{11}] & [f_{12}] & \cdots & [f_{1M}] \\ \cdot & \cdot & & \cdot \\ \cdot & \cdot & & \cdot \\ \cdot & \cdot & & \cdot \\ [f_{M1}] & [f_{M2}] & & [f_{MM}] \end{bmatrix} + \begin{bmatrix} [f'_{11}] & & \\ & [f'_{22}] & \\ & & \\ 0 & & [f'_{MM}] \end{bmatrix}
\qquad (4.4.6)
$$

where the first matrix on the right is the flexibility matrix of the released box-girder and $[f'_{11}]$ is the flexibility matrix of the stiffening frame at section I. $[f'_{11}]$ is zero if the frame is assumed to be rigid or if the support is unyielding. The elements of the first matrix are also found by using the basic finite strip method described in Section 4.3 and those of the second matrix by usual structural analysis.

At a stiffened section such as that seen in Fig. 4.4.1(c) the restraining forces must be in a state of self-equilibrium, meaning that there are $N-3$ independent unknown forces acting on the frame or diaphragm. Thus, if the forces at node 1 (see Fig. 4.4.2(b)) are chosen as being the dependent forces it is found that

$$
\begin{aligned}
H_1 &= -H_2 - H_3 \ldots - H_N \\
V_1 &= -V_2 - V_3 \ldots - V_N \\
M_1 &= H_2 \bar{y}_2 - V_2 \bar{x}_2 - M_2 + H_3 \bar{y}_3 \ldots - M_N
\end{aligned}
\qquad (4.4.7)
$$

where $\bar{x}_n$ and $\bar{y}_n$ are the coordinates of node n with the origin at node 1. Hence eqn (4.4.1) can be redefined as

$$
[r]_{\mathrm{I}} = [M]_{\mathrm{I}} [R]_{\mathrm{I}}
\qquad (4.4.8)
$$

where $[R]_{\mathrm{I}}$ is a column vector of the independent unknown forces $H_2, V_2, M_2 \ldots M_N$ and

$$
[M]_{\mathrm{I}} = \begin{bmatrix}
-1 & 0 & 0 & -1 & 0 & 0 & \cdots & -1 & 0 & 0 \\
0 & -1 & 0 & 0 & -1 & 0 & \cdots & 0 & -1 & 0 \\
\bar{y}_2 & -\bar{x}_2 & -1 & \bar{y}_3 & -\bar{x}_3 & -1 & & \bar{y}_N & -\bar{x}_N & -1 \\
1 & & & & & & & & & \\
& 1 & & & & & 0 & & & \\
& & 1 & & & & & & 1 & \\
& 0 & & & & & & & & 1 \\
& & & & & & & & & 1
\end{bmatrix}
\qquad (4.4.9)
$$

From the contragradient law (eqns (4.2.7) and (4.2.8)) a set of relative displacements corresponding to $[R]_I$ can be obtained as

$$[\Delta]_I = [M]_I^T[\delta]_I. \tag{4.4.10}$$

When all of the restrained sections are considered the equivalents of eqns (4.4.8) and (4.4.10) for the complete structure can be assembled, that is,

$$[r] = [M][R] \tag{4.4.11}$$

$$[\Delta] = [M]^T[\delta] \tag{4.4.12}$$

where $[R]$ contains all of the independent unknown redundant forces, $[\Delta]$ are the corresponding displacements and

$$[M] = \begin{bmatrix} [M]_1 & & 0 \\ & [M]_2 & \\ 0 & & [M]_M \end{bmatrix}. \tag{4.4.13}$$

When a section is supported $[M]_I$ is a unit matrix because the reactions become the dependent forces in eqn (4.4.8). Substituting eqns (4.4.11) and (4.4.12) into eqn (4.4.3) yields

$$[F][R] = -[\Delta] \tag{4.4.14}$$

in which

$$[F] = [M]^T[f][M]. \tag{4.4.15}$$

After obtaining $[R]$ from eqn (4.4.14) the redundant forces $[r]$ are found by back-substituting in eqn (4.4.11). The final solution is obtained by superimposing the redundant forces and the applied loads on the released structure by using the basic finite strip method.

Loo (1976) has used the technique described here to analyse some box girder bridges and some of the results are illustrated in Figs. 4.4.3 to 4.4.7. Figure 4.4.3 refers to a model of a two-span continuous box beam with framing at B and a rigid diaphragm at C. Results are compared with those from a program (FINPLA) by Meyer and Scordelis (1970). Figures 4.4.4 to 4.4.7 refer to a two-span continuous box bridge under two standard AASHO trucks. The loadings are shown in Fig. 4.4.4 and Fig. 4.4.5 shows the finite strip simulations, Fig. 4.4.6 shows the spanwise distribution of deflections and in Fig. 4.4.7 the distributions of longitudinal in-plane stress resultants (N mm^{-1}) are again compared with results from the FINPLA program of Meyer and Scordelis. The computing time is much less (about one-third) in the case of the finite strip method.

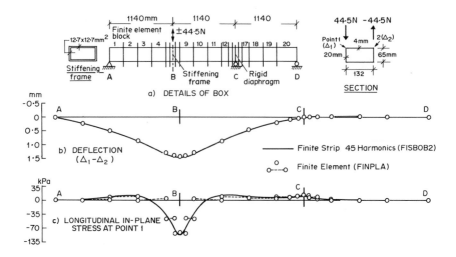

a) DETAILS OF BOX

b) DEFLECTION
($\Delta_1 - \Delta_2$)

—— Finite Strip 45 Harmonics (FISBOB2)

o——o Finite Element (FINPLA)

c) LONGITUDINAL IN-PLANE
STRESS AT POINT 1

Fig. 4.4.3. Spanwise distributions of deflections and stresses for a model box-girder (Loo 1976).

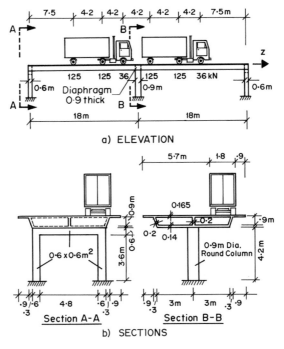

a) ELEVATION

Section A-A

Section B-B

b) SECTIONS

Fig. 4.4.4. Details of two-span continuous box-girder bridge (Loo 1976).

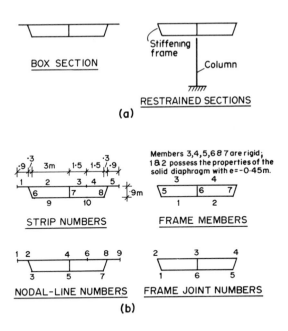

Fig. 4.4.5. Finite strip simulations of the bridge illustrated in Fig. 4.4.4 (Loo 1976).

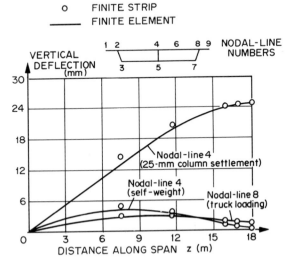

Fig. 4.4.6. Spanwise distributions of deflections of bridge illustrated in Fig. 4.4.4 (Loo 1976).

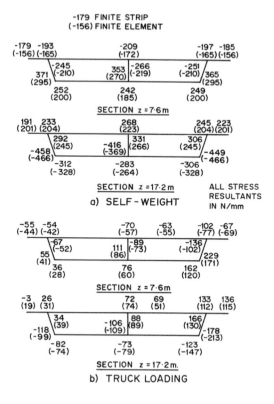

Fig. 4.4.7. Distributions of longitudinal in-plane stress resultants of bridge illustrated in Fig. 4.4.4 (Loo 1976).

4.5. CRITICAL LOADS OF STIFFENED PLATES AND OTHER FOLDED PLATE STRUCTURES WITH AXIAL STRESS

In this section it is assumed that each finite strip (Fig. 4.5.1) carries a uniform axial stress. It is perfectly flat up to the point just before buckling with the uniformly applied stress is $\bar{\sigma}_z$ and in this condition the strip is defined as having zero total potential energy. This is simply an arbitrary datum from which changes in the total potential energy can be measured. When the strip is displaced into its buckled form an arbitrary point (x, z) is displaced (u, w, v). At this point there will be changes in the in-plane stress during buckling as follows:

—in the x direction from 0 to σ_x;
—in the z direction from $\bar{\sigma}_z$ to $\bar{\sigma}_z + \sigma_z$;
—the shear stress from 0 to τ_{xz}.

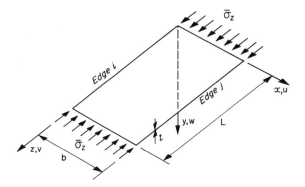

Fig. 4.5.1. Finite strip element with axial stress.

The strain components change as follows:

—in the x-direction the strain increases by ε_x;
—in the z-direction the strain increases by ε_z;
—the shear strain increases by γ_{xz}.

The increase in total potential energy of a small element $dx \times dz$ during buckling is the sum of:

(a) the work done by the stresses on its boundary, i.e.

$$\tfrac{1}{2}t[\sigma_x\varepsilon_x + \sigma_z\varepsilon_z + \tau_{xz}\gamma_{xz}]\,dx\,dz$$

(b) the work done by the bending and twisting moments on its boundary, i.e.

$$-\frac{1}{2}\left[M_X\frac{\partial^2 w}{\partial x^2} + M_z\frac{\partial^2 w}{\partial z^2} - 2M_{xz}\frac{\partial^2 w}{\partial x\,\partial z}\right]dx\,dz,\text{ and}$$

(c) the work done by $\bar{\sigma}_z$ acting through an apparent strain $\bar{\varepsilon}_z$ which makes allowance for the rotation of the element about axes parallel to the x- and z-directions (see below). The positive senses of the bending and twisting moments are shown in Fig. 4.5.2.
The total potential energy of the strip is therefore

$$U = \frac{t}{2}\int_0^L \int_0^b [\sigma_x\varepsilon_x + \sigma_z\varepsilon_z + \tau_{xz}\gamma_{xz}]\,dx\,dz$$

$$-\frac{1}{2}\int_0^L \int_0^b \left[M_X\frac{\partial^2 w}{\partial x^2} + M_z\frac{\partial^2 w}{\partial z^2} - 2M_{xz}\frac{\partial^2 w}{\partial x\,\partial z}\right]dx\,dz$$

$$+ t\int_0^L \int_0^b \bar{\sigma}_z\bar{\varepsilon}_z\,dx\,dz \quad (4.5.1)$$

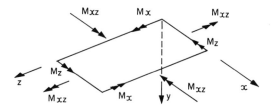

Fig. 4.5.2. Positive plate bending and twisting moments. (Left-hand screw rule is used.)

Most of the analysis is identical to that given in Section 4.3 with the surface and edge loads X_m, Z_m, X_{im}, etc., now assumed to be zero. Indeed those parts of the stiffness matrix which arise from the first two integrals are the previously obtained in-plane and out-of-plane stiffness matrices S_{pm} and S_{bm}, where the suffix m again indicates that the mth Fourier component is being considered. This part of the stiffness matrix is obtained from eqns (4.3.25) and (4.3.38)

$$
S_m^{(0)} = \begin{bmatrix}
\overset{1}{S_1+S_2} & \overset{2}{} & \overset{3}{} & \overset{4}{} & \overset{5}{} & \overset{6}{} & \overset{7}{} & \overset{8}{} \\
S_3-S_4 & S_5+S_6 & & & \text{symmetrical} & & & \\
 & & S_7 & & & & & \\
 & & S_8 & S_{11} & & & & \\
-S_1+S_2/2 & -S_3-S_4 & & & S_1+S_2 & & & \\
S_3+S_4 & S_5/2-S_6 & & & -S_3+S_4 & S_5+S_6 & & \\
 & & S_9 & -S_{10} & & & S_7 & \\
 & & S_{10} & S_{12} & & & -S_8 & S_{11}
\end{bmatrix}
$$

(4.5.2)

If the displacements of a fibre of length dz were small its strain would be $\partial v/\partial z$. However, because the fibre rotates about axes parallel to the x- and y-directions there is an additional (apparent) strain (Fig. 4.5.3). The total apparent strain is therefore

$$
\bar{\varepsilon}_z = \frac{\partial v}{\partial z} - \frac{1}{2}\left[\frac{\partial u}{\partial z}\right]^2 - \frac{1}{2}\left[\frac{\partial w}{\partial z}\right]^2
$$

(4.5.3)

The first term is the usual expression for strain while the last two terms are derived by considering the apparent shortening in the z-direction due to the rotation of the element. For example in Fig. 4.5.3(b) the rotation of the element dz about an axis parallel to the y-axis is $\partial u/\partial z$. Hence

$$
BD = \frac{\partial u}{\partial z}\,dz
$$

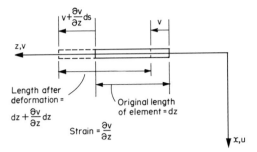

(a) First term of equation (4.5.3)

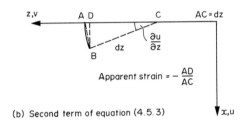

(b) Second term of equation (4.5.3)

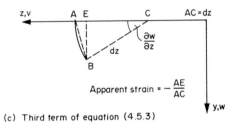

(c) Third term of equation (4.5.3)

Fig. 4.5.3. The apparent strain is the sum of the axial strain and that due to rotations of an element about the y- and x-axes.

and the apparent axial shortening

$$AD = \left(\frac{1}{2}\frac{\partial u}{\partial z}\right)\left(\frac{\partial u}{\partial z}dz\right) = \frac{1}{2}\left(\frac{\partial u}{\partial z}\right)^2 dz,$$

i.e. the apparent strain is $-\frac{1}{2}(\partial u/\partial z)^2$. Similarly, a rotation about an axis parallel to the x-axis gives an additional apparent strain of $-\frac{1}{2}(\partial w/\partial z)^2$.

On substituting from eqns (4.3.4) and (4.3.29) into eqns (4.5.3) and

(4.5.1) the following *additional* stiffness matrix is obtained

$$S_m^{(g)} = \frac{tLk_m^2}{2}$$

	1	2	3	4	5	6	7	8
	$b/3$				$b/6$			
			$\frac{13}{35}b$	$\frac{11}{210}b^2$			$\frac{9}{70}b$	$-\frac{13}{420}b^2$
				$\frac{1}{105}b^3$			$\frac{13}{420}b^2$	$-\frac{1}{140}b^3$
					$b/3$			
symmetrical								
							$\frac{13}{35}b$	$-\frac{11}{210}b^2$
								$\frac{1}{105}b^3$

(4.5.4)

where $S_m^{(g)}$ is called the *geometric stiffness matrix*. At the buckling stress the total stiffness is zero, i.e.

$$S_m^{(0)} + \bar{\sigma}_z S_m^{(g)} = 0. \tag{4.5.5}$$

When this equation is compared with that for a simple column buckling, that is,

$$EI\frac{d^2y}{dx^2} + Py = 0$$

it is seen that the terms correspond. The first term arises from the linear elastic theory of a simple beam, while P is the axial load and y is the deflection or geometry of the distortion. After multiplying eqn (4.5.5) by the deflection vector similar meanings can be given to its corresponding terms.

So far only an isolated finite strip has been considered. Before assembling the strips it is necessary to transform their coordinate axes from local to global coordinates by a rotation through angle α. The displacement vector in local coordinates (δ') is related to that in global coordinates (δ) by the following equation (see eqn (4.3.44) et seq.)

$$\delta' = C\delta \tag{4.5.6}$$

where

$$C = \begin{bmatrix} c & & -s & & & & & \\ & 1 & & & & & & \\ s & & c & & & & & \\ & & & 1 & & & & \\ & & & & c & & -s & \\ & & & & & 1 & & \\ & & & & s & & c & \\ & & & & & & & 1 \end{bmatrix} \tag{4.5.7}$$

and $c = \cos \alpha$ and $s = \sin \alpha$.

The element stiffness eqn (4.5.5) transforms to global coordinates by using eqn (4.2.9), thus

$$C^T[S_m^{(0)} + \bar{\sigma}_z S_m^{(g)}]C = 0. \tag{4.5.8}$$

Each element of the structure has an equation of this form and they can be assembled to form the global stiffness matrix S_m. For stiffened plates it is found that provided the nodes are numbered in an orderly way from, say, left to right, S_m will be banded. Advantage is taken of the banded nature of S_m in selecting the method of solution by computer. Some details of a suitable method are described by Murray and Thierauf (1981). For a structure with N nodal points there will be $4N$ equations in S_m so there are up to $4N$ buckling modes and $4N$ critical values of $\bar{\sigma}_z$. It is only the lowest value which is of interest; this lowest value is called σ_{cr}. The length of the buckle strongly influences σ_{cr}. One way of proceeding would be to select a large value for L and then let $m = 1, 2, 3$, etc., until the buckling length with the lowest σ_{cr} is found. This is physically the same as introducing restraints first only at the ends, secondly at the ends and the mid-point, and so on. However, a more convenient procedure is simply to let $m = 1$ and to increase L in steps from a very small to a very large value. At each value of L the critical stress σ_{cr} is calculated by solving the assembled form of eqn (4.5.8) as an eigenvalue problem.

When this method is used to find the critical stresses of various stiffened plates at least four different types of buckling plots (σ_{cr}–L) are obtained. They are illustrated in Fig. 4.5.4. In Fig. 4.5.4(a) it is seen that there is no minimum critical stress. A cross-section in this class buckles

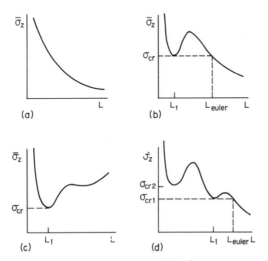

Fig. 4.5.4. Typical buckling plots.

like an Euler column. A flat plate which is free in all directions along its edges has this kind of σ_{cr}–L curve. A strut which buckles by torsional instability may also have a curve of this type.

Figure 4.5.4(b) is the most common type of curve. The minimum σ_{cr} which occurs at a small wavelength L_1 is associated with a local buckle. Usually L_1 is of the same order as the spacing of the stiffeners. The drooping part of the curve which occurs with large L represents Euler buckling of the stiffened plate as a long column, i.e. without distortion of the cross-section. The curve plotted in Fig. 4.5.4(b) is that obtained from a computer using the method described, i.e. $m = 1$ and L is varied. If the length of a real stiffened plate is made $2L_1$ it is self-evident that instead of one buckle of this wavelength forming, two local buckles each of wavelength L_1 will form at a lower critical stress. The same reasoning applies if the length of the stiffened plate is made $3L_1$ and so on. In other words the σ_{cr}–L curve should be that shown by the full line in Fig. 4.5.5. An example of a stiffened plate with this curve is one with almost any significant stiffening and whose longitudinal edges are free in all directions. Also a wide plate with many stiffeners and restrained (pinned or fixed) edges behaves in this way because edge restraint is only felt locally. These multiple local buckles are not drawn on the σ_{cr}–L curves for the sake of clarity.

In Fig. 4.5.4(c) there is no Euler buckling, but local buckling occurs at a wavelength of L_1. A long narrow plate with only one, two, or three stiffeners and with pinned or fixed sides behaves in this manner. Such a plate cannot buckle as an Euler strut. Again the other minima at $2L_1$, $3L_1$, etc., should be shown on the graph.

The curve in Fig. 4.5.4(d) exhibits two minima instead of one as in Fig. 4.5.4(b). This is because some of the more complicated cross-sections have more than one mode of local buckling. For example a trough-stiffened plate may exhibit both the symmetric and the anti-symmetric buckling mode.

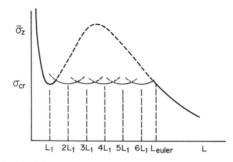

Fig. 4.5.5. Multiple local buckles develop in a stiffened plate whose length L is $L_1 < L < L_{\text{euler}}$.

The tables derived by Murray and Thierauf (1981) list the usual cross-section properties (area of cross-section, second moment of area, etc.) and σ_{cr}, L_1, and L_{euler} for about 4000 different cross-sections. The stiffener shapes treated are thin rectangles, L, V, and troughs. By using elementary dimensional analysis the results can be scaled for different sizes of panel. The value of L_{euler} is the recommended maximum spacing of transverse stiffeners. If their spacing exceeds L_{euler} it is likely that a stiffened panel will behave rather like an Euler column but the Euler formula

$$\frac{\sigma_{cr}}{E} = \frac{\pi^2}{(L/r)^2} \tag{4.5.9}$$

where r is the radius of gyration, cannot always be used to determine the buckling stress because there is usually significant interaction of critical loads in this region. As an example, Fig. 4.5.6 shows a σ_{cr}–L plot and the Euler buckling curve obtained from eqn (4.5.9). It is seen that in this case a designer who tries to use the Euler formula in the region where $L > L_{euler}$ would grossly overestimate the buckling stress. This is because for this panel there is severe interaction of the critical loads for local, Euler column, and torsional column buckling. Therefore it is recommended that the spacing of the transverse stiffeners should always be $\leqslant L_{euler}$.

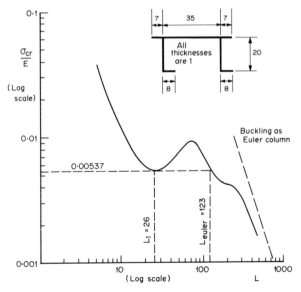

Fig. 4.5.6. Buckling plot of a stiffened plate showing severe interaction of buckling loads in the region of L_{euler}.

Figure 4.5.7 shows some other typical cases. In some unusual cases it is possible that the longitudinal edges of the plate are prevented from moving laterally, i.e. $u = 0$ when $x = b$. At the buckling stress σ_{cr} there must be a lateral compressive stress $\sigma_x = \nu\sigma_{cr}$. The effect of this stress is to reduce the critical stress, in some cases by a large amount. In most practical situations these stresses do not occur because it is difficult to restrain a long stiffened plate except locally where the transverse stiffeners are attached.

Stiffened plates have been analysed by other methods. Design tables of critical stresses of a large number of stiffened panels of practical interest have been calculated by Klöppel and Scheer (1960) and Klöppel and Möller (1968) who used an energy (Rayleigh–Ritz) approach. Wittrick (1968) and Williams and Wittrick (1969) first set up the governing equation of a perfect isolated plate and obtained closed-form solutions for specific loading cases by use of complex variables (see Chapter 5). By

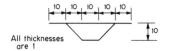

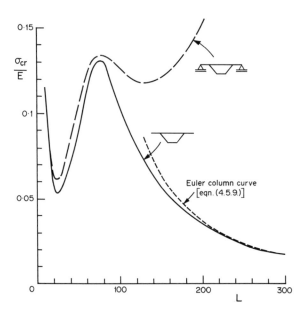

Fig. 4.5.7. Buckling plots of stiffened plate showing the effect of edge restraint.

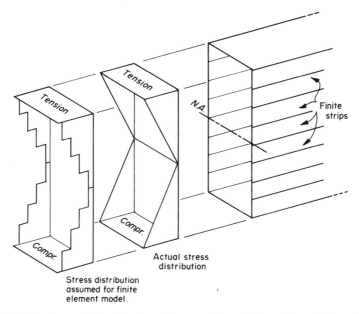

Fig. 4.5.8. Bending analysis of a box-girder may be carried out by replacing the non-uniform stress by one with a stepped distribution.

combining these solutions for two or more plates they were able to obtain the critical stresses of stiffened plates.

The finite strip method may also be used to find the critical loads of box-girders with and without diaphragms. In the former case it is necessary to use more strips and apply an equivalent uniform axial stress to each strip (Fig. 4.5.8). In this case the theory presented in this section may be applied directly. When diaphragms are present Loo's analysis (Section 4.4) must be represented as an eigenvalue problem by including the geometric stiffness matrix. It has the same form as that presented in this section.

The case of torsional buckling may also be treated by the finite strip method, but for this problem it is necessary to allow for the phase angle of the buckle because the nodal lines ($w = 0$) of the buckles are no longer straight and lying exactly in the transverse direction.

Another interesting buckling problem has been solved by Thierauf. The cross-section shown in Fig. 4.5.9(a) is used as a strut or a beam and it was required to evaluate the critical loads for different proportions of the cross-section. The finite strip method described here proved to be a most suitable way of solving this problem. Furthermore, it was found that local buckling of the webs could be suppressed by indenting the cross-section as shown in Fig. 4.5.9(b) and the critical stress was greatly increased. The

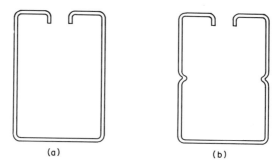

Fig. 4.5.9. The buckling analysis of cold-rolled sections may be carried out by the finite strip method. Results show that the critical stress may often be increased substantially by using longitudinal indentations.

finite strip method was used by Manko (1979) to analyse orthotropic steel deck plates of bridges under transverse loading only. Stability of the structure was not considered. The stiffeners were either open or closed profile. The results from a number of experiments on two models were compared with the finite strip analyses and agreement was found to be excellent. The loading pattern generally represented that of the wheel loads specified in the Polish Design Code for bridge decks. The advantage of the finite strip over the analysis which assumes that the deckplate is an orthotropic plate (e.g. AISC (1963), Pelikan and Esslinger (1957)) is that the detailed stress distribution can be obtained. Manko's paper shows that local wheel loads can produce very high local stresses. These stresses do not usually lead to collapse when the load is applied as a static load (as was shown by Pelikan and Esslinger) but they can give rise to serious fatigue problems. The analysis also shows that panels with stiffeners of closed profiles spread the load so that it is shared by more stiffeners than is the case of stiffeners with open profiles. In the latter case nearly all of the load is carried by the longitudinal strip of the panel which lies under the wheel load. In the next chapter some of the classical and more recent numerical methods for treating isolated plates are presented. Although isolated plates can be analysed by the finite strip method it is necessary for analysts to have a yardstick against which their methods can be gauged and these methods provide such a yardstick. The methods treated in the next chapter are also used to study plates with initial imperfections which can profoundly affect behaviour patterns of stiffened and un-stiffened plates. Finally these powerful numerical methods are useful for determining the buckling stresses of, say, a panel which is a part of a box-girder, when the calculated boundary stresses are distributed in a complicated way.

EXERCISES

4.1 Prove that the work done by the applied loads acting through the associated displacements in the local and in the global coordinate systems are equal. (Hint: Use eqns (4.2.7) and (4.2.8.))

4.2 The simple spring illustrated is inclined to the vertical at an angle α and its stiffness is S_1. The local coordinate axes are x' and y' and the global coordinate axes x and y.

(a) Show that the local stiffness matrix is

$$S' = \begin{bmatrix} S_1 & 0 \\ 0 & 0 \end{bmatrix}$$

(b) Show that the transformation matrix (see eqn (4.3.44)) is

$$C = \begin{bmatrix} \cos \alpha & -\sin \alpha \\ \sin \alpha & \cos \alpha \end{bmatrix}$$

(c) Use eqn (4.3.48) to derive the stiffness matrix in global coordinates.
(d) Check the result by using the laws of statics.

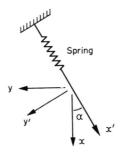

Ex. 4.2

5

ELASTIC BUCKLING OF ISOLATED PLATES AND STIFFENED PLATES WITH IMPERFECTIONS

5.1. INTRODUCTION

In Chapter 4 of this book the elastic buckling behaviour of perfectly-flat stiffened and unstiffened plates with in-plane loading has been analysed by one method only, that is, the finite strip method. For these problems first-order (linear) buckling theory has been used and the buckling stress is obtained as the solution of an eigenvalue problem. From this analysis one could gain the impression that, as a perfectly flat plate buckles and its deflections increase, the applied stress remains constant in the same way as it does for an Euler strut. However, laboratory tests show that, for example, a nearly flat plate with pinned sides and which is loaded with in-plane forces remains nearly flat until the buckling stress is reached after which out-of-plane deflections occur. As these deflections increase the distribution of in-plane stress may change dramatically and in many cases it is found that substantial increases in the in-plane loading are required in order to increase the magnitude of the buckle. In other words a plate may not buckle at a constant in-plane load.

Laboratory tests also show that an initial imperfection in the form of an initial dishing of the plate can have a great influence upon its behaviour. Furthermore some plates have to carry simultaneously in-plane loads and loads which act normal to the plane of the plate. Tests suggest that the effect of the normal loads on the post-buckling behaviour of a plate is similar to that of some initial dishing.

In Chapter 6 the mechanism of failure of plates is studied but it is first necessary to understand their elastic post-buckling behaviour. The first purpose of this chapter is to investigate the elastic post-buckling behaviour of isolated plates whose principal loading is in-plane. The plates may have initial imperfections and they may carry loads which act in a direction normal to the plane of the plate (the y-direction). The second purpose is to study the post-buckling behaviour of a stiffened plate which may be perfect or may have initial imperfections. To solve these problems second-order stability analyses are used and it is necessary to solve non-linear partial differential equations by approximate, iterative and numerical methods.

It is well-known that a simple Euler strut buckles at a critical load and

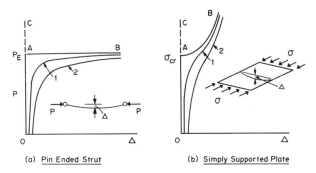

Fig. 5.1.1. Buckling curves compared.

the plot of load against deflection is as shown in Fig. 5.1.1(a). In this diagram the behaviour of a perfectly straight strut is represented by the graph OAB. An initially deformed strut behaves according to Curve 1 and one with larger initial imperfections deforms along Curve 2. Such structures deflect at almost constant load. A simply supported square or nearly square plate which is loaded axially in one direction [Fig. 5.1.1(b)] exhibits a somewhat different load–deflection characteristic. When it is perfectly flat and the in-plane loading is increased the deflection is zero until the critical stress σ_{cr} is reached, after which a very small disturbance causes it to follow curve AB. A similar plate with initial dishing will follow Curve 1, and one with larger initial dishing follows Curve 2. It is seen that as these plates buckle their in-plane loads must be increased. This effect arises from the membrane stresses acting in the plane of the plate in the x-direction which is at right angles to the loading direction, the z-direction. These membrane forces partially restrain the plate against further deformation as it deflects. Figure 5.1.2(a) is a simple analogue of this behaviour but the two lateral springs can be replaced by one in the plane of the two linkages so Fig. 5.1.2(b) represents a simple two-dimensional analogue. Such models have been used to study qualitatively the post-buckling behaviour of plates (Walker and Murray 1975).

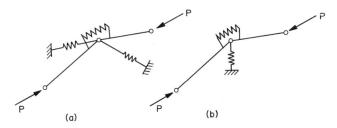

Fig. 5.1.2. Analogue of plate.

However, when one of the longitudinal edges is completely free of restraint, i.e. free in every direction, and the other three edges are pinned these lateral membrane effects cannot develop. In this case the load–deflection curve has a similar shape to that shown in Fig. 5.1.1(a). Another way of looking at this type of plate is to imagine it as a number of longitudinal strips each of which is an Euler column with a small amount of twisting as it buckles. It is easy to imagine that such a combination of Euler columns would have a load–deflection curve with the shape shown in Fig. 5.1.1(a). Thus it is seen that the conditions of support along the boundaries can have a strong influence upon post-buckling behaviour.

It is pointed out in Section 4.5 that in the case of a stiffened plate the assumed length of the buckle has a great influence upon the calculated buckling stress. It is found that in a long, simply supported, stiffened plate the lowest critical stress may occur when several buckles appear within the length of the plate. A similar effect has been thoroughly studied in the case of isolated flat plates. A well-known result (Timoshenko and Gere 1961) of a Rayleigh–Ritz analysis is that a long rectangular plate, which is simply supported along its edges and loaded axially in the longitudinal direction, behaves in a similar manner to a square plate except that it buckles into almost square panels. The buckling mode of a rectangular plate with a 3:1 aspect ratio is shown in Fig. 5.1.3. This result is also easily obtained by a finite strip analysis described in Chapter 4. Thus the critical stress of a long rectangular plate is almost the same as that of a square plate whose sides are equal in length to the width of the rectangular plate. The more general case of a perfect simply supported plate loaded by a stress which varies linearly across its width has also been solved by the Rayleigh–Ritz and the finite strip methods. The results of

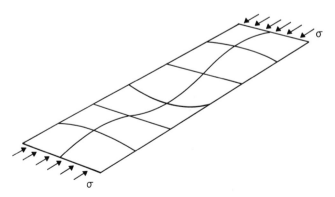

Fig. 5.1.3. Buckled form of plate with 3:1 aspect ratio and sides simply supported.

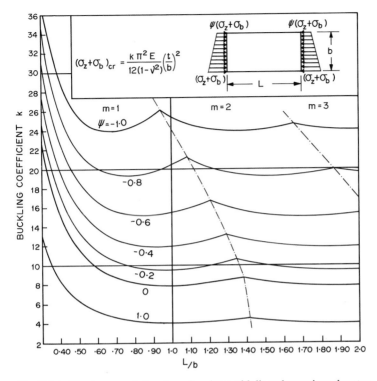

Fig. 5.1.4. Simply supported rectangular plates with linearly varying edge stress.

these analyses are presented in Fig. 5.1.4 which can be used to evaluate the critical stress of isolated plates with this type of loading.

In Section 4.5 it was shown that a thin plate stiffened in the longitudinal direction has several possible buckling modes. For b/t ratios of 30 or more, one of the important modes to consider is local plate buckling. Thin stiffeners of open profile have very little torsional rigidity so their main influence is to provide a nodal line down the sides of each panel. The plate then undergoes local buckling with a sympathetic twisting of the stiffener in the manner indicated in Fig. 5.1.5. Figure 4.1.2 also shows this type of buckling as it is seen in a laboratory test. For this type of

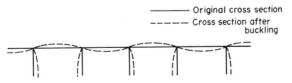

Fig. 5.1.5. Deformation of the cross-section of a stiffened plate due to local buckling.

buckling the axial stress across the plate close to where it meets the
stiffener must be greater than the stresses in the plate midway between
the stiffeners. This is because at the edges the plate is kept straight by the
stiffeners whereas in the central strip the plate is free to deflect up and
down so its axial membrane strain and therefore its membrane stress must
be less. It is this non-uniformity of in-plane stress which has led to the
concepts of effective-width referred to in Section 6.5 and Chapter 7. As
the in-plane load increases and the buckles grow in amplitude both the
effective stiffness of the plate and the distribution of stress across its width
change. These effects are also a function of the magnitude of the initial
imperfections. The theory presented in the next section allows the influence of all of these parameters to be studied quantitatively for an
isolated plate.

5.2. DERIVATION OF MARGUERRE'S EQUATIONS FOR AN INITIALLY DISHED ELASTIC PLATE

The following theory enables the behaviour of an isolated rectangular
plate which is initially deformed out of its plane and which carries
in-plane load and a normal load Y per unit area to be studied. The plate
is treated like a shallow shell with initial shape given by

$$y = y(x, z). \tag{5.2.1}$$

It is assumed that y is small enough to make the usual approximations for
small slopes. The deflection w of a point in the middle surface of the plate
is measured from its initial position to its final position in a direction
parallel to that of the y-axis. Thus, the distance between a general point
(x, z) in the deformed plate and the x–z plane after deformation is $y + w$.
In a plate of thickness t, the stress resultants (per unit width of plate) are
as follows, where integrations are taken from $-t/2$ to $t/2$. For consistency
with most other authors in this field tensile stresses and tensile loads are
taken as positive throughout this chapter. Thus the critical stresses and
end loads will be compressive and therefore have negative values.

$$N_x = \int \sigma_x \, dh$$

$$N_z = \int \sigma_z \, dh$$

$$N_{xz} = N_{zx} = \int \tau_{xz} \, dh$$

$$Q_x = \int \tau_{xy} \, dh \tag{5.2.2}$$

$$Q_z = \int \tau_{yz} \, dh$$

$$M_x = \int \sigma_x h \, dh$$

$$M_z = \int \sigma_z h \, dh$$

$$M_{xz} = M_{zx} = \int \tau_{xz} h \, dh.$$

The equilibrium equations for an element in its deformed position (Fig. 5.2.1) are

$$N_z' + N_{xz}^{\cdot} = 0, \qquad N_{xz}' + N_x^{\cdot} = 0$$
$$Q_z' + Q_x^{\cdot} + Y = 0$$
$$M_z' + M_{xz}^{\cdot} + N_z(y+w)' + N_{xz}(y+w)^{\cdot} - Q_z = 0 \qquad\qquad (5.2.3)$$
$$M_{xz}' + M_x^{\cdot} + N_x(y+w)^{\cdot} + N_{xz}(y+w)' - Q_x = 0$$

where the dash and dot indicate differentiation with respect to z and x, respectively.

The following algebraic manipulations are now carried out. Q_z and Q_x are eliminated from the last three of eqns (5.2.3), the first two of eqns (5.2.3) are introduced and finally, the following substitutions which satisfy the first two of eqns (5.2.3) are made.

$$\Phi^{\cdot\cdot} = N_z; \qquad \Phi'' = N_x; \qquad \Phi^{\cdot\prime} = -N_{xz} \qquad\qquad (5.2.4)$$

where Φ is a stress function. After ignoring certain terms which are considered to be small this leads to the following equation

$$M_z'' + 2M_{xz}^{\cdot\prime} + M_x^{\cdot\cdot} + \Phi^{\cdot\cdot}(y+w)'' - 2\Phi^{\cdot\prime}(y+w)'^{\cdot} + \Phi''(y+w)^{\cdot\cdot}(y+w)^{\cdot\cdot}$$
$$+ Y = 0. \qquad (5.2.5)$$

The stress–strain relationships have their usual form (Timoshenko and

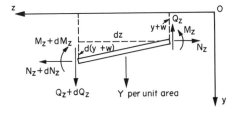

Fig. 5.2.1. Forces acting on an element of a plate in its deformed position (side view only).

Woinowsky-Krieger 1959), thus,

$$N_z = \frac{Et}{1-\nu^2}(\varepsilon_z + \nu\varepsilon_x); \qquad M_z = \frac{Et^3}{12(1-\nu^2)}(\chi_z + \nu\chi_x);$$

$$N_x = \frac{Et}{1-\nu^2}(\varepsilon_x + \nu\varepsilon_z); \qquad M_x = \frac{Et^3}{12(1-\nu^2)}(\chi_x + \nu\chi_z); \qquad (5.2.6)$$

$$N_{xz} = \frac{Et\gamma_{xz}}{2(1+\nu)}; \qquad M_{xz} = \frac{Et^3}{12(1+\nu)}\chi_{xz}$$

where ε and γ are strains and χ is a curvature or twist.

The compatibility conditions are obtained by considering the geometrical changes as the element is stretched and deformed (Fig. 5.2.2). By

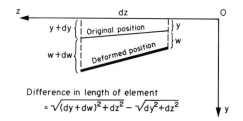

Fig. 5.2.2. Displacement w causes stretching of the element.

using Pythagoras' Theorem and a truncated binomial expansion it can be shown that the strains and curvatures are

strains: $\varepsilon_z = v' + y'w' + \dfrac{w'^2}{2}$

$$\varepsilon_x = u^{\cdot} + y^{\cdot}w^{\cdot} + \frac{w^{\cdot 2}}{2}$$

$$\gamma_{xy} = v^{\cdot} + u' + y'w^{\cdot} + y^{\cdot}w' + w'w^{\cdot} \qquad (5.2.7)$$

curvatures and twist: $\chi_z = -w''; \qquad \chi_x = -w^{\cdot\cdot}$

$$\chi_{xz} = -w'^{\cdot},$$

By eliminating u and v from these equations the compatibility equation is obtained, i.e.

$$\varepsilon_z^{\cdot\cdot} + \varepsilon_x'' - \gamma_{xz}'^{\cdot} + y''w^{\cdot\cdot} - 2y'^{\cdot}w'^{\cdot} + y^{\cdot\cdot}w'' - w'^{\cdot 2} + w''w^{\cdot\cdot} = 0. \qquad (5.2.8)$$

After rearranging eqns (5.2.6) so as to express the strains and curvatures as functions of the stress resultants and by using eqns (5.2.4) the compatibility equation (5.2.8) becomes

$$\nabla^4\Phi + Et(y''w^{\cdot\cdot} - 2y'^{\cdot}w'^{\cdot} + y^{\cdot\cdot}w'' + w''w^{\cdot\cdot} - w'^{\cdot 2}) = 0. \qquad (5.2.9)$$

By similar algebraic steps it is possible to write the equilibrium eqn (5.2.5) in terms of w and Φ, thus

$$D\nabla^4 w - [\Phi^{\cdot\cdot}(y+w)'' - 2\Phi^{\cdot\prime}(y+w)'^{\cdot} + \Phi''(y+w)^{\cdot\cdot}] - Y = 0. \qquad (5.2.10)$$

Equations (5.2.9) and (5.2.10) are Marguerre's (1938) simultaneous non-linear partial differential equations. Except for the assumption that the material is isotropic they are quite general, covering both dished plates $(y \neq 0)$ and flat plates $(y = 0)$. Some methods used for solving them are described in the next section.

5.3. SOLUTION OF MARGUERRE'S EQUATIONS

5.3.1. Introduction

Marguerre's equations (eqns (5.2.9) and (5.2.10)) can be used to study a number of cases of practical interest. The simplest cases involve plates which are initially flat (i.e. $y = 0$), carry uniformly distributed in-plane loads (i.e. Φ is quadratic in x and z) and which carry no transverse load (i.e. $Y = 0$). For this class of problem $\nabla^4\Phi = 0$ and eqns (5.2.9) and (5.2.10) can be satisfied by the solution $w = 0$ when the applied load is less than the critical value. The critical load and buckling mode can be found by solving eqn (5.2.10). For the case when $Y = 0$ it is only necessary to solve the following simplified linear homogeneous form of eqn (5.2.10) which is now defining an eigenvalue problem.

$$D\nabla^4 w - N_z w'' - 2N_{xz}w'^{\cdot} - N_x w^{\cdot\cdot} = 0 \qquad (5.3.1)$$

where N_x and N_z are tensile loads per unit width or length of plate. In other words the critical value of N_z will have a negative value.

A closed-form solution of eqn (5.3.1) when $N_x = 0$ has been obtained by Wittrick (1968) and he has developed stiffness matrices for plates with longitudinal and shear in-plane loading. Williams and Wittrick (1969) showed how these matrices can be used to obtain the critical stresses of assemblies of plates, e.g. stiffened plates. Their work is outlined in Section 5.3.2. When some of the above restrictions are relaxed (i.e. N_z is no longer uniform and N_x is no longer zero) the equations become too complicated to solve in closed form but numerical techniques are available. Michelutti (1976) has used the Runge–Kutta method to obtain the critical stresses and buckling modes for a variety of non-uniform stress conditions acting along the boundaries of both isolated plates and plates stiffened with thin rectangular stiffeners. The finite difference method offers an alternative and very powerful way of solving eqn (5.3.1) and the work of Kollbrunner and Meister (1958) and of Fok (1980) is also described in Section 5.3.2.

So far in this section only perfectly flat plates have been considered. It should be noted that the governing equation (5.3.1) is linear and that is why the methods mentioned above can be used. To study the behaviour of a plate with initial imperfection $(y \neq 0)$ it is necessary to solve the simultaneous non-linear Marguerre equations (5.2.9) and (5.2.10). There appear to be no exact solutions but they have been solved for isolated plates by approximate analytical methods (e.g. Coan 1951, Yamaki 1959, and Walker 1969) and by numerical methods (e.g. an iterative Runge–Kutta method and the finite difference method). These methods are described in Sections 5.3.3 and 5.3.4, respectively.

It appears that so far there have been few successful attempts to analyse theoretically the elastic behaviour of an imperfect stiffened plate. A number of solutions were obtained by Bilstein (1974) who used an iterative Runge–Kutta method. These numerical methods all require enormous amounts of computing time so it is unlikely that comprehensive tables or graphs for the design and analysis of stiffened plates with imperfections will ever become available.

Another problem which arises from these investigations is that the amount of information derived even for a single plate is very large. It is, of course, desirable to condense research results so that they are presented to designers in their most compact form. The several ways of doing this have been reviewed by Walker (1980) and one method is described in Section 5.3.5. Section 5.4 is a brief review of some literature on the elastic buckling of plates.

5.3.2. Buckling solution for flat isolated and stiffened plates

In this section perfectly flat plates are considered, the governing equation being eqn (5.3.1). The analytical solution of Wittrick (1968), the Runge–Kutta method used by Michelutti (1976) and the finite difference method are described. In each case a representative section of an infinitely long plate is considered (Fig. 5.3.1). Below the buckling load the solution of eqn (5.3.1) is trivial, i.e. $w = 0$. Let λ_1 be the wavelength of the edge forces along $x = 0$ and $x = b$. When shear stress τ_{xz} is present the post-buckling deflection is not a simple function of z but varies in phase across the width of the plate (Fig. 5.3.1). Thus it is necessary to assume a sinusoidal deflection pattern for w as the real part of a complex function, i.e.

$$w(x, z) = \operatorname{Re} W(X)e^{iZ} \tag{5.3.2}$$

where

$$X = \frac{\pi x}{b} \quad \text{and} \quad Z = \frac{\pi z}{\lambda_1}. \tag{5.3.3}$$

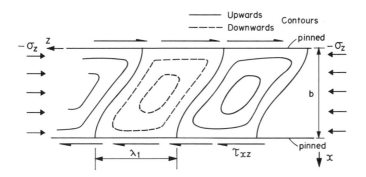

Fig. 5.3.1. Buckling pattern of a plate under combined axial and shear stress.

If $N_{xz} \neq 0$, $W(X)$ is complex and eqn (5.3.1) becomes

$$W^{\cdots\cdot} + (K_x - 2\Omega^2)W^{\cdot\cdot} + 2i\Omega H W^{\cdot} + \Omega^2(\Omega^2 - K_z)W = 0 \qquad (5.3.4)$$

where

$$\Omega = \frac{b}{\lambda_1}, \qquad H = -\frac{b^2 N_{xz}}{\pi^2 D}, \qquad K_x = -\frac{b^2 N_x}{\pi^2 D}, \qquad K_z = -\frac{b^2 N_z}{\pi^2 D}. \qquad (5.3.5)$$

For the case when $N_x = 0$ eqn (5.3.4) is assumed to have a solution of the form (m is complex)

$$W = A e^{mX} \qquad (5.3.6)$$

leading to the following quartic auxiliary equation

$$\left(\frac{m}{\Omega}\right)^4 - 2\left(\frac{m}{\Omega}\right)^2 + 2i\frac{H}{\Omega^2}\left(\frac{m}{\Omega}\right) + 1 - \frac{K_z}{\Omega^2} = 0. \qquad (5.3.7)$$

Just as the solution of the simple Euler strut problem is expressed in sines and cosines when the axial load is compressive and sinhs and coshs when it is tensile, so the nature of the solution of the plate buckling problem depends upon where the roots of eqn (5.3.7) lie in the H/Ω^2, K_z/Ω^2 plane. This plane is divided into regions by curves along which there are repeated roots. These curves are found by equating the discriminant (Sokolnikoff 1941) Δ of eqn (5.3.7) to zero where

$$\Delta = \left(\frac{4}{3} - \frac{K_z}{\Omega^2}\right)^3 - 27\left(\frac{H^2}{4\Omega^4} + \frac{K_z}{3\Omega^2} - \frac{8}{27}\right)^2. \qquad (5.3.8)$$

The remainder of Wittrick's paper (1968) is concerned with the establishment of the stiffness matrices, which relate the edge forces to the edge rotations, for the different regions in the H/Ω^2, K_z/Ω^2 plane. These stiffness matrices are listed in his paper and are too lengthy to reproduce here.

The way in which these stiffness matrices can be used to obtain the buckling load and mode of an assembly of plates is described by Williams and Wittrick (1969). It involves matching equilibrium and compatibility conditions along each edge (i.e. at the join of two or more plates). Because the length λ_1 of the buckle is not known at the start it is necessary to devise a strategy for obtaining the numerical value of λ_1 which minimises the buckling stress. The technique is the same as that described previously in Section 4.5. Figure 5.3.2 shows a typical graph of calculated buckling stress against assumed values of λ_1. The right-hand branch represents Euler buckling while the left-hand branch represents combined plate/stiffener buckling (i.e. local buckling).

Attention is now turned to the problem of a flat plate carrying in-plane stresses N_x, N_z and N_{xz} which are no longer constant. A linearly distributed N_z is, for example, of interest in the design of the webs in a box-girder. Equation (5.2.9) can be solved with $y = w = 0$ by finding the stress function Φ which is biharmonic (i.e. $\nabla^4\Phi = 0$) and satisfies the known boundary conditions, i.e. it is necessary to solve only a simple plane stress problem. The buckling stress and mode are obtained from eqn (5.3.4). This equation is split into its real and imaginary parts by first writing W as a complex function.

$$W(x) = W_1 + iW_2. \tag{5.3.9}$$

On substitution in eqn (5.3.4) a complex equation is obtained. It can only be satisfied for all x if both the real and imaginary parts are zero. Hence the following simultaneous ordinary differential equations must be solved.

$$W_1'''' + (K_x - 2\Omega^2)W_1'' - 2\Omega H W_2' + \Omega^2(\Omega^2 - K_z)W_1 = 0 \tag{5.3.10}$$

$$W_2'''' + (K_x - 2\Omega^2)W_2'' + 2\Omega H W_1' + \Omega^2(\Omega^2 - K_z)W_2 = 0. \tag{5.3.11}$$

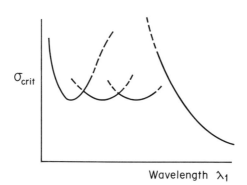

σ_{crit}

Wavelength λ_1

Fig. 5.3.2. Buckling stress of a given stiffened plate as the assumed buckling wavelength is varied.

These coupled equations are linear and can be solved numerically by the Runge–Kutta technique (Piaggio 1954). The variables w_1 to w_8 are introduced and defined as follows

$$w_1 = W_1$$
$$w_2 = W_1'$$
$$w_3 = W_1''$$
$$w_4 = W_1''' \qquad\qquad (5.3.12)$$
$$w_5 = W_2$$
$$w_6 = W_2'$$
$$w_7 = W_2''$$
$$w_8 = W_2'''$$

where each of the symbols w_1 to w_8 represent a function of x. Equations (5.3.10) and (5.3.11) are now replaced by the following differential equations of first order.

$$\frac{dw_1}{dx} = w_2$$

$$\frac{dw_2}{dx} = w_3$$

$$\frac{dw_3}{dx} = w_4$$

$$\frac{dw_4}{dx} = -(K_x - 2\Omega^2)w_3 + 2\Omega H w_6 - \Omega^2(\Omega^2 - K_z)w_1 \qquad\qquad (5.3.13)$$

$$\frac{dw_5}{dx} = w_6$$

$$\frac{dw_6}{dx} = w_7$$

$$\frac{dw_7}{dx} = w_8$$

$$\frac{dw_8}{dx} = -(K_x - 2\Omega^2)w_7 - 2\Omega H w_2 - \Omega^2(\Omega^2 - K_z)w_5.$$

These equations are in a form to which a standard Runge–Kutta computer subroutine can be applied. The Runge–Kutta method is a numerical integration technique which is carried out in a stepwise manner across the domain of interest. The domain, which in the present case is the width b of the plate, is first divided into equal steps by grid points. The boundary conditions are now considered, and, for the purposes of illustration, the case of a plate whose boundaries $x = 0$ and b are simply supported is treated. For this case w and $w\ddot{}$ are both zero at $x = 0$ and b. From eqns (5.3.12) and (5.3.9) the following eight conditions must apply.

$$w_1(0) = w_5(0) = w_3(0) = w_7(0) = w_1(b) = w_5(b) = w_3(b) = w_7(b) = 0 \tag{5.3.14}$$

The Runge–Kutta method generates the solution by integrating in a stepwise manner between $x = 0$ and $x = b$ using known boundary conditions at the origin. Unfortunately only four of the boundary conditions are known at the origin, the other four being known at $x = b$, and it is necessary to adopt the following device. Eight sets of linearly independent but otherwise arbitrary boundary conditions are selected and used as starting values for the Runge–Kutta subroutine. Two such typical sets of boundary conditions are

$$w_1(0) = 1, \qquad w_2(0) = 0, \qquad w_3(0) = 0 \ldots w_8(0) = 0$$
$$w_1(0) = 0, \qquad w_2(0) = 1, \qquad w_3(0) = 0 \ldots w_8(0) = 0. \tag{5.3.15}$$

These boundary conditions, being linearly independent of one another, lead to eight linearly independent solutions of eqn (5.3.13). These solutions are called complementary functions and they are stored in the computer in 'book' form with each 'page' being a table. For example, one 'page' will contain all of the value of w_2 [$= W_1'$ from eqn (5.3.12)], i.e. the values of w_2 for all of the eight complementary functions at all of the grid points. The notation w_{265} is the value of w_2 at the sixth grid point for the fifth complementary function. Figure 5.3.3 is, for example, a graphical representation of all of the information which relates to w_2 and is stored as one 'page' in the computer.

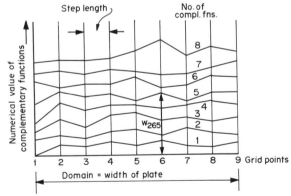

Fig. 5.3.3. Eight complementary functions for w_2 generated by the Runge–Kutta method from eight linearly independent boundary conditions at grid point 1.

To obtain the complete solution for $w_1, w_2, \ldots, w_8$ it is necessary to multiply each complementary function by a corresponding integration constant A_1 to A_8. Thus all of the values of w_{mn1} ($m = 1$ to 8, $n = 1$ to 9), have to be multiplied by A_1, w_{mn2} by A_2 and so on up to A_8, and these values are added to obtain w_1 to w_8. This is similar to the normal procedure used to solve a differential equation, such as $\ddot{y} + a^2 y = 0$, where the complementary functions are $\cos ax$ and $\sin ax$ and the complete solution is $y = A_1 \cos ax + A_2 \sin ax$. The only difference is that with the Runge–Kutta method all of the derivatives e.g. $w_2 (= W_1^{\cdot})$, $w_3 (= W_1^{\cdot\cdot})$ are also recorded.

To evaluate the integration constants A_1 to A_8 it is necessary to introduce the actual boundary conditions of the plate. For example, the eight conditions given in eqn (5.3.14) lead to the following relationships between A_1 to A_8, respectively.

$$A_1 w_{111} + A_2 w_{112} + A_3 w_{113} + \ldots + A_8 w_{118} = 0$$

$$A_1 w_{511} + A_2 w_{512} + A_3 w_{513} + \ldots + A_8 w_{518} = 0$$

and so on . . . (5.3.16)

$$A_1 w_{791} + A_2 w_{792} + A_3 w_{793} + \ldots + A_8 w_{798} = 0$$

This set of homogeneous linear algebraic equations can be written in matrix form, thus

$$[W] \quad [A] = [0]$$
$$(8 \times 8) \quad (8 \times 1) = (8 \times 1)$$

(5.3.17)

Apart from the trivial solution $[A] = [0]$ there is another of practical interest when $[W]$ is singular. This occurs when the value of axial stress is large enough to make the determinant $|W|$ zero. As the axial stress increases $|W|$ diminishes until it reaches zero, at which point the plate has reached the buckling stress but care must be exercised to ensure that the lowest buckling stress is obtained and not one of the many larger buckling stresses associated with higher buckling modes.

The buckling mode is next found. It is represented by the real part of eqn (5.3.2). As in all structural eigenvalue problems the magnitude of the buckles cannot be determined but the shape can be found. This is done by dividing each of the equations in (5.3.17) by A_1, then transferring the first term of each equation to the right-hand side and solving for the unknowns A_2/A_1, A_3/A_1, etc. For the pin-sided plate considered previously the equations to be solved are built up from the known boundary conditions (eqns (5.3.14)).

A number of buckling problems with isolated plates have been treated by Michelutti (1976) in the above manner and the results are illustrated and compared with some classical results (Bulson 1970) in Table 5.3.1.

Table 5.3.1. Results of single plate buckling analyses by the Runge–Kutta method (Michelutti 1976).

No.	Boundary conditions	b/t	L/b	ψ	σ_x/σ_z	$\sigma_{z\,cr}$ Computed	(MPa) Classical (Bulson 1970)	k	Notation
1		56	1	1	0	−236	−238	4.00	
2		56	1	1	0.5	−159		2.68	
3		56	0.66	1	0	−415	−414	6.97	clamped
4		56	1	1	1	−227		3.82	simply-supported
5		56	1.63	1	0	−75.9	−73.0	1.25	free
6		56	5	1	0	−27.9	−26.8	0.46	$\sigma_{cr} = \dfrac{-k\pi^2 E}{12(1-\nu^2)}\left(\dfrac{t}{b}\right)^2$
7		56	0.79	1	0	−320	−323	5.41	
8		56	1	0.5	0	−107		1.80	
9		56	1	0.5	0	−154		2.59	
10		56	1	0.5	0.5	−367		6.18	

The phenomenon of local buckling of assemblies of thin plates (e.g. stiffened plates) whose joint lines remain straight can be analysed by a slight extension of the above method. It is only necessary to ensure rotational compatibility and moment equilibrium at each joint line. A set of five stiffened plates (Fig. 5.3.4) was analysed for buckling stress and mode by Michelutti (1976) and the results are presented in Table 5.3.2 while two representative buckling modes are illustrated in Fig. 5.3.5. It should be noted that no attempt was made to vary the length of the buckle in order to find the minimum local buckling stress as is suggested by Fig. 5.3.2.

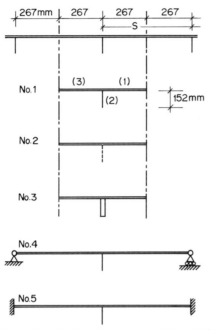

Fig. 5.3.4. Stiffened plates analysed for buckling stress by Michelutti (1976) [see Table 5.3.2 for results].

However, the Runge–Kutta method can be employed in such a search.

Equation (5.3.1) can also be solved by the finite difference method (Shaw 1953). The next few paragraphs are a brief explanation of the finite difference method. It will be seen that a linear differential equation can be replaced by an equivalent set of linear simultaneous algebraic equations which can then be solved in a variety of ways, e.g. by matrix inversion, by the relaxation method, and so on.

Table 5.3.2. Critical stress for stiffened plates (refer Fig. 5.3.4).

No.	b/t for plates			L/S	$\sigma_{z\,cr}$(MPa)	
	1	2	3		Computed	Theory
1	28.0	20.7	28.0	1.00	−253	
2	28.0	207	28.0	1.00	−240	−238[†]
3	28.0	2.07	28.0	0.66	−417	−414[‡]
4	56.0	20.7	56.0	1.00	−245	
5	56.0	20.7	56.0	0.80	−331	

[†] Single plate with simply supported edges.
[‡] Single plate with clamped edges.

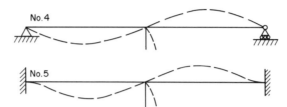

Fig. 5.3.5. Buckled shape of cross-section shown in Fig. 5.3.4 and Table 5.3.2 (Michelutti, 1976).

The first, second, third, and so on derivatives of a continuous function $w = f(x)$ (Fig. 5.3.6) can be approximated in the following way. The domain over which the derivatives are required is divided into steps of length h.

The first derivative at point n is approximately $(w_{n+1} - w_{n-1})/2h$. The second derivative at point n is the rate at which the first derivative changes, i.e.

$$\frac{1}{h}\left[\frac{(w_{n+1} - w_n)}{h} - \frac{(w_n - w_{n-1})}{h}\right].$$

The third derivative at point n is the rate at which the second derivative changes and so on. Hence a 'table' of derivatives can be drawn up as follows:

1st derivative $= (w_{n+1} - w_{n-1})/2h$

2nd derivative $= (w_{n+1} - 2w_n + w_{n-1})/h^2$

3rd derivative $= (w_{n+2} - 2w_{n+1} + 2w_{n-1} - w_{n-2})/2h^3$

4th derivative $= (w_{n+2} - 4w_{n+1} + 6w_n - 4w_{n-1} + w_{n-2})/h^4.$ (5.3.18)

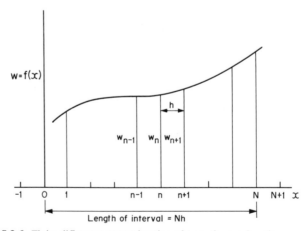

Fig. 5.3.6. Finite difference approximation of a continuous function, $w = f(x)$.

The extension to higher orders is obvious. Equation (5.3.18) can be used to replace an ordinary linear differential equation by an equivalent set of simultaneous linear algebraic equations with $w_0, w_1, \ldots w_N$ as the unknowns when it is applied at each point between 0 and N (Fig. 5.3.6). However, two features related to the boundaries upset this simple pattern. Firstly, it may be necessary to introduce so-called 'fictitious' nodes outside of the required domain (e.g. at -1 and $N+1$) and, secondly, the given boundary conditions must also be written in finite difference form and included in the set of equations to make the problem complete and solvable.

Partial differential equations can be set up in finite difference form by extending these ideas to two or more dimensions. For example, at point (m, n) (Fig. 5.3.7) the Laplacian of w is

$$\nabla^2 w = \frac{\partial^2 w}{\partial x^2} + \frac{\partial^2 w}{\partial z^2}$$

$$= \frac{w_{m+1,n} - 2w_{m,n} + w_{m-1,n}}{h^2} + \frac{w_{m,n+1} - 2w_{m,n} + w_{m,n-1}}{h^2}$$

$$= \frac{(w_{m+1,n} + w_{m-1,n} + w_{m,n+1} + w_{m,n-1} - 4w_{m,n})}{h^2} \qquad (5.3.19(a))$$

The twist of a surface can be expressed by using the same technique, thus

$$\left(\frac{\partial^2 w}{\partial x \, \partial z}\right)_{\text{at } m,n} = (w_{m+1,n+1} - w_{m+1,n-1} - w_{m-1,n+1} + w_{m-1,n-1})/4h^2$$

$$(5.3.19(b))$$

Operators such as ∇^2 and $\partial^2/\partial x \, \partial z$ can be arranged as stencils (Fig. 5.3.8), a number of which may be found in standard textbooks on numerical methods (e.g. Shaw 1953).

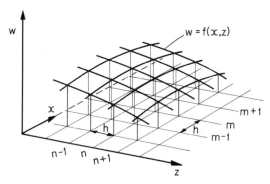

Fig. 5.3.7. Finite difference approximation of a continuous function, $w = f(x, z)$.

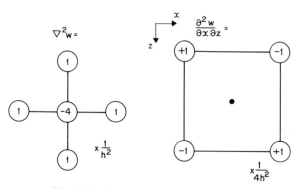

Fig. 5.3.8. Stencils of two differential operators.

Equation (5.3.1) is written in the following finite difference form at the point (i, k).

$$w_{i+2,k} + 2w_{i+1,k+1} + 2w_{i+1,k-1} - 8w_{i+1,k} - 8w_{i-1,k}$$
$$+ w_{i-2,k} + 20w_{i,k} + w_{i,k+2} + 2w_{i-1,k+1}$$
$$+ 2w_{i-1,k-1} - 8w_{i,k+1} - 8w_{i,k-1}$$
$$+ w_{i,k-2} = -\frac{h^2}{D}[-N_x(w_{i+1,k} - 2w_{i,k} + w_{i-1,k})$$
$$- \tfrac{1}{2}N_{xz}(w_{i+1,k+1} + w_{i-1,k-1} - w_{i+1,k-1} - w_{i-1,k+1})$$
$$- N_z(w_{i,k+1} - 2w_{i,k} + w_{i,k-1})]. \qquad (5.3.20)$$

This equation applies at each nodal point inside of the boundary and is more conveniently written in the pictorial form of a stencil, thus,

$$
\begin{array}{ccccc}
\cdot & \cdot & +1 & \cdot & \cdot \\
\cdot & +2 & -8 & +2 & \cdot \\
+1 & -8 & +20 & -8 & +1 \\
\cdot & +2 & -8 & +2 & \cdot \\
\cdot & \cdot & +1 & \cdot & \cdot
\end{array}
\cdot w = -\frac{h^2}{D}
\left\{
-N_x
\begin{array}{ccc}
\cdot & \cdot & \cdot \\
+1 & -2 & +1 \\
\cdot & \cdot & \cdot
\end{array}
\cdot w
\right.
$$

$$
\left.
-\frac{N_{xz}}{2}
\begin{array}{ccc}
+1 & \cdot & -1 \\
\cdot & \cdot & \cdot \\
-1 & \cdot & +1
\end{array}
\cdot w - N_z
\begin{array}{ccc}
\cdot & +1 & \cdot \\
\cdot & -2 & \cdot \\
\cdot & +1 & \cdot
\end{array}
\cdot w
\right\}
\qquad (5.3.21)
$$

As an example a pin-sided square plate which carried a load N_z in the z-direction and has $N_{xz} = N_x = 0$ is now considered. The displacements w

along the boundaries are zero. Such a plate can be thought of as one element of a chequerboard of similar elements (Fig. 5.3.9(a)), half of which buckle upwards (say the black squares) and the other half downwards (say the white squares). Figure 5.3.9(b) shows a coarse finite difference net $(h = b/2)$ and on it are the unknown deflections after allowing for symmetry. Thus, for example, the node at A actually lies outside of the plate being studied. It lies at the centre of an (imaginary) adjoining plate whose deflection is in the opposite direction so its deflection is shown as $-w_{1,1}$. By first making N_x and $N_{xz} = 0$ in eqn (5.3.21) and then using the boundary condition $w_{0,0} = w_{0,1} = 0$ the following equation is obtained.

$$-w_{1,1} - w_{1,1} + (20 - 2K)w_{1,1} - w_{1,1} - w_{1,1} = 0 \qquad (5.3.22)$$

where for convenience

$$K = -\frac{b^2 N_z}{4D} = -\frac{b^2 \sigma_z t}{4D}. \qquad (5.3.23)$$

Equation (5.3.22) has a trivial solution, that is, $w_{1,1} = 0$. This of course means that the plate remains flat. (Equilibrium and compatibility are obviously satisfied by this solution.) However, a non-trivial solution also exists, that is, when $K = 8$. Plate buckling solutions are usually written in the following form

$$\sigma_{cr} = -k \frac{\pi^2 E}{12(1 - v^2)} \left(\frac{t}{b}\right)^2 \qquad (5.3.24)$$

where t is the plate thickness. For the present problem, therefore,

$$k = 3.24. \qquad (5.3.25)$$

The exact solution is known (Fig. 5.1.4) to be $k = 4$, indicating that even this very crude finite difference net has given a solution within 20% of the correct one. A finer net which divides the side into three equal intervals (Fig. 5.3.9(c)), can also be used. This problem is left as an exercise for the reader. Its solution is

$$K = \left(-\frac{b^2 N_z}{9D}\right) = 4,$$

i.e. $k = 3.65$ \qquad (5.3.26)

which is within 10% of the correct value.

When the net is further decreased in size it is found that a number of simultaneous homogeneous equations (i.e. the right-hand sides are all zero) have to be satisfied. Apart from the trivial solution, that is, $w = 0$ everywhere, there is a buckled shape which will satisfy these equations. It

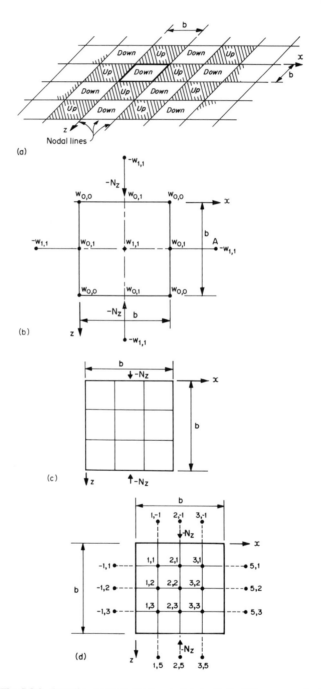

Fig. 5.3.9. Buckling analysis of square plate by finite difference method.

does this when the determinant of the left-hand side is zero. The following example illustrates how the buckling load and the buckling mode may be found by this method.

Example 5.3.1

Use the finite difference net shown in Fig. 5.3.9(d) to determine the critical value of the applied load N_z of a square plate $b \times b$ of thickness t and which is simply supported along all boundaries. Also define the shape of the buckling mode.

In eqn (5.3.21) $N_{xz} = 0$ and let $K = -h^2 N_z/D$ where $h = b/4$. From the conditions of symmetry it is only necessary to consider one-quarter of the plate but we simplify the solution by assuming that the buckled form of the plate is also symmetrical about the two diagonals. In other words since

$$w_{1,1} = w_{3,1} = w_{1,3} = w_{3,3}, \qquad w_{1,2} = w_{2,1} = w_{2,3} = w_{3,2}$$

the only deflections to be used in the analysis are $w_{1,1}$, $w_{2,1}$ and $w_{2,2}$.

For a simply supported boundary the deflection at a fictitious node is equal in magnitude but opposite in sign to that at the node which is its mirror image inside of the boundary. Thus these boundary conditions are represented by

$$w_{1,-1} = w_{-1,1} = -w_{1,1}, \qquad w_{2,-1} = -w_{2,1}.$$

In addition it is required that at all of the boundary nodes $w = 0$. When these conditions are inserted into eqn (5.3.21) and it is applied to nodes 1,1, 2,1 and 2,2 the following simultaneous equations are obtained, respectively

$$(20 - 2K)w_{1,1} - (16 - K)w_{2,1} + 2w_{2,2} = 0$$
$$-16w_{1,1} + (24 - 2K)w_{2,1} - (8 - K)w_{2,2} = 0$$
$$8w_{1,1} - (32 - 2K)w_{2,1} + (20 - 2K)w_{2,2} = 0$$

The determinant of these equations is made zero by changing K on a trial-and-error basis. It is found that when $K = 2.343$ the determinant is close to zero. Hence

$$(N_z)_{cr} = -\frac{KD}{h^2} = -2.343 \frac{Et^3}{12(1 - v^2)} \left(\frac{4}{b}\right)^2,$$

i.e. $\sigma_{cr} = -3.800 \dfrac{\pi^2 E}{12(1 - v^2)} \left(\dfrac{t}{b}\right)^2.$

(By classical theory the coefficient should be 4, a difference of only 5%.) To find the buckling mode it is convenient to let $w_{2,2} = 1$ and substitute

$K = 2.343$ in the first two of the simultaneous equations, whence

$$15.314w_{1,1} - 13.657w_{2,1} = -2$$
$$-16w_{1,1} + 19.314w_{2,1} = 5.657.$$

The solution of these equations is $w_{1,1} = 0.5002$, $w_{2,1} = 0.7073$. If the shape of the buckle had been assumed as a double sine curve of unit central height, i.e. $w = \sin\dfrac{\pi x}{b}\sin\dfrac{\pi z}{b}$ these values would then be $w_{1,1} = 0.5$, $w_{2,1} = 0.7071$ which are in good agreement with the results of the above numerical analysis.

The boundary conditions for a plate with simply supported, built-in, or free edges are now considered. In the examples just treated the problem was avoided by simply taking account of the symmetry of the problem. The well-known relationship between moment and curvature of a plate (Timoshenko and Woinowsky-Krieger 1959) is conveniently written as a matrix equation

$$\begin{bmatrix} M_x \\ M_z \\ M_{xz} \end{bmatrix} = -D \begin{bmatrix} 1 & \nu & 0 \\ \nu & 1 & 0 \\ 0 & 0 & -(1-\nu) \end{bmatrix} \begin{bmatrix} \dfrac{\partial^2 w}{\partial x^2} \\ \dfrac{\partial^2 w}{\partial z^2} \\ \dfrac{\partial^2 w}{\partial x\,\partial z} \end{bmatrix}. \tag{5.3.27}$$

For an edge the local x-axis can be taken in the direction of the unit normal $\bar{n}$ and the z-axis in the direction of the unit tangent $\bar{t}$ (Fig. 5.3.10).

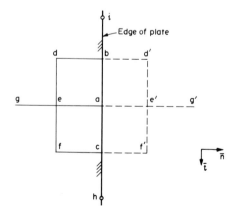

Fig. 5.3.10. Boundary conditions of a plate are obtained in finite difference form by using this net.

For a simply supported edge

$$\frac{\partial^2 w}{\partial n^2} = w = 0. \tag{5.3.28}$$

For a built-in edge

$$\frac{\partial w}{\partial n} = w = 0. \tag{5.3.29}$$

For a free edge the moment and shear force are both zero.
    From eqn (5.3.27) the moment is zero when

$$\frac{\partial^2 w}{\partial n^2} + \nu \frac{\partial^2 w}{\partial t^2} = 0. \tag{5.3.230}$$

From standard texts on plate theory (Timoshenko and Woinowsky-Krieger 1959) the shear force is zero when

$$\frac{\partial^3 w}{\partial n^3} + (2 - \nu) \frac{\partial^3 w}{\partial n \, \partial t^2} = 0. \tag{5.3.31}$$

It is a simple matter to write eqns (5.3.28)–(5.3.31) as equivalent finite difference equations. Reference to Fig. 5.3.10 and eqns (5.3.28) and (5.3.29) shows that for
    (a) a simply supported edge:

$$w_a = w_b = w_c = 0 \quad \text{and} \quad w'_e = -w_e \tag{5.3.32}$$

    (b) a built-in edge:

$$w_a = w_b = w_c = 0 \quad \text{and} \quad w'_e = w_e \tag{5.3.33}$$

    (c) for a free edge:

w is not zero, in general, and hence from eqn (5.3.30)

$$\begin{aligned}
w'_e &= 2w_a(1+\nu) - \nu(w_b + w_c) - w_e \\
w'_d &= 2w_b(1+\nu) - \nu(w_i + w_a) - w_d \\
w'_f &= 2w_c(1+\nu) - \nu(w_a + w_h) - w_f.
\end{aligned} \tag{5.3.34}$$

Applying eqns (5.3.31) and (5.3.18) to node a

$$w'_g = -2w_e(3-\nu) + 2w'_e(3-\nu) + (2-\nu)(w_d + w_f - w'_d - w'_f) + w_g. \tag{5.3.35}$$

On substituting eqns (5.3.34) into (5.3.35) the following explicit expression for w'_g in terms of w for the internal nodes is obtained.

$$\begin{aligned}
w'_g = {} & -2w_e(3-\nu) + 2[2w_a(1+\nu) - \nu(w_c + w_b) - w_e](3-\nu) \\
& + (2-\nu)[2w_f + 2w_d - 2w_c(1+\nu) + \nu(w_h + w_a) \\
& - 2w_b(1+\nu) + \nu(w_a + w_i)] + w_g.
\end{aligned} \tag{5.3.36}$$

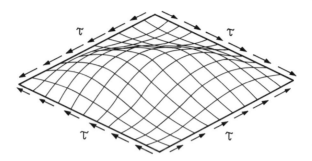

Fig. 5.3.11. Buckling mode of square plate in shear; $\tau_{cr} = 9.72 \dfrac{\pi^2 E}{12(1-\nu^2)} \left(\dfrac{t}{b}\right)^2$.

Equation (5.3.21) is applied at each of the internal nodes but it is found that when this is done, for example, at node e (Fig. 5.3.10), the unknown w'_e is included in the equation. However, w'_e is easily eliminated by again using the first of eqns (5.3.34). Special provisions also have to be made at corners.

Thus the governing equation for a flat plate (eqn (5.3.1)) has now been replaced by an equivalent finite difference equation (eqn (5.3.21)) and a technique for eliminating all of the displacements of the external (fictitious) nodes from the set of simultaneous linear algebraic (finite difference) equations has been developed. The technique has been applied by Fok (1980) to a wide variety of isolated plates with various combinations of edge restraint and in-plane loading and their buckling loads and modes have been determined. Whenever possible his results have been checked against existing theories and it was found that results agreed to within a few percent. Figures 5.3.11 and 5.3.12 show two typical buckling patterns and Table 5.3.3 is a summary of some analyses of plates with uniform and non-uniform stresses.

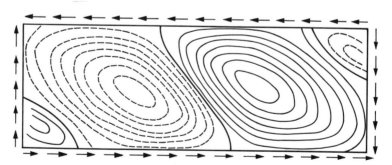

Fig. 5.3.12. Buckling mode of a rectangular 1:3 plate in pure shear. With 200 nodes
$$\tau_{cr} = \frac{6.01\pi^2 E}{12(1-\nu^2)} \left(\frac{t}{b}\right)^2.$$ (By an energy method $k = 5.9$.)

Table 5.3.3. Finite difference evaluation of k factor for various plates (Fok 1980).

Case	L/b	b/t	c/b	No. of buckles	k
1.	1	102	0.5	1	10.12
	1	102	0.33	1	10.00
	3	115	0.33	4	7.40
2.	3	115	0.5	4	5.60
	3	84	0.5	4	5.60
3.	1	102	0.5	1	5.40
	1	68	0.5	1	5.40
	3	84	0.5	4	5.40
	3	115	0.5	4	5.40
4.	3	84	0.5	4	6.98
	3	115	0.5	4	6.98
	3	84	0.425	4	6.92
	3	115	0.425	4	6.92
	3	115	0.33	4	6.75
5.	3	84	0.5	3	4.00
	3	115	0.5	3	4.00
	1	102	0.5	1	4.00
	1	68	0.5	1	4.00
	3	140	0.5	3	4.00
	3	84	0.5	3	4.00
	3	115	0.425	3	3.95
	1	102	0.425	1	3.95
	1	68	0.425	1	3.95
	3	84	0.33	3	3.88
	3	115	0.33	3	3.88
	1	102	0.33	1	3.89
	1	68	0.33	1	3.89
	3	140	0.33	3	3.88

Case	L/b	b/t	c/b	No. of buckles	k
6.	1	102	0.5	3	6.85
	3	84	0.5	4	4.28
	3	115	0.5	4	4.28
	3	140	0.5	4	4.28
	1	102	0.33	1	6.78
	3	115	0.33	4	4.27
	3	140	0.33	4	4.27
7.	3	115	0.5	1	0.530
	3	84	0.5	1	0.530
	1	68	0.5	1	1.400
	2	102	0.5	1	0.675
	1	84	0.425	1	0.680
	1	114	0.425	1	0.680
8.	1	68	0.33	1	2.35
	2	102	0.33	1	1.23
	3	140	0.575	1	0.719
	2	102	0.5	1	1.954
	3	84	0.5	1	0.850
9.	1	127	0.5	1	2.52
	3	140	0.5	1	0.850
	3	84	0.425	1	1.016
	2	102	0.33	1	2.390
	3	115	0.5	2	1.25
	1	68	0.5	1	1.64
	3	140	0.5	2	1.25
	3	115	0.425	2	1.52
	3	140	0.425	2	1.52
	1	68	0.33	1	3.86

/////// built-in
——— simply supported
- - - free

Note: The end loads are distributed linearly with resultant distance c from the left-hand side.

Fok has also used the finite difference method to determine the critical load for local buckling of a number of plate assemblies, i.e. it was assumed that the joint lines remain straight during buckling. A few results are shown here in Figs. 5.3.13 and 5.3.14. It was shown in Chapter 4 that the finite strip method can be used to derive the buckling stress of

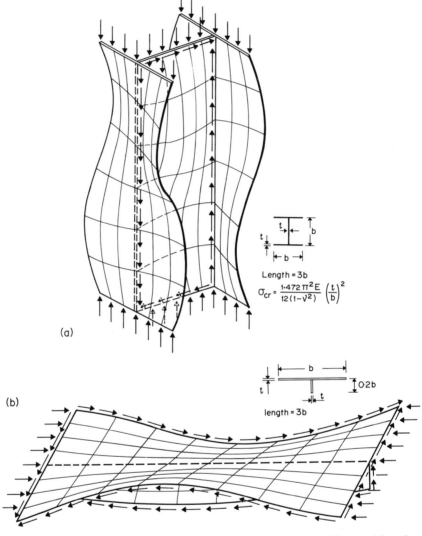

Length = 3b

$$\sigma_{cr} = \frac{1.472\,\pi^2 E}{12(1-\nu^2)} \left(\frac{t}{b}\right)^2$$

(a)

(b)

length = 3b

Fig. 5.3.13. (a) Buckling mode of an I-column; (b) buckling mode of a T-beam subjected to normal and shearing stresses. With 84 nodes $\sigma_{cr} = -\dfrac{1.78\,\pi^2 E}{12(1-\nu^2)} \left(\dfrac{t}{b}\right)^2$. (Fok 1980.)

stiffened plates and it was seen that it can take both local and global buckling into account simultaneously, i.e. if interaction of buckling modes occurs it is taken into account. In the way they are used in this section the Runge–Kutta and the finite difference methods can only be used to study local buckling. This is because it is assumed that the edges of each plate element do not move in space. Both the finite difference and Runge–

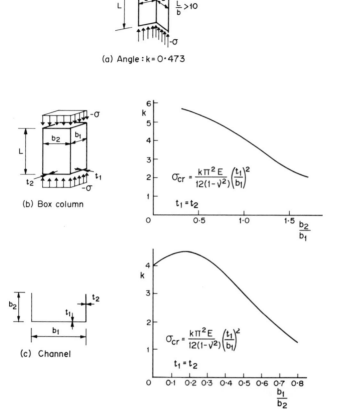

(a) Angle : k = 0·473

(b) Box column

$$\sigma_{cr} = \frac{k\pi^2 E}{12(1-\nu^2)}\left(\frac{t_1}{b_1}\right)^2$$

$t_1 = t_2$

(c) Channel

$$\sigma_{cr} = \frac{k\pi^2 E}{12(1-\nu^2)}\left(\frac{t_1}{b_1}\right)^2$$

$t_1 = t_2$

Fig. 5.3.14. Critical stress factors for local buckling obtained by the finite difference method. (Fok 1980.)

Kutta methods can be used to solve Marguerre's equations (Sections 5.3.4 and 5.4) and in this application both local and global buckling are analysed.

Since both the finite strip and Runge–Kutta methods assume that the deflected shape is a sine curve in the z-direction they cannot cope with built-in ends for example. The finite difference method is not restricted in this way.

The theory contained in this section is concerned only with an eigen-value problem and therefore no second-order effects and no post-buckling behaviour are considered. They are taken into account in the section which follows.

5.3.3. Approximate analytical solution of Marguerre's equations for a rectangular plate with initial dishing

In this section a simply supported rectangular plate (Fig. 5.3.15) $b \times L \times t$ and with assumed initial dishing

$$y = y_0 \cos \beta x \cos \lambda z \qquad (5.3.37)$$

is considered where

$$\beta = \pi/b \quad \text{and} \quad \lambda = \pi/L. \qquad (5.3.38)$$

For convenience the origin is taken here at the centre of the plate.

The purpose of the following analysis is to obtain expressions for the central deflection of the plate and to study the distribution of both membrane and bending stresses as the axial stress σ_z changes. The results enable the effects of the magnitudes of the initial imperfection and of the aspect ratio b/L to be studied. Also the influence of boundary conditions along the longitudinal edges $x = \pm b/2$ can be analysed. Yamaki (1959) and Abdel-Sayed (1969) have solved Marguerre's equations for various boundary conditions. Their analyses cover three cases (Fig. 5.3.15) all of which have pinned sides.

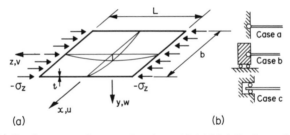

Fig. 5.3.15. (a) Simply supported rectangular plate with initial deflection and in-plane stress σ_z; (b) boundary conditions along $x = \pm b/2$.

Case (a):

The edges $x = \pm b/2$ are kept straight and fixed ($u = 0$). In practice this represents a plate which is restrained against lateral movement by massive framing of other members.

Case (b):

The edges $x = \pm b/2$ are kept straight ($u = \text{constant}$) but the total lateral load is zero, i.e.

$$\int_{-L/2}^{L/2} N_x \, dz = 0. \qquad (5.3.39)$$

This case represents a plate panel in, say, a wide stiffened plate with axial loading. The edges of the plate panel are constrained to remain straight by the surrounding plate panels.

Case (c):
The edges $x = \pm b/2$ are stress free so they can develop in-plane deflections as the plate buckles ($u \neq$ constant). An example of this kind of buckling is a plate panel in a narrow stiffened plate, a square box column or the web of a channel or I-column because in these cases the surrounding plate panels cannot prevent in-plane distortions.

For each of these three cases the edges $z = \pm L/2$ are constrained to remain straight during buckling and for all edges $w = 0$ and $\tau_{xz} = 0$.

As a solution of Marguerre's equations (eqns (5.2.9) and (5.2.10)) the following form of the deflection w is assumed

$$w = w_0 \cos \beta x \cos \lambda z \qquad (5.3.40)$$

so after the deformation the total distance of a point in the middle plane of the plate from the x–z plane is $w + y$. It is seen that although eqn (5.2.9) is non-linear with respect to w it is linear in Φ so its solution is relatively easy to obtain. It is found that by substituting eqns (5.3.37) and (5.3.40) into eqn (5.2.9) a solution for Φ which covers all three boundary conditions can be found. Φ has the form

$$\frac{\Phi}{Et} = -\frac{w_0(w_0 + 2y_0)}{16} \left[\frac{\beta^2}{\lambda^2} \cos^2 \lambda z + \frac{\lambda^2}{\beta^2} \cos^2 \beta x \right] + \frac{px^2}{2}$$

$$+ K_1 \left[\frac{w_0(w_0 + 2y_0)}{16} \beta^2 + \frac{vp}{2} \right] z^2$$

$$+ K_2 [C_1 \cosh 2\lambda x + C_2 x \sinh 2\lambda x] \cos 2\lambda z \qquad (5.3.41)$$

where

$$p = (N_z)_{av}/Et \qquad (5.3.42)$$

and for

Case (a): $K_1 = 1$, $K_2 = 0$;

Case (b): $K_1 = 0$, $K_2 = 0$; $\qquad (5.3.43)$

Case (c): $K_1 = 0$, $K_2 = 1$.

The terms C_1 and C_2 therefore exist only for Case (c). They are evaluated by using the conditions $\tau_{xz} = 0$ along $x = \pm b/2$ and $z = \pm L/2$, whence

$$C_1 = -\frac{\beta^2 w_0(w_0 + 2y_0)(b \coth \lambda b + 1/\lambda)}{32\lambda^2[b \sinh \lambda b - b \cosh \lambda b \coth \lambda b - (\cosh \lambda b)/\lambda]}$$

$$\qquad (5.3.44)$$

$$C_2 = \frac{\beta^2 w_0(w_0 + 2y_0)}{16\lambda^2[b \sinh \lambda b - b \cosh \lambda b \coth \lambda b - (\cosh \lambda b)/\lambda]}$$

Equation (5.2.10) cannot be solved exactly for the assumed deflection pattern given by eqn (5.3.40) but a good approximation for w_0 can be found by a simple application of the 'principle of virtual loads'. The method about to be described is sometimes called Galerkin's method. It is a general method used in theoretical physics for obtaining approximate solutions of differential equations. In the case of a dished plate it will be seen that the quantities involved and the steps taken in the calculation have physical significance.

The assumed solution for w (eqn (5.3.40)) is only an approximation to the true shape of the deflection pattern. Therefore when the values of w and Φ from eqns (5.3.40) and (5.3.41) are substituted into eqn (5.2.10) the left-hand side will not be zero. This error will be a force per unit of area acting normal to the x–z plane because the other terms in eqn (5.2.10) are also quantities of this nature. Thus if the error or residual, as it is often called, is denoted by the symbol R the residual force acting on an element of dimensions $dx \times dz$ is $R \, dx \, dz$. There are many possible ways of minimizing this residual force but the Galerkin method minimizes the 'total complementary energy'. Another way of describing the method is to say that the residual forces are multiplied by the deflection $w(x, z)$, which is used as a 'weight function'. The product is integrated over the entire area of the plate and the integral equated to zero, i.e.

$$\iint_{\text{whole plate}} Rw \, dx \, dz = 0. \tag{5.3.45}$$

This integration is long and tedious and it is difficult to find a simple form of eqn (5.3.45). The following equation has been obtained

$$\frac{bL}{32}\left[-\frac{2Dw_0(\beta^2+\lambda^2)^2}{Et} - 2p(w_0+y_0)(\lambda^2+\lambda\beta^2 K_1) \right.$$

$$\left. -\frac{w_0(w_0+y_0)(w_0+2y_0)}{8}\{\beta^4(2K_1+1)+\lambda^4\} \right]$$

$$+\tfrac{1}{8}L\beta^2\lambda C_1 K_2(w_0+y_0)\sinh\lambda b$$

$$+\tfrac{1}{8}L\lambda^2 C_2 K_2(w_0+y_0)(\beta^2-\lambda^2)\left[\frac{\lambda b}{2}\left(\frac{1}{\lambda^2}-\frac{1}{\beta^2+\lambda^2}\right)\cosh\lambda b \right.$$

$$\left. +\left(\frac{\lambda^2-\beta^2}{2(\beta^2+\lambda^2)^2}-\frac{1}{2\lambda^2}\right)\sinh\lambda b \right]$$

$$+\frac{LC_2 K_2\lambda^3\beta(w_0+y_0)}{4(\beta^2+\lambda^2)}\left[\frac{\beta b}{2}\cosh\beta b - \frac{\beta b \sinh\beta b}{\beta^2+\lambda^2} \right]=0. \tag{5.3.46}$$

Since C_1 and C_2 are quadratic functions of w_0 it is seen that eqn (5.3.46) is a cubic equation of the form

$$B_1 w_0^3 + B_2 w_0^2 + B_3 w_0 + B_4 = 0 \qquad (5.3.47)$$

which expresses the relationship between the axial load parameter p and the central deflection w_0. This equation is used in the following way. The plate dimensions are given and an initial central deflection y_0 and an actual central deflection w_0 are assumed. All terms except p in eqn (5.3.46) are then evaluated directly with a computer and it is easy to find p. The results of an analysis of a family of square plates $(60 \times 60 \times 1)$ with different magnitudes of initial deformation are presented in Fig. 5.3.16(a) and Fig. 5.3.17 compares the central deflection curves of various plates with different geometry and in-plane boundary conditions.

The critical value of p, that is, p_{cr}, is found by assuming y_0 is zero in eqn (5.3.46), in which case B_2 and B_4 in eqn (5.3.47) are found to be zero. Hence this case is represented by the vertical axis ($w_0 = 0$) or the parabola

$$B_1 w_0^2 + B_3 = 0. \qquad (5.3.48)$$

This parabola is seen as the uppermost curve in each of the load–deflection diagrams of Figs. 5.3.16(a) and 5.3.17 and it meets the vertical axis at the critical stress.

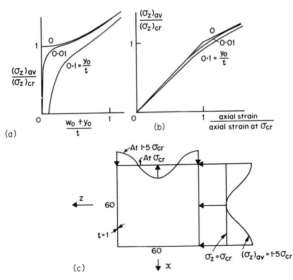

Fig. 5.3.16. Theoretical behaviour of a family of square plates $(60 \times 60 \times 1$, case (c) type supports); (a) vertical deflection at centre of plate; (b) axial strain; (c) stress distribution for $y_0/t = 0$.

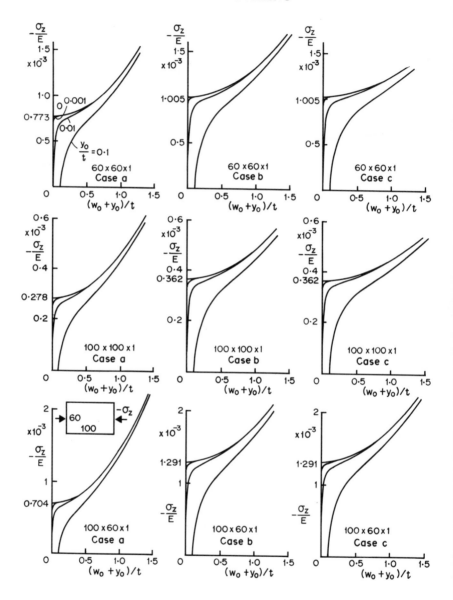

Fig. 5.3.17. Graphs of stress σ_z against central deflection for simply supported plates with various aspect ratios, sizes, initial imperfections, and in-plane constraints along the unloaded sides.

The results of this analysis are as follows
Case (a):

$$p_{cr} = \frac{\sigma_{cr}}{E} = -\frac{\pi^2 \left(\frac{b}{L} + \frac{L}{b}\right)^2}{12(1-\nu)^2 \left(1 + \nu\left(\frac{L}{b}\right)^2\right)} \left(\frac{t}{b}\right)^2$$

Cases (b) and (c):   (5.3.49)

$$p_{cr} = \frac{\sigma_{cr}}{E} = -\frac{\pi^2 \left(\frac{b}{L} + \frac{L}{b}\right)^2}{12(1-\nu^2)} \left(\frac{t}{b}\right)^2$$

It is seen that for Case (a) p_{cr} is less than that for Cases (b) and (c). For a square steel plate, p_{cr} for Case (a) is 77% of that for Cases (b) and (c). This is because for Case (a) the lateral expansion due to the Poisson's ratio effect is prevented and this induces a lateral compressive stress which reduces the critical stress.

In using these equations to evaluate the critical stresses of simply supported perfect plates it is necessary to recognise that the same problem exists as that discussed in connection with stiffened plates and described by Fig. 4.5.5. In other words, if L is greater than b it may be necessary to see whether buckles of length $L/2$, $L/3$, $L/4$, and so on, give a smaller critical stress. This is because a long panel will buckle into a number of waves at a lower stress than if it is forced to buckle into a single wave. This problem arises only because at the outset it was assumed that the panel's deflections were of the form of a single sine function in z given by eqns (5.3.37) and (5.3.40). Yamaki (1959), Coan (1951) and Walker (1969) have solved the same problem but assumed that the deflected shape in the z-direction was an infinite series of double trigonometric functions. Solutions to this problem are also obtained by using the perturbation technique described in Section 1.3.3. The general conclusions of such studies appear to agree with those reached by using the simpler analysis described above.

Expressions for the membrane stresses are derived from eqn (5.2.4).

$$\frac{\sigma_z}{E} = \frac{1}{Et}\frac{\partial^2\Phi}{\partial x^2} = \frac{w_0(w_0 + 2y_0)\lambda^2 \cos 2\beta x}{8} + p + K_2[4C_1\lambda^2 \cosh 2\lambda x$$
$$+ C_2(4\lambda^2 x \sinh 2\lambda x + 4\lambda \cosh 2\lambda x)] \cos 2\lambda z$$

$$\frac{\sigma_x}{E} = \frac{1}{Et}\frac{\partial^2\Phi}{\partial z^2} = \frac{\beta^2 w_0(w_0 + 2y_0)(K_1 + \cos 2\lambda z)}{8} - 4\lambda^2 K_2[C_1 \cosh 2\lambda x$$
$$+ C_2 x \sinh 2\lambda x] \cos 2\lambda z + K_1 \nu p$$ (5.3.50)

$$\frac{\tau_{xz}}{E} = -\frac{1}{Et}\frac{\partial^2\Phi}{\partial x\,\partial z} = 2\lambda K_2[2\lambda C_1 \sinh 2\lambda x + C_2(2\lambda x \cosh 2\lambda x$$
$$+ \sinh 2\lambda x)] \sin 2\lambda z.$$

The stresses due to bending are derived from the expressions (5.3.27) for the bending moments, that is

$$M_z = -D\left(\frac{\partial^2 w}{\partial z^2} + v\frac{\partial^2 w}{\partial x^2}\right), \quad M_x = -D\left(\frac{\partial^2 w}{\partial x^2} + v\frac{\partial^2 w}{\partial z^2}\right). \tag{5.3.51}$$

Thus the bending stresses at the outer fibres are

$$\sigma_{z\,max} = \frac{6M_z}{t^2} = \pm\frac{Ew_0t(\lambda^2 + v\beta^2)}{2(1-v^2)}\cos\lambda z\cos\beta x$$

$$\sigma_{x\,max} = \frac{6M_x}{t^2} = \pm\frac{Ew_0t(\beta^2 + v\lambda^2)}{2(1-v^2)}\cos\lambda z\cos\beta x. \tag{5.3.52}$$

The development of membrane stresses in a square plate with case (c) boundary conditions may be studied by referring to Fig. 5.3.16(c). It is seen that for a perfectly flat plate the stresses remain uniformly distributed until the critical stress is reached. Thereafter the membrane stresses σ_z became concentrated towards the outer edges ($x = \pm b/2$) of the plate and at the centre of the plate they will decrease and may even become tensile. However, while the bending stresses are always zero at the edges, because the edges remain straight, they increase rapidly at the centre of the plate after buckling. It is therefore not possible to see by inspection where the greatest stress occurs, but it will be either at the mid-point of the longitudinal edge or at the mid-point of the plate. Which of these two points carries the greatest stress can be found by using eqn (5.3.50) and (5.3.52). It will be shown in Chapter 7 that some codes of practice require the evaluation of the maximum stress in plate panels at working loads as a check so that yielding of the steel does not occur.

The change of axial length of the panel is obtained by evaluating the axial deflection v at $z = \pm L/2$. From eqn (5.2.7)

$$v' = \varepsilon_z - y'w' - \tfrac{1}{2}w'^2 \tag{5.3.53}$$

and from the usual stress–strain relationship

$$\varepsilon_z = \frac{1}{E}(\sigma_z - v\sigma_x) = \frac{1}{Et}\left(\frac{\partial^2\Phi}{\partial x^2} - v\frac{\partial^2\Phi}{\partial z^2}\right). \tag{5.3.54}$$

By integration of eqn (5.3.53) and using (5.3.54)

$$v = \frac{1}{Et}\int_0^z \frac{\partial^2\Phi}{\partial x^2}dz - \frac{v}{Et}\frac{\partial\Phi}{\partial z} - \int_0^z \frac{\partial y}{\partial z}\frac{\partial w}{\partial z}dz - \frac{1}{2}\int_0^z \left(\frac{\partial w}{\partial z}\right)^2 dz. \tag{5.3.55}$$

The general expression for v is obtained by substituting for Φ, y and w from eqns (5.3.41), (5.3.37) and (5.3.40). The displacement at the end of the panel ($z = L/2$) which is, of course, one-half of its total axial extension is

$$v = \tfrac{1}{2}pL(1 - K_1 v) - \frac{Lw_0(w_0 + 2y_0)}{16}(\lambda^2 + vK_1\beta^2) \qquad (5.3.56)$$

which shows that the ends of the plate remain straight during buckling. Curves for the axial strain of nine perfectly flat plates with different edge conditions are shown in Fig. 5.3.18. One important feature of these

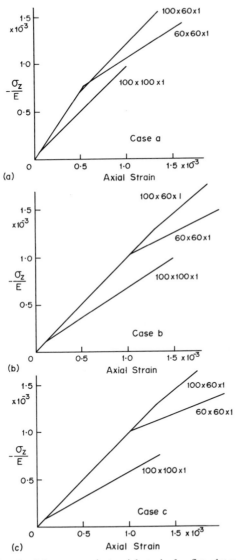

Fig. 5.3.18. Graphs of axial stress against axial strain for flat plates with different sizes, shapes, and boundary conditions

curves is the sudden reduction of the in-plane stiffness of a perfectly flat plate after the critical stress is reached. The graphs indicate that the reduction in stiffness can be as high as 60%. For a plate with initial imperfection the axial stiffness, which is proportional to the slope of the graph, varies continuously. If one plate panel, i.e. a segment between two stiffeners, in a stiffened plate has initial imperfections it will be less stiff than surrounding plate panels. Therefore it will not carry as much of the applied axial load as the other plate panels. Designers need to have some way of estimating this so-called *tangent stiffness* K_{bt} of such a plate panel so that the buckling behaviour of stiffened panels can be studied. The foregoing theory enables this to be done but it is necessary to resort to the use of a computer and this is not always a convenient approach to designers. Therefore this important aspect of plate behaviour can now be analysed by using design curves (Merrison rules 1973) which have been derived from the theory presented here. The application of these curves is studied in Chapter 7. Other sets of useful curves for initially deformed plates have been presented by Keays and Williams (1975) and Williams and Aalami (1979).

The analytical method just described is difficult to apply to plates with other than simply supported boundaries. For these cases it is necessary to resort to numerical methods for solving Marguerre's equations and some of these are described in the next section.

5.3.4. Solution of Marguerre's equations by numerical methods

The numerical analysis of initially deformed plates of rectangular shape has been carried out by several authors (e.g. Basu and Chapman 1966; Aalami and Chapman 1969; Williams 1971) using the finite difference method described briefly in Section 5.3.2. Because of the non-linear nature of the Marguerre equations their equivalent finite difference equations are also non-linear and their solution is found by using iterative methods such as dynamic relaxation (Otter, Cassel, and Hobbs 1966; Rushton 1969). All of these iterative methods converge if the non-linear terms do not have a dominating influence. The linear terms are those which are constant or those which are the product of a constant and an unknown displacement, e.g. $c_m w_m$ where w_m is the displacement at the mth node. The non-linear terms are those which involve the product of two such displacements, e.g. $c_{mn} w_m w_n$. The following simple technique converges if the non-linear terms are not dominant.

The finite difference equivalent of eqns (5.2.9) and (5.2.10) can be written in the following matrix notation

$$[A][\Phi] + [B(w)][W] = 0 \qquad\qquad (5.3.57)$$

$$[C][W] + [D(w)][\Phi] = 0 \qquad\qquad (5.3.58)$$

where [A] and [C] are matrices whose elements are constants, [B(w)] and [D(w)] are matrices whose elements are constants or linear functions of the as yet unknown displacements w at the nodal points and [Φ] and [W] are column vectors whose elements are the unknown stress functions and displacements, respectively, at the nodal points. It is assumed that the coefficients in [B(w)] and [D(w)] are not very large so that the non-linearity is fairly small.

To start the process it is assumed that the plate is unloaded but, of course, has initial displacements. This starting point is the origin where [Φ] = [W] = 0. If the displacements are increased by small amounts, at the same time holding [Φ] = 0, errors will be introduced into eqns (5.3.57) and (5.3.58) which can now be written as follows

$$[A][\Phi_0] + [B(w_1)][W_1] = [E_1^{(1)}] \tag{5.3.59}$$

$$[C](W_1) + [D(w_1)][\Phi_0] = [E_2^{(1)}]. \tag{5.3.60}$$

The suffix 0 and 1 indicate that the original value of [Φ] applies whereas the first increment in [W] has been made. To reduce the error on the right-hand side of eqn (5.3.59) to zero Φ_0 is increased by an amount $\Delta\Phi_0$ where

$$[\Delta\Phi_0] = -[A]^{-1}[E_1^{(1)}] \tag{5.3.61}$$

and this increment is then added to [Φ_0]. Hence the new value of [Φ] is

$$[\Phi_1] = [\Phi_0] + [\Delta\Phi_0]. \tag{5.3.62}$$

This establishes new values [Φ_1] and [W_1] which can be substituted into eqn (5.3.60) to find the new error $E_2^{(2)}$, thus

$$[C][W_1] + [D(w_1)][\Phi_1] = [E_2^{(2)}]. \tag{5.3.63}$$

This error can be reduced to zero by the same device as that used to reduce $E_1^{(1)}$ to zero. Thus, [W_1] is incremented by

$$[\Delta W_1] = -[C]^{-1}[E_2^{(2)}] \tag{5.3.64}$$

and this increment is used to establish a new value [W_2] of the displacements, thus

$$[W_2] = [W_1] + [\Delta W_1]. \tag{5.3.65}$$

This process can be continued until the errors become sufficiently small. The displacements can then be increased to find the next point on the curve. This technique is a convenient and straightforward way to solve small non-linear problems but for solving large problems there are more sophisticated methods available.

Fok (1980) analysed the elastic post-buckling behaviour of seventy-three isolated plates by the finite difference method and confirmed the results experimentally by the Moire fringe technique and with electrical

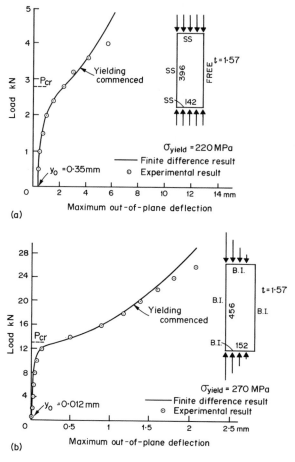

Fig. 5.3.19. Comparison of theoretical and experimental results for two plates, typical of results obtained by Fok (1980).

resistance strain gauges. The plates had a variety of aspect ratios, thicknesses, boundary conditions and deliberately-imposed initial deflections of various patterns. The end loads (in the z-direction) could be applied eccentrically through rigid plattens which in some cases were allowed to rotate in the plane of the plate specimen. Figure 5.3.19 shows some typical comparisons of experimental results with those from a finite difference solution of Marguerre's equations. From these and other comparisons it is concluded that the finite difference method can be used to solve Marguerre's equations with confidence. The method has the drawback that it is iterative and can be rather costly in computer time which has led some research in the direction indicated in the next section.

5.3.5. Application of perturbation technique to isolated plate problems and presentation of design data

It was shown in previous sections that the post-buckling behaviour of isolated plates can be studied by solving Marguerre's equations (eqns (5.2.9) and (5.2.10)) and that solutions can be obtained by using Galerkin's method or the finite difference method. From these studies it was seen that a typical perfectly flat plate buckles in the manner indicated by the full line AB shown in Fig. 5.1.1(b) and that above the bifurcation point A at the critical stress there is an unstable arm AC which is shown as a broken line. The determination of the critical stress is a relatively easy problem compared with the derivation of the post-buckling curve AB because it requires only the solution of the linear buckling eqn (5.3.1) whereas the post-buckling curve is obtained by solving simultaneously the non-linear Marguerre equations with $y = 0$.

The perturbation technique takes advantage of the fact that A is relatively easy to locate and once it is established solutions can be obtained as infinite series expanded about A. These series solutions can be truncated after a few terms so that any desired level of accuracy can be achieved. For a given plate any of the required properties such as central deflection, stress at a given point in the plate, bending moments, and so on are derived as truncated series. Thus not only does the perturbation technique enable quite difficult linear and non-linear buckling problems to be solved but there is another important advantage which was first recognised by Williams and Walker (1975). One way of presenting design data would be to publish extensive tables or graphs to show the load–deflection, load–stress, and load–bending-moment curves for various points in a plate. Walker and Williams showed that it is only necessary to list the coefficients of the truncated series thereby condensing the results enormously and also having them in a more convenient form for further machine computation. Their method is especially advantageous when it is applied to plates with initial imperfections and with boundaries which were not just simply supported because then they were able to establish the coefficients by the finite difference method as an aid to the perturbation method. This combination of methods also lead to large savings in computer time. It is the aim of this section to show how Marguerre's equations may be solved by the perturbation technique and to explain how the results may be presented in condensed form.

It is worth noting that the perturbation method has only recently been used extensively to solve many difficult stability problems. This is surprising because Koiter (1945) used the method of asymptotic expansion of the potential energy to solve problems of elastic stability in 1945. Thompson (1962, 1964) applied the method to study snap-through buckling of shells and thereafter followed many papers and a book (Thompson and Hunt 1973) which demonstrate the great potential of the perturbation technique for solving many and varied stability problems. Stein (1959) was the first to apply the technique to plate buckling problems. He derived the post-buckling behaviour of a perfectly flat plate (i.e. $y = 0$) in the following manner.

The governing equations of a perfectly flat plate with in-plane loading can be taken in the form of eqn (5.2.10) with $y = Y = 0$ and the first two of eqns (5.2.3). It will be recalled that in these equations N_x and N_z are positive when they are tensile. By using eqns (5.2.6) and (5.2.7) they can be written in terms of the displacement components u, v and w. In the perturbation technique it is assumed that these variables can be expanded in terms of an arbitrary perturbation parameter ε about the point of buckling, at which point $\varepsilon = 0$. ε is a function of the applied load and the central deflection. It is not a function of the coordinates x and z. Thus

$$u = \sum_{n=0,2}^{\infty} u^{(n)} \varepsilon^n, \qquad v = \sum_{n=0,2}^{\infty} v^{(n)} \varepsilon^n, \qquad w = \sum_{n=1,3}^{\infty} w^{(n)} \varepsilon^n \qquad (5.3.66)$$

where the coefficients $u^{(n)}$, $v^{(n)}$ and $w^{(n)}$ are functions of x and z only. For flat plates which are just on the point of buckling u and v are not zero but w is. Thus for small positive values of ε (i.e. just above buckling) u and v start with zero power of ε and that for w with a non-zero power. The odd powers of u and v and the even powers of w have been omitted because of the symmetry of the problem.

In making the above substitutions the following relationships are obtained

$$N_z = \sum_{n=0,2}^{\infty} N_z^{(n)} \varepsilon^n + \sum_{m=1,3}^{\infty} \sum_{n=1,3}^{\infty} N_z^{(mn)} \varepsilon^{(m+n)}$$

$$N_x = \sum_{n=0,2}^{\infty} N_x^{(n)} \varepsilon^n + \sum_{m=1,3}^{\infty} \sum_{n=1,3}^{\infty} N_x^{(mn)} \varepsilon^{(m+n)} \qquad (5.3.67)$$

$$N_{xz} = \sum_{n=0,2}^{\infty} N_{xz}^{(n)} \varepsilon^n + \sum_{m=1,3}^{\infty} \sum_{n=1,3}^{\infty} N_{xz}^{(mn)} \varepsilon^{(m+n)}$$

where

$$N_z^{(n)} = \frac{Et}{1-\nu^2} (v'^{(n)} + \nu u^{\cdot(n)})$$

$$N_x^{(n)} = \frac{Et}{1-\nu^2} (u^{\cdot(n)} + \nu v'^{(n)})$$

$$N_{xz}^{(n)} = \frac{Et}{2(1+\nu)} (v^{\cdot(n)} + u'^{(n)})$$

$$N_z^{(mn)} = \frac{Et}{2(1-\nu^2)} (w'^{(m)} w'^{(n)} + \nu w^{\cdot(m)} w^{\cdot(n)}) \qquad (5.3.68)$$

$$N_x^{(mn)} = \frac{Et}{2(1-\nu^2)}(w^{\cdot(m)}w^{\cdot(n)} + \nu w'^{(m)}w'^{(n)})$$

$$N_{xz}^{(mn)} = \frac{Et}{2(1+\nu)}w'^{(m)}w^{\cdot(n)}.$$

Since ε is an arbitrary parameter each coefficient of the power series must vanish when eqns (5.3.66), (5.3.67) and (5.2.4) are substituted into eqn (5.2.10) and the first two of (5.2.3). This leads to the following infinite set of equations

$$N_z'^{(0)} + N_{xz}^{\cdot(0)} = 0$$
$$N_{xz}'^{(0)} + N_x^{\cdot(0)} = 0 \tag{5.3.69}$$

$$D\nabla^4 w^{(1)} - N_z^{(0)}w''^{(1)} - 2N_{xz}^{(0)}w'^{\cdot(1)} - N_x^{(0)}w^{\cdot\cdot(1)} = 0 \tag{5.3.70}$$

$$N_z'^{(2)} + N_{xz}^{\cdot(2)} = -(N_z'^{(11)} + N_{xz}^{\cdot(11)})$$
$$N_{xz}'^{(2)} + N_x^{\cdot(2)} = -(N_{xz}'^{(11)} + N_x^{\cdot(11)}) \tag{5.3.71}$$

$$D\nabla^4 w^{(3)} - N_z^{(0)}w''^{(3)} - 2N_{xz}^{(0)}w'^{\cdot(3)} - N_x^{(0)}w^{\cdot\cdot(3)}$$
$$= (N^{(2)} + N_z^{(11)})w''^{(1)} + (N_x^{(2)} + N_x^{(11)})w^{\cdot\cdot(1)} + 2(N_{xz}^{(2)} + N_{xz}^{(1)})w'^{\cdot(1)} \tag{5.3.72}$$

and so on for higher powers of ε.

As an example of the use of these equations a pin-sided rectangular plate $(L \times b)$ with the origin at one corner (Fig. 5.3.20) and loaded in the z-direction with an in-plane load $-P$ (i.e. compressive) is considered.

The out-of-plane boundary conditions are:
  zero deflection
  $w(0, z) = w(b, z) = w(x, 0) = w(x, L) = 0$
  zero moment
  $w^{\cdot\cdot}(0, z) = w^{\cdot\cdot}(b, z) = w''(x, 0) = w''(x, L) = 0$

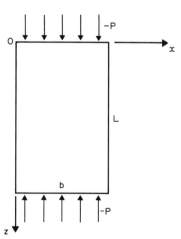

Fig. 5.3.20. Rectangular plate with simply supported sides analysed by Stein's (1959) perturbation method.

and the in-plane boundary conditons are assumed to be,
  constant displacement
  $$u^{\cdot}(0, z) = u^{\cdot}(b, z) = v'(x, 0) = v'(x, L) = 0$$
  zero shear stress
  $$u'(0, z) = u'(b, z) = v^{\cdot}(x, 0) = v^{\cdot}(x, L) = 0$$
  loaded edges
  $$\int_0^b (N_z)_{z=0,L}\, dx = -P$$
  unloaded edges
  $$\int_0^L (N_x)_{x=0,b}\, dz = 0 \tag{5.3.73}$$

When the expressions for u, v and w in eqns (5.3.66) are substituted into the first four boundary conditions it is seen that every value $u^{(n)}$, $v^{(n)}$, and $w^{(n)}$ must satisfy these boundary conditions. Substituting from the first of eqns (5.3.67) into the conditions on the loaded edges gives

$$P = \sum_{n=0,2} P^{(n)} \varepsilon^n \tag{5.3.74}$$

where

$$P^{(0)} = -\int_0^b (N_z^{(0)})_{z=0,L}\, dx$$
$$P^{(2)} = -\int_0^b (N_z^{(2)} + N_0^{(11)})_{z=0,L}\, dx \tag{5.3.75}$$

and so on for the higher-order coefficients.
  Similarly for the unloaded edge from eqn (5.3.67)

$$\int_0^L (N_x^{(0)})_{x=0,b}\, dz = 0$$
$$\int_0^L (N_x^{(2)} + N_x^{(11)})_{x=0,b}\, dz = 0 \tag{5.3.76}$$

and so on for the higher-order coefficients.
  Equations (5.3.69) can be written in terms of $u^{(0)}$ and $v^{(0)}$ by making use of eqns (5.3.68).

$$v''^{(0)} + \frac{1-v}{2} v^{\cdot\cdot(0)} + \frac{1+v}{2} u'^{\cdot(0)} = 0$$
$$\frac{1+v}{2} v'^{\cdot(0)} + u^{\cdot\cdot(0)} + \frac{1-v}{2} u''^{(0)} = 0. \tag{5.3.77}$$

By inspection of the in-plane boundary conditions and of the forms of these equations it is seen that $u^{(0)}$ and $v^{(0)}$ are linear functions of x and z. Thus these equations and the boundary conditions are satisfied if

$$u^{(0)} = \frac{\nu P^{(0)}}{Etb}\left(x - \frac{b}{2}\right)$$

$$v^{(0)} = -\frac{P^{(0)}}{Etb}\left(z - \frac{L}{2}\right).$$

(5.3.78)

Hence from eqn (5.3.68)

$$N_x^{(0)} = N_{xz}^{(0)} = 0, \quad \text{and} \quad N_z^{(0)} = -\frac{P^{(0)}}{b}.$$

(5.3.79)

The boundary conditions and eqn (5.3.70) can be satisfied by taking $w^{(1)}$ in the following form

$$w^{(1)} = w_1 \sin \frac{m\pi x}{b} \sin \frac{n\pi z}{L}$$

(5.3.80)

where w_1 has yet to be determined. After substituting from eqns (5.3.79) and (5.3.80) into eqn (5.3.70) it is found that

$$P^{(0)} = Db\left[\left(\frac{m\pi}{b}\right)^2 + \left(\frac{n\pi}{L}\right)^2\right]^2 \Big/ \left(\frac{n\pi}{L}\right)^2.$$

(5.3.81)

So far the perturbation technique has yielded the small-deflection solution, and, just as in that theory, the central displacement w_1 of the plate is indeterminate as yet. To obtain w_1 it is necessary to proceed further. The values of $N_x^{(11)}$, $N_z^{(11)}$, and $N_{xz}^{(11)}$ are found by substituting into eqns (5.3.68) and then eqn (5.3.71) may be solved. Solutions which satisfy the boundary conditions are

$$v^{(2)} = -\left[\frac{P^{(2)}}{Etb} + \frac{w_1^2}{8}\left(\frac{n\pi}{L}\right)^2\right]\left(z - \frac{L}{2}\right) - \frac{w_1^2}{16}\left[\frac{\left(\frac{n\pi}{L}\right)^2 - \nu\left(\frac{m\pi}{b}\right)^2}{\frac{n\pi}{L}}\sin\frac{2n\pi z}{L}\right.$$

$$\left. -\frac{n\pi}{L}\sin\frac{2n\pi z}{L}\cos\frac{2m\pi x}{b}\right]$$

(5.3.82)

$$u^{(2)} = \left[\frac{\nu P^{(2)}}{Etb} - \frac{w_1^2}{8}\left(\frac{m\pi}{b}\right)^2\right]\left(x - \frac{b}{2}\right) - \frac{w_1^2}{16}\left[\frac{\left(\frac{m\pi}{b}\right)^2 - \nu\left(\frac{n\pi}{L}\right)^2}{\frac{m\pi}{b}}\sin\frac{2m\pi x}{b}\right.$$

$$\left. -\frac{m\pi}{b}\cos\frac{2n\pi z}{L}\sin\frac{2m\pi x}{b}\right].$$

The next step is to obtain a solution of $w^{(3)}$ which satisfies eqn (5.3.72) and the boundary conditions. After substituting from eqns (5.3.82), (5.3.68) and (5.3.75) into eqn (5.3.72) it becomes

$$
\begin{aligned}
D\nabla^4 w^{(3)} + \frac{P^{(0)}}{b} w'''^{(3)} = & w_1 \left[\left\{ \frac{P^{(2)}}{b} \left(\frac{n\pi}{L} \right)^2 - \frac{Etw_1^2}{16} \right. \right. \\
& \left. \times \left[\left(\frac{n\pi}{L} \right)^4 + \left(\frac{m\pi}{b} \right)^4 \right] \right\} \sin \frac{m\pi x}{b} \sin \frac{n\pi z}{L} \\
& + \frac{Etw_1^2}{16} \left(\frac{n\pi}{L} \right)^4 \sin \frac{n\pi z}{L} \sin \frac{3m\pi x}{b} \\
& + \frac{Etw_1^2}{16} \left(\frac{m\pi}{b} \right)^4 \sin \frac{3n\pi z}{L} \sin \frac{m\pi x}{b} \right].
\end{aligned}
\tag{5.3.83}
$$

Since $\sin(m\pi z/b) \sin(n\pi z/L)$ is a complementary function of this equation its coefficient must be zero (see footnote†) and this enables a relationship between w_1 and $P^{(2)}$ to be established.

$$
w_1^2 = \frac{16 P^{(2)}}{Etb} \frac{\left(\dfrac{n\pi}{L} \right)^2}{\left(\dfrac{n\pi}{L} \right)^4 + \left(\dfrac{m\pi}{b} \right)^4}.
\tag{5.3.84}
$$

This procedure can be carried forward, as indeed it was by Stein, to obtain further relationships between the coefficients of the various series. Here the process is now terminated. So far the parameter ε has remained arbitrary but now it is related to some function of the load P or of the central deflection or of the axial shortening or of some other convenient parameter. The final solution will not depend upon this choice. Stein chose the following simple relationship

$$
\varepsilon^2 = (P - P_{cr})/P_{cr}
\tag{5.3.85}
$$

with the understanding that $P \geqslant P_{cr}$. From eqn (5.2.81) it is seen that P_{cr} may be replaced by $P^{(0)}$ and hence

$$
P = P^{(0)} + \varepsilon^2 P^{(0)}.
\tag{5.3.86}
$$

Thus from eqn (5.3.74) it is seen that

$$
P^{(2)} = P^{(0)}, \quad \text{and} \quad P^{(n)} = 0 \quad \text{for all } n \geqslant 4
\tag{5.3.87}
$$

---

† This principle is easily demonstrated by considering the ordinary differential equation $\ddot{y} + a^2 y = A \cos ax$, where the right-hand side is a complementary function. In trying a solution of the form $B \cos ax$ it is found that A must be zero.

Hence first approximations for the following parameters are

$$v = -\frac{\pi^2}{3(1-\nu^2)}\frac{t^2 L}{b^2}\left\{\frac{Pb}{4D\pi^2}\left(\frac{z}{L}-\frac{1}{2}\right)+\delta^2\left[\frac{\beta^2}{2}\left(\frac{z}{L}-\frac{1}{2}\right)\right.\right. \tag{5.3.88}$$

$$+\frac{\beta^2-\nu m^2}{4n\pi}\sin\frac{2n\pi z}{L}-\frac{\beta^2}{4n\pi}\sin\frac{2n\pi z}{L}\cos\frac{2m\pi x}{b}\Bigg]\Bigg\}$$

$$u = -\frac{\pi^2}{3(1-\nu^2)}\frac{t^2}{b}\left\{-\frac{\nu Pb}{4D\pi^2}\left(\frac{x}{b}-\frac{1}{2}\right)+\delta^2\left[\frac{m^2}{2}\left(\frac{x}{b}-\frac{1}{2}\right)\right.\right. \tag{5.3.89}$$

$$+\frac{m^2-\nu\beta^2}{4m\pi}\sin\frac{2m\pi x}{b}-\frac{m}{4\pi}\cos\frac{2n\pi z}{L}\sin\frac{2m\pi x}{b}\Bigg]\Bigg\} \tag{5.3.90}$$

$$w = \frac{2t\delta}{[3(1-\nu^2)]^{\frac{1}{2}}}\sin\frac{n\pi z}{L}\sin\frac{m\pi x}{b} \tag{5.3.91}$$

$$N_z = -\frac{P}{b}-\frac{2D\pi^2}{b^2}\beta^2\delta^2\cos\frac{2m\pi x}{b} \tag{5.3.92}$$

$$N_x = -\frac{2D\pi^2 m^2\delta^2}{b^2}\cos\frac{2n\pi z}{L} \tag{5.3.93}$$

$$N_{xz} = 0$$

where

$$\left.\begin{array}{l}\delta^2 = \dfrac{3(1-\nu^2)}{4}\dfrac{\varepsilon^2 w_1^2}{t^2}\\[3mm]\beta = \dfrac{nb}{L}.\end{array}\right\} \tag{5.3.94}$$

Stein's analysis of a flat square plate is presented in Fig. 5.3.21. The initial part of the graph represents the pre-buckling region in which the

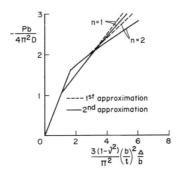

Fig. 5.3.21. Non-dimensional graph of load P against axial shortening Δ for a simply-supported square plate. (Stein 1959.)

plate remains flat. The curve labelled $n = 1$ gives the axial shortening as a function of P when the plate buckles into a shape given by the first Fourier component ($\sin \pi x/b \sin \pi z/L$) while that labelled $n = 2$ shows the axial shortening for the Fourier component ($\sin \pi x/b \sin 2\pi z/L$). It is seen that at about twice the critical load the plate has a tendency to change its buckling pattern from a symmetrical to an antisymmetrical mode. These results were confirmed by Stein for plates with other ratios of b/L. Furthermore he increased the accuracy by increasing the number of terms in each series and he also observed experimentally the tendency of plates to change buckling modes in the post-buckling region. The graphs shown in Fig. 5.3.21 suggest that the lower-order analysis described here is sufficiently accurate until the deflections are very large.

It is seen that the perturbation technique is a recursive method and that it enables a non-linear problem to be transformed into a succession of linear problems (e.g. eqns (5.3.69) and (5.3.72) are all linear). While the perturbation technique increases the understanding of post-buckling phenomena it is difficult to see how it could be extended to plate assemblies such as stiffened plates because of the algebraic complexity.

Walker (1969) also studied the post-buckling behaviour of square plates by using the perturbation technique and his results were used in an ingenious way by Williams and Walker (1975). Williams (1971) published some results of finite difference analyses he had carried out on initially deformed plates. His computer program (PLAP) used dynamic relaxation. Although PLAP is an efficient program it must be rerun every time the plate geometry, or applied stress, or the initial imperfection is changed. However, this costly procedure is avoided by using the perturbation technique. To calculate the response of a family of plates with given length, width, and thickness but with varying in-plane stresses and initial imperfections (such as the set of curves shown in Fig. 5.1.1(b)) it may be necessary to run the program twenty to thirty times. As is shown below, by using the perturbation technique it is only necessary to run the program twice to obtain all of the response curves of the given family. Furthermore, it is not necessary to store the information as a series of curves but simply to record two coefficients for every parameter (e.g. central deflection, bending moment at a point, stress at a point).

To illustrate Walker's method (1969) it is applied here to a flat square plate $(b \times b \times t)$ which is simply supported and loaded in the z-direction. The origin is located at the centre and the following boundary conditions are treated.

At loaded edges $z = \pm b/2$

$$\text{in-plane} \quad v = \text{constant}, \frac{\partial^2 \Phi}{\partial x\, \partial z} = 0$$

out-of-plane $\quad w = \dfrac{\partial^2 w}{\partial z^2} = 0$

At loaded edges $x = \pm b/2$

in-plane $\quad \dfrac{\partial^2 \Phi}{\partial z^2} = \dfrac{\partial^2 \Phi}{\partial x \, \partial z} = 0$

out-of-plane $\quad w = \dfrac{\partial^2 w}{\partial x^2} = 0.$ $\hfill (5.3.95)$

Thus the plate is simply supported and the edges undergo uniform displacement, which is case (b) in Section 5.3.3. The shape of the buckled plate is assumed as

$$w = \sum_{m=1,3}^{\infty} \sum_{n=1,3}^{\infty} w_{mn} \cos \frac{m\pi x}{b} \cos \frac{n\pi z}{b}. \qquad (5.3.96)$$

On substituting into eqn (5.2.9) with $y = 0$ it is found that the resulting equation in Φ has a solution

$$\Phi = -\frac{Px^2}{2} + \sum_{p=0,2}^{\infty} \sum_{q=0,2}^{\infty} (b_{pq} - b_{0(pq)}) \cos \frac{p\pi z}{b} \cos \frac{q\pi x}{b} \qquad (5.3.97)$$

where

$$b_{pq} = \frac{Et}{4(p^2 + q^2)} \sum B_n \qquad (5.3.98)$$

and the coefficients B_n have been listed by Coan (1951).

The equations for the unknown coefficients w_{mn} are found by using Galerkin's method which requires that

$$\int_{b/2}^{b/2} \int_{-b/2}^{b/2} (D\nabla^4 w - \Phi^{\cdot\cdot} w'' + 2\Phi' w'^{\cdot} - \Phi'' w^{\cdot\cdot}) w \, dx \, dz = 0. \qquad (5.3.99)$$

On substituting from eqns (5.3.96) and (5.3.97) into this equation there follows a set of simultaneous cubic algebraic equations in the unknowns w_{mn}. Walker considered only three terms in the series for w so there are three simultaneous equations in w_{11}, w_{13} and w_{31} as follows.

$$400 w_{11} t^2 \left[\frac{1}{3(1 - \nu^2)} + \frac{\bar{\lambda}}{\pi^2} \right] = 50 w_{11}^3 - 25 w_{13} w_{11}^2 - 25 w_{31} w_{11}^2 - 50 w_{11}^2 w_{13}$$

$$- 50 w_{11}^2 w_{31}$$

$$+ 426 w_{11} w_{13}^2 + 426 w_{11} w_{31}^2$$

$$+ 200 w_{11} w_{13} w_{31} - 650 w_{13}^2 w_{31} - 650 w_{13} w_{31}^2$$

$$400w_{13}t^2\left[\frac{25}{3(1-\nu^2)}+\frac{\bar{\lambda}}{\pi^2}\right]=-25w_{11}^3+225w_{11}^2w_{13}+201w_{11}^2w_{13}$$

$$+100w_{11}^2w_{31}+130w_{11}w_{31}w_{13}-2950w_{13}w_{31}^2$$

$$+650w_{11}w_{31}^2-2050w_{13}^3$$

$$400w_{31}t^2\left[\frac{25}{3(1-\nu^2)}+\frac{9\bar{\lambda}}{\pi^2}\right]=-25w_{11}^3+100w_{11}^2w_{13}+650w_{11}w_{13}^2$$

$$+225w_{11}^2w_{31}+1300w_{11}w_{13}w_{31}-2950w_{13}^2w_{31}$$

$$-2050w_{31}^3 \qquad\qquad (5.3.100)$$

where $\bar{\lambda}$ is a non-dimensional form of P, i.e.

$$\bar{\lambda}=-\frac{bP}{Et^3}. \qquad\qquad (5.3.101)$$

Equations (5.3.100) can be written in tensor notation as

$$(A_i+B_i\bar{\lambda})w_i=C_{ijkl}w_jw_kw_l. \qquad\qquad (5.3.102)$$

Following Stein's suggestion (see eqn (5.3.66)) each of the parameters w_{11}, w_{13}, and w_{31} expressed here as w_i can be expanded in terms of a perturbation parameter ε, thus

$$w_i=\sum_{n=1,3}^{\infty}w_i^{(n)}\varepsilon^n \qquad\qquad (5.3.103)$$

and ε is given by eqn (5.3.85).

After truncating this series at $n=3$ and substituting into eqn (5.3.102) terms in ε and ε^3 are obtained. Because ε is arbitrary their coefficients must be zero. The first of these yields the solution of the linear buckling problem, as it did in Stein's method, that is, $\bar{\lambda}_{cr}=-A_i/B_i$ which gives the lowest critical stress as $-\pi^2Et^2/3(1-\nu^2)b^2$. From the remaining equations it is found that

$$w_{11}/t=\left[\frac{8}{3(1-\nu^2)}\right]^{\frac{1}{2}}\varepsilon+\frac{5}{128}\left[\frac{8}{3(1-\nu^2)}\right]^{\frac{1}{2}}\varepsilon^3$$

$$w_{13}/t=-\frac{1}{48}\left[\frac{8}{3(1-\nu^2)}\right]^{\frac{1}{2}}\varepsilon^3 \qquad\qquad (5.3.104)$$

$$w_{31}/t=-\frac{1}{32}\left[\frac{8}{3(1-\nu^2)}\right]^{\frac{1}{2}}\varepsilon^3.$$

On substituting these coefficients into eqn (5.3.96) the following expression for the central deflection is obtained

$$w_{max}/t=1.71\varepsilon-0.02\varepsilon^3 \qquad\qquad (5.3.105)$$

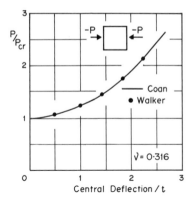

Fig. 5.3.22. Comparison of Coan's (1951) and Walker's (1969) analyses of a square simply supported plate (loaded sides remain straight in-plane, other sides remain straight with zero total load.)

Figure 5.3.22 shows a comparison between this result and that due to Coan (1951) who considered more terms. The good agreement indicates that truncation of the series for w (eqn (5.3.96)) and of that for w_{11}, w_{13} and w_{31} (eqn (5.3.103)) after three terms is satisfactory. This is also concluded from Stein's analysis.

This perturbation method is easily extended to initially imperfect plates by using the findings of Dawson (1971). He discovered that by replacing ε by a new parameter ψ, defined as follows

$$\psi = \left[\frac{P}{P_{cr}} - 1 + \frac{y}{w+y}\right]^{\frac{1}{2}} \tag{5.3.106}$$

the coefficients in the series do not change. Equation (5.3.105) is replaced by

$$\{[(w+y)/t]^2 - (y/t)^2\}^{\frac{1}{2}} = 1.71\psi - 0.02\psi^3 \tag{5.3.107}$$

where $w + y$ is the sum of the actual and the initial deflections at a point in the plate. Again results agree with those of Yamaki (1959) and Coan (1951). Equation (5.3.107) is a remarkably simple equation which enables a whole family of curves (e.g. Fig. 5.1.1(b)) to be plotted.

The difficulty with the perturbation method as described here is that the coefficients of the series need to be obtained analytically and this requires a considerable effort. Also the boundary conditions can present enormous problems if they are other than simply supported. Williams and Walker (1975) overcame this difficulty by making use of the fact that two coefficients (see for example eqn (5.3.107)) are sufficient to describe the response of an initially deformed plate to in-plane loading. Thus they reasoned that, by using their existing finite difference program (PLAP),

Table 5.3.4. Explicit solutions for imperfect plates obtained by Williams and Walker (1975).

(a) Two-term series coefficients for 1:1 plate with all sides simply supported

Variable	Location	Equation (below)	Load case							
			(a)		(b)		(c)		(d)	
			A	B	A	B	A	B	A	B
$w+y$	4	1	2.157	0.010	1.840	-0.259	1.761	-0.082	1.790	-0.126
v		2	0.341	0.013	0.325	0.019	0.954	0.032	0.584	0.023
u		3					0.954	0.032	2.564	0.070
σ_z	1	4	0.055	0.004	0.331	0.023	0.658	0.041	0.495	0.032
	2	4	0.628	0.010	0.338	-0.001	0.665	0.009	0.506	0.000
	3	4	-0.269	-0.004	-0.331	-0.012	-0.664	-0.017	-0.498	-0.014
	4	4	-0.383	0.011	-0.329	0.007	-0.658	0.019	-0.488	0.008
σ_x	1	5	0	0	0.361	-0.004	0.658	0.041	1.023	0.048
	2	5	0	0	-0.301	-0.031	-0.664	-0.017	-0.968	-0.035
	3	5	0.209	-0.006	0.365	-0.021	0.665	0.009	1.028	0.003
	4	5	-0.201	0.016	-0.299	-0.011	-0.658	0.019	-0.962	0.023
M_z	4	6	1.178	-0.066	1.195	-0.080	1.174	-0.062	1.175	-0.064
M_x	4	7	1.174	-0.074	1.168	-0.055	1.174	-0.062	1.172	-0.059
M_{xz}	1	8	0.603	0.066	0.605	0.054	0.610	0.050	0.607	0.051

(b) Two-term series coefficients for 2:1 plate with all sides simply supported

Variable	Location	Equation (below)	Load case							
			(a)		(b)		(c)		(d)	
			A	B	A	B	A	B	A	B
$w+y$	4	1	1.430	0.203	1.408	0.168	1.409	0.113	1.427	0.093
v		2	0.881	0.116	0.887	0.095	0.415	0.103	0.162	0.105
u	1	3	0	0	0.204	0.138	1.545	0.247	3.366	0.531
σ_z	2	4	0.198	−0.047	0.176	−0.001	0.301	0.186	0.150	0.157
	3	4	0	0	−0.267	−0.061	0.271	−0.062	0.135	−0.016
	4	4	−0.195	0.045	−0.236	0.075	−0.246	0.070	−0.199	−0.085
σ_x	1	5	0.652	0.046	0.896	0.098	1.084	0.204	1.322	0.278
	2	5	−0.820	0.097	−0.829	0.084	−1.048	0.132	−1.227	0.180
	3	5	1.100	0.197	0.870	0.101	1.081	0.158	1.337	0.194
	4	5	−0.822	0.137	−0.816	0.124	−1.035	0.207	−1.212	0.286
M_z	4	6	0.505	−0.094	0.502	−0.088	0.533	−0.127	0.564	−0.161
M_x	4	7	0.962	−0.026	0.971	−0.028	0.969	−0.033	0.988	−0.050
M_{xz}	1	8	0.285	0.098	0.293	0.083	0.264	0.119	0.238	0.148

(c) Two-term series coefficients for 3:1 plate with all sides simply supported

Variable	Location	Equation (below)	Load case							
			(a)		(b)		(c)		(d)	
			A	B	A	B	A	B	A	B
$w+y$	4	1	1.221	0.397	1.220	0.356	1.205	0.415	1.204	0.380
v		2					0.201	0.163	0.091	0.092
u		3	1.150	0.326	1.153	0.310	1.814	0.693	3.458	1.712
σ_z	1	4	0	0	−2.047	1.463	0.162	0.312	0.081	0.248
σ_z	2	4	0.096	−0.067	−2.100	1.186	0.105	−0.088	0.072	−0.086
σ_z	3	4	0	0	−2.351	1.039	−0.170	−0.288	−0.081	−0.251
σ_z	4	4	−0.093	0.064	−2.292	1.312	0.112	0.131	−0.068	0.076
σ_x	1	5	1.019	0.150	3.349	−0.922	1.283	0.486	1.385	0.710
σ_x	2	5	−0.976	0.315	1.211	−0.938	−1.104	0.506	−1.221	0.669
σ_x	3	5	1.269	0.521	3.320	−0.923	1.251	0.505	1.386	0.650
σ_x	4	5	−0.981	0.331	1.204	−0.918	−1.110	0.541	−1.226	0.704
M_z	4	6	0.340	−0.055	0.340	−0.055	0.377	−0.099	0.379	−0.105
M_x	4	7	0.924	−0.020	0.909	−0.010	0.930	−0.032	0.937	−0.036
M_{xz}	1	8	0.150	0.172	0.161	0.158	0.101	0.229	0.032	0.309

(d) Two-term series coefficients for 0.66 : 1 $(L : b)$ plate with loaded sides $(z = 0, L)$ simply supported and unloaded sides $(x = 0, b)$ clamped.

Variable	Location	Equation (below)	Load case (a)		(b)	
			A	B	A	B
$w + y$	4	1	1.846	−0.101	1.812	−0.097
v	1	2	0.357	0.016	0.354	0.015
σ_z	2	4	0.229	0.009	0.345	0.017
	3	4	0.486	0.025	0.370	0.011
	4	4	−0.456	0.007	−0.472	0.006
		4	−0.515	0.024	−0.506	0.021
σ_x	1	5	0	0	0.131	0.017
	2	5	0	0	−0.133	−0.012
	3	5	0.234	−0.013	0.251	−0.008
	4	5	−0.206	0.024	−0.227	0.010
M_z	4	6	2.448	−0.094	2.440	−0.089
M_x	2	7	−3.190	−0.442	−3.163	−0.476
	4	7	2.009	−0.188	2.011	−0.183
M_{xz}	1	8	0.026	0.017	0.039	−0.002

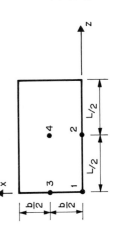

Locations referred to in table
f means in-plane stress normal to edge = 0
s means edge is straight and total load = 0.

Summary of general form of explicit solutions

(a) Deflection normal to plate

$$\{[(w+y)/t]^2 - (y/t)^2\}^{\frac{1}{2}} = A\psi + B\psi^3 \tag{1}$$

(b) End shortening

$$\frac{v}{v_{cr}} - \frac{\sigma_{z\,av}}{\sigma_{z\,cr}} = A\eta + B\eta^2 \tag{2}$$

$$\frac{u}{u_{cr}} - \frac{\sigma_{x\,av}}{\sigma_{x\,cr}} = A\eta + B\eta^2 \tag{3}$$

(c) Membrane stress

$$\frac{\sigma_z}{\sigma_{z\,cr}} - \frac{\sigma_{z\,av}}{\sigma_{z\,cr}} = A\eta + B\eta^2 \tag{4}$$

$$\frac{\sigma_x}{\sigma_{x\,cr}} - \frac{\sigma_{x\,av}}{\sigma_{x\,cr}} = A\eta - B\eta^2 \tag{5}$$

(d) Bending moment

$$M_z\left(\frac{b^2}{Et^4}\right) = A\Omega + B\xi \tag{6}$$

$$M_x\left(\frac{b^2}{Et^4}\right) = A\Omega + B\xi \tag{7}$$

(e) Twisting moment

$$M_{xz}\left(\frac{b^2}{Et^4}\right) = A\Omega + B\xi \tag{8}$$

where

$$\psi = \left[\frac{P}{P_{cr}} - 1 + \frac{y}{w+y}\right]^{\frac{1}{2}}$$

$$\Omega = w/t$$

$$\eta = [(w+y)/t]^2 - (y/t)^2$$

$$\xi = [(w+y)/t]^3 - (y/t)^3$$

$$v_{cr} = \left[L\left(1 - v\frac{\sigma_{x\,av}}{\sigma_{z\,av}}\right)\Big/E\right]\sigma_{z\,cr}$$

$$u_{cr} = \left[b\left(1 - v\frac{\sigma_{z\,av}}{\sigma_{x\,av}}\right)\Big/E\right]\sigma_{x\,cr}$$

$$P/P_{cr} = \frac{\sigma_{z\,av}}{\sigma_{z\,cr}} \text{ when main load is in the } z\text{-direction}$$

or

$$P/P_{cr} = \frac{\sigma_{x\,av}}{\sigma_{x\,cr}} \text{ when main load is in the } x\text{-direction.}$$

they need only calculate the response of two plates in a given family to obtain the response curves of the whole family. The general equation used is

$$\{[(w+y)/t]^2 - (y/t)^2\}^{\frac{1}{2}} = A\psi + B\psi^3. \tag{5.3.108}$$

As an example, suppose for a plate with $y_0/t = 0.5$ the computer program gives the following solutions at the centre of the plate.

(a) At $P/P_{cr} = 0.505$ $(w_0 + y_0)/t = 0.842$

(b) At $P/P_{cr} = 1.023$ $(w_0 + y_0)/t = 1.416$.

On substitution into eqn (5.3.106) $\psi = 0.314$ and 0.613, respectively, and on substitution into eqn (5.3.108)

$$[(0.842)^2 - (0.5)^2]^{\frac{1}{2}} = A(0.314) + B(0.314)^3$$
$$[(1.416)^2 - (0.5)^2]^{\frac{1}{2}} = A(0.613) + B(0.613)^3.$$

On solving, $A = 2.157$ and $B = 0.010$, hence

$$\{[(w_0 + y_0)/t]^2 - (y_0/t)^2\}^{\frac{1}{2}} = 2.157\psi + 0.010\psi^3. \tag{5.3.109}$$

This expression, which is called an *explicit function*, can now be used to plot the whole family of curves of load against central deflection for geometrically similar plates but whose initial deflection varies. Comparisons of the explicit functions with curves obtained by other methods show that they are sufficiently accurate for design purposes. Obviously terms of higher order could be added to the right-hand side of eqn (5.3.108) to increase accuracy if this became necessary. Fok (1980) developed other explicit functions which not only improve accuracy but also avoid those cases when ψ is imaginary. The stresses, bending moments, axial shortening, and so on can also be expressed as explicit functions and Table 5.3.4 summarizes some results of Williams and Walker's analyses. Their paper goes on to explain how this information may be used for design purposes.

5.4. A BRIEF REVIEW OF SOME LITERATURE ON THE ELASTIC BUCKLING OF THIN-WALLED STRUCTURES

In the earlier sections of this chapter the Marguerre equations were taken as the starting point for the analysis of both perfect and imperfect plates. In this section the historical background and further developments after Marguerre are reviewed. The history of the governing equations is first described and then that of various solutions and applications is traced.

The membrane theory of plates was first studied by Euler (1766) and the flexural theory by Bernoulli (1789) and Navier (1823). The theory for combined membrane and flexural effects was developed by Kirchhoff

(1877) and Saint Venant (1883). At this stage the governing equation for thin isotropic plates loaded transversely with Y per unit area and in-plane forces N_x, N_z and N_{xz} per unit length was

$$D\left[\frac{\partial^4 w}{\partial x^4}+2\frac{\partial^4 w}{\partial x^2 \partial z^2}+\frac{\partial^4 w}{\partial z^4}\right]=Y+N_x\frac{\partial^2 w}{\partial x^2}+2N_{xz}\frac{\partial^2 w}{\partial x \partial z}+N_z\frac{\partial^2 w}{\partial z^2} \qquad (5.4.1)$$

By comparison with the previously presented derivation of Marguerre's equations it is seen that this is the same as the equilibrium equation (5.2.10) with the initial imperfection $y = 0$. Equation (5.4.1) has been used in Section 5.3.1 (eqn (5.3.1)) with $Y = 0$ to study the buckling of perfectly flat plates. In other words at the turn of this century the equation governing the buckling of flat plates was available and it was known that it forms the basis of an eigenvalue problem. At that time it was not recognised that as the plate buckles the values of N_x, N_z and N_{xz} at a given point would vary because of the stretching of the plate.

The next development which overcame this deficiency was due to Föppl (1907) who introduced the stress function Φ and paved the way for von Karman (1910) to derive the governing equations for perfectly flat plates, that is,

$$D\left[\frac{\partial^4 w}{\partial x^4}+2\frac{\partial^4 w}{\partial x^2 \partial z^2}+\frac{\partial^4 w}{\partial z^4}\right]=\frac{\partial^2 \Phi}{\partial z^2}\frac{\partial^2 w}{\partial x^2}+\frac{\partial^2 \Phi}{\partial x^2}\frac{\partial^2 w}{\partial z^2}-2\frac{\partial^2 \Phi}{\partial x \partial z}\frac{\partial^2 w}{\partial x \partial z}+Y$$

$$\frac{\partial^4 \Phi}{\partial x^4}+2+\frac{\partial^4 \Phi}{\partial x^2 \partial z^2}+\frac{\partial^4 \Phi}{\partial z^4}=Et\left[\left(\frac{\partial^2 w}{\partial x \partial z}\right)^2-\frac{\partial^2 w}{\partial x^2}\frac{\partial^2 w}{\partial z^2}\right]. \qquad (5.4.2)$$

These equations enabled the post-buckling behaviour of perfectly flat plates to be studied and it will be seen later that they were used to carry out the large-deflection analysis of flat plates with transverse loads Y only (i.e. N_x, N_z, and N_{xz} are zero at the boundaries).

Marguerre's equations for imperfect plates were derived in 1938 and have been used extensively since then. They are essentially the governing equations for shallow shells.

Flat plates which are stiffened to different degrees in orthogonal directions behave like orthotropic plates, the theory of which was developed by Gehring (1860) and Boussinesq (1879). For a transverse load Y the governing equation

$$D_x\frac{\partial^4 w}{\partial x^4}+2H\frac{\partial^4 w}{\partial x^2 \partial z^2}+D_z\frac{\partial^4 w}{\partial z^4}=Y \qquad (5.4.3)$$

was developed by Huber (1914) and is now known as 'Huber's equation'. The section properties are defined as follows.

$$D_x=(EI)_x/(1-\nu_x\nu_z)=\text{average flexural rigidity of the stiffened}$$
$$\text{plate under bending moments } M_x \quad (5.4.4)$$

$D_z = (EI)_z/(1 - \nu_x\nu_z) =$ average flexural rigidity of the stiffened plate under bending moments M_z (5.4.5)

$$H = \tfrac{1}{2}(\nu_x D_z + \nu_z D_x) + 2(GI)_{xz} \tag{5.4.6}$$

$2(GI)_{xz} = \dfrac{\partial^2 w}{\partial x\,\partial z}\bigg/ M_{xz} =$ average torsional rigidity $= G_{xz}t^3/12$ (5.4.7)

$\nu_x, \nu_z =$ Poisson's ratio in the x- and z-directions

$$E = \sqrt{(E_x E_z)} = \text{modified Young's modulus} \tag{5.4.8}$$

$$G_{xz} \simeq \frac{E}{2[1 + \sqrt{(\nu_x\nu_z)}]} = \text{modified shear modulus.} \tag{5.4.9}$$

Huber (1914) applied these equations to the analysis of a reinforced concrete slab stiffened by orthogonal ribs. The von Karman large-deflection equations for flat isotropic plates with in-plane loading were modified to account for anisotropy by Rostovtsev (1940) and later the effects of initial imperfections were included resulting in the following simultaneous equations.

$$D_x \frac{\partial^4 w}{\partial x^4} + 2H \frac{\partial^4 w}{\partial x^2\,\partial z^2} + D_z \frac{\partial^4 w}{\partial z^4} = \frac{\partial^2 \Phi}{\partial z^2} \frac{\partial^2 (y + w)}{\partial x^2}$$

$$+ \frac{\partial^2 \Phi}{\partial x^2} \frac{\partial^2 (y + w)}{\partial z^2} - 2 \frac{\partial^2 \Phi}{\partial x\,\partial z} \frac{\partial^2 (y + w)}{\partial x\,\partial z} + Y$$

$$\frac{1}{t_z E_z} \frac{\partial^4 \Phi}{\partial x^4} + 2\left(\frac{1}{K_{xz}} - \frac{\nu_x}{t_x E_x} - \frac{\nu_z}{t_z E_z}\right) \frac{\partial^4 \Phi}{\partial x^2\,\partial z^2} + \frac{1}{t_x E_x} \frac{\partial^4 \Phi}{\partial z^4}$$

$$= -\frac{\partial^2 y}{\partial z^2} \frac{\partial^2 w}{\partial x^2} + 2 \frac{\partial^2 y}{\partial x\,\partial z} \frac{\partial^2 w}{\partial x\,\partial z} - \frac{\partial^2 y}{\partial x^2} \frac{\partial^2 w}{\partial z^2} - \frac{\partial^2 w}{\partial z^2} \frac{\partial^2 w}{\partial x^2} + \left(\frac{\partial^2 w}{\partial x\,\partial z}\right)^2$$

  (5.4.10)

where t_x and t_z are the equivalent thicknesses in the x- and z-directions and K_{xz} is the shearing rigidity of the equivalent plate.

These appear to be the most general equations currently available for solving plate buckling problems. When the plate is isotropic eqns (5.4.10) degenerate into eqns (5.2.9) and (5.2.10) which in turn simplify into eqn (5.3.1) when the plate is isotropic and initially flat. Attention is now turned to some of the solutions obtained to plate buckling problems and this survey begins with the buckling of a flat plate.

Bryan (1891) was the first to solve the problem of a simply supported rectangular plate with two opposite sides carrying uniform compressive loads. What is especially interesting is that he was the first to derive the solution of a plate buckling problem by using energy principles. The same problem was later solved by Timoshenko (1907) by integrating eqn (5.3.1). He also (1907, 1910, 1913) analysed many other flat plates with different boundary conditions by these two methods. Following the advent of the finite difference and relaxation techniques and later with the increasing use of electronic computers and finite elements it has become relatively easy to solve eqn (5.3.1) for a wide variety of plate shapes and stress distributions. The results of such analyses have been published as design data in books such as those by Williams and Aalami (1979), Bulson (1970), Klöppel and Scheer (1960), Klöppel and Möller (1968), and C.R.C. of Japan (1971), but designers will find many gaps in the available design data. However, with the techniques described in earlier sections of this chapter a designer should be able to obtain new solutions for himself.

As seen in earlier sections of this chapter the large-deflection theory of isolated plates is much more complicated because of the non-linearity of the governing equations (eqns (5.4.2) and (5.4.10)) and it is only fairly recently that design data has become available. The non-linearity arises because as the deflections increase the membrane stresses are redistributed and begin to dominate over flexural action. This problem occurs for example when a plate carries normal forces $Y(x, z)$ which induce deflections of the order of the thickness or more. Marguerre (1938) and Way (1938) obtained solutions of problems of this kind by energy methods and Levy (1942a, b, c, and d) and Levy and Greenman (1942) used truncated double Fourier series to obtain solutions. In all of these cases the edges of the plates were constrained to remain straight with or without translation. In other words, their solution could be applied to an internal plate panel of a wide stiffened plate but not to an edge panel or to a box column, a plate girder and so on.

To study the post-buckling behaviour of plates Coan (1951) solved Marguerre's equations by assuming the deflected shape of a rectangular simply supported plate as a double Fourier series and overcame the restriction on Levy's solution referred to above. Edge pull-in of three kinds were allowed for by adding further complementary functions to the expressions for the stress function Φ. The theory was applied to a square plate with the central initial deflection $y_0 = 0.1t$ and results compared favourably with experimental values. Similar problems were analysed by Yamaki (1959) and Walker (1969); the essential difference between these papers lies in the way they deal with the non-linearity of the equations. As is described in Section 5.3.5 Walker solved the simultaneous cubic equations by the perturbation technique. Each of these workers assumed simply supported boundaries and a double sine series for the shape of both the initial and the final deflection so the results are applicable to a very limited range of problems. The main value of these papers is that they have become a yardstick for judging the accuracy of other methods, especially numerical methods. Rhodes and Harvey (1971) also analysed post-buckling behaviour. They assumed that the stress function Φ was a combination of trigonometric and hyperbolic functions.

Kaiser (1936) had earlier used the finite difference and relaxation techniques to solve von Karman's equations for the cases where edge pull-in is allowed.

Green and Southwell (1946) also solved von Karman's equations by these techniques but first modified the equations so that they were expressed in terms of three unknown displacements, u, v, and w. Wang (1948a, b) and later Basu and Chapman (1966) applied the finite difference technique directly to von Karman's equations to solve, with the aid of a computer, isolated plate problems.

A considerable number of practical solutions of Marguerre's equations (for imperfect plates) have been obtained by Williams (1971, 1973), Aalami and Chapman (1969), and Keays and Williams (1975). Much of this work is consolidated in a design manual on thin plates with in-plane loading by Williams and Aalami (1979). In that volume extensive graphs are presented to enable designers to predict the behaviour of imperfect rectangular plates. An earlier book by Aalami and Williams (1975) dealt with loading which is normal to the plane of the plate. The work of Williams and Walker (1975, 1977) which enables these graphs to be expressed as explicit functions has already been presented in Section 5.3.5.

This review now turns its attention to the elastic buckling of stiffened plates. There have been hundreds of attempts to apply orthotropic plate theory to study the buckling of stiffened plates. It is beyond the scope of this book to review them in depth and the reader is referred to Troitsky's book (1976) for a fuller coverage of this branch of thin-walled structures. It appears that most of the writers in this field have been concerned with the buckling analysis of stiffened bridge decks. Pelikan and Esslinger (1957) developed a practical method for bridge-deck design and it was used by the American Institute of Steel Construction for their design manual for orthotropic steel plate deck bridges (1963). They assume that the stiffened plate is an orthotropic plate with rigid supports at the main girders and elastic supports at the lighter transverse floor beams.

The main value in applying orthotropic plate theory to stiffened plate problems is that it provides designers with an alternative approach. On the whole the results of the theory are fairly easy to apply. For example, there are many well-established formulae for the critical stress of rectangular orthotropic plates with a variety of boundary conditions and axial loading. Falconer and Chapman (1953), Williams (1971) and Troitsky (1976) are typical of these source of information. Although the theory is valid for a plate of uniform thickness which is made from an anisotropic material such as a fibre-reinforced plastic it should, however, be recognised that there are many assumptions which are difficult to justify when the theory is applied to stiffened plates. Thus the neutral planes in the two orthogonal directions usually do not coincide, the theory 'smears' the stiffeners over the plates and the way this should be done is not universally agreed. Also the theory cannot take into account stiffners which are unevenly spaced. The most serious deficiency is that local buckling (see Chapter 4) cannot be treated and this is usually the governing criterion of stiffened plate design.

The alternative approach to the elastic buckling analysis of stiffened plates is to divide the cross-section into its elements, i.e. sections of plate and sections of stiffeners as was done in Chapter 4 and Section 5.3 where the finite strip method and work of Wittrick and Williams (1974), Michelutti (1976) and Fok (1980) in solving von Karman's equations was described. Many valuable graphs and charts for the buckling stresses of stiffened plates of practical interest were obtained by using energy principles and brought out in book form by Klöppel and Scheer (1960) and Klöppel and Möller (1968). Designers will find these books, which are the fruits of many productive years of research of Professor Klöppel and his team at Darmstadt University, very useful indeed.

All of the papers and books discussed in the last paragraph deal with the elastic analysis of stiffened plates without initial imperfections. Comparatively little attention has been paid to stiffened plates with initial imperfections. The Marguerre equations were, however, applied to isolated plates with initial imperfections by Bilstein (1974), also of Darmstadt. He also assumed that deflections, bending moments and stresses were simply trigonometric functions of the z-coordinate. This reduced the Marguerre equations to non-linear ordinary differential equations in x, which he solved by an iterative Runge–Kutta method. After obtaining good agreement of his results for isolated imperfect plates with those of other authors Bilstein then applied the same method to stiffened plates with initial imperfections. The following describes only two of the imperfect stiffened plates studied by Bilstein. The plate shown in Fig. 5.4.1 has three simply supported sides and one side which is stiffened by a rectangular stiffener attached with a small eccentricity. The post-buckling behaviour of this perfect plate was analysed first by Klöppel and Unger (1969) by using an energy method and agreement between the two methods is reasonably good up to $\sigma_{av}/\sigma_{cr} = 1.5$. Some results of Bilstein's analysis of this plate with three types of initial imperfection are shown in Fig. 5.4.2, where for 'load-case I' the ends $z = 0$ and L remain straight and for 'load-case II' the applied stress distribution is maintained throughout the study and the ends can undergo in-plane distortion. It is seen that the most noticeable features of these graphs are, firstly, the large difference between the load–deflection curves for load-cases I and II

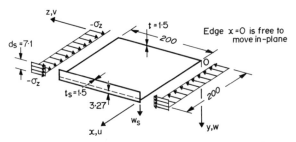

Fig. 5.4.1. Plate simply supported on three sides and stiffened on the unsupported side analysed by Bilstein (1974).

for a perfect panel. In load-case I the stiffness of the central strip of the panel falls to a very small value as the panel buckles but the outer edges can carry increasing loads because their stiffness is still positive after the onset of buckling. This is reflected in the steepness of the graphs for case I shown in Fig. 5.4.2. For load-case II the buckling strength of the panel is largely determined by that of the central strip. Increases in the applied load result in the large deflections indicated in Fig. 5.4.2. Secondly, a 'type 1' initial deflection, in which the plate is imperfect but the stiffener

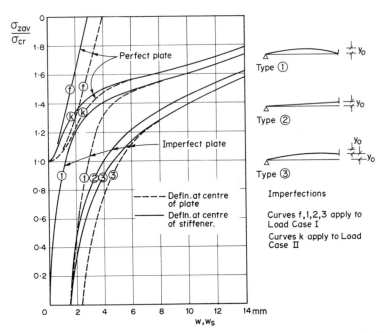

Fig. 5.4.2. Graphs showing effects of load type and imperfection type upon the buckling behaviour of the plate shown in Fig. 5.4.1.

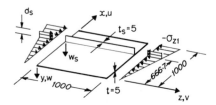

Fig. 5.4.3. Plate simply supported on all four sides and stiffened at the third point analysed by Bilstein (1974).

is perfectly straight, results in smaller deformations of the panel than either 'type 2 or 3' initial deflections.

Another of Bilstein's many studies of stiffened plates with initial deformation is illustrated in Fig. 5.4.3. This is a typical panel from the web of a box-girder where the in-plane stress varies linearly across the ends and a longitudinal stiffener is supposed to stabilise the most highly stressed part of the plate. Figure 5.4.4 shows a typical deflection pattern and Fig. 5.4.5 shows two of many curves of maximum stress. The latter curves show that small initial imperfections can greatly increase the magnitude of the maximum stress. In other words a stiffened plate with small initial imperfections will reach the point of first yielding at a much lower load than one which is perfect and this has important implications for designers who may be required to carry out checks to ensure that no part of the panel yields at working loads (serviceability checks) and for fabricators who could be asked to build such a panel to within specified tolerances (e.g. by the Merrison rules and BS5400). The problem of the local buckling of an imperfect plated structure which was solved by Bilstein with the Runge–Kutta technique can also be solved by the finite difference method. Fok (1980) has demonstrated this by analysing an imperfect box-column and an imperfect channel-column. In the latter case computer results were compared with experimental results obtained by Khoo (1979) and found to give very good agreement.

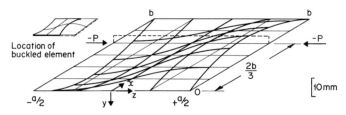

Fig. 5.4.4. Buckled shape of plate shown in Fig. 5.4.3 when $d_s = 40$, $P/P_{cr} = 0.8052$ ($P = -346.28$ kN). (Bilstein 1974.)

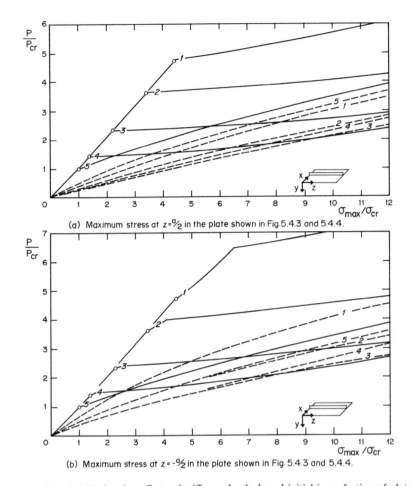

(a) Maximum stress at $z = \frac{d}{2}$ in the plate shown in Fig.5.4.3 and 5.4.4.

(b) Maximum stress at $z = -\frac{d}{2}$ in the plate shown in Fig. 5.4.3 and 5.4.4.

Fig. 5.4.5. Graphs showing effects of stiffener depth d_s and initial imperfection of plate on maximum stress at two cross-sections.
Data: Triangular loading (Fig. 5.4.3). ——— zero imperfection; ---- $y_0/t = 1.5$. Curve parameters: 1, $d_s = 50$; 2, $d_s = 40$; 3, $d_s = 30$; 4, $d_s = 20$; 5, $d_s = 0$. (Bilstein 1974.)

Although there have been many papers published about experiments concentrating on the collapse load of thin-walled structures (see Chapter 6) there have been relatively few specifically devoted to their elastic response. Before reviewing some typical papers on experimental studies of the elastic behaviour of thin-walled structures the experimental techniques available are briefly described and discussed. These techniques are divided into those aimed at evaluating in-plane stresses and strains and those aimed at recording the out-of-plane effects, e.g. deflections, curvatures, and bending moments. The most satisfactory way of obtaining in-plane stresses and strains in a plated structure appears to be to use electrical resistance strain gauges, of which there is a vast array now available. For many plated structures the directions of the principal stresses at points in the plate are known beforehand so it is necessary to use only two single gauges oriented in these directions at a point on each surface of the plate. An example of this is the deck of a large box-girder bridge where the principal directions are parallel to the longitudinal axis of the bridge and at right angles to it. Membrane stresses and strains are found by averaging the readings on the two surfaces and bending effects are derived by taking their differences, but the latter method is usually not very accurate. This same problem occurs even more acutely when the directions of the principal stresses are not known and it is necessary to use strain gauge rosettes on each surface. In these cases the accuracy of the results of a bending analysis will be even less than when the principal directions are known. Perhaps the most satisfactory way of overcoming these difficulties is to use the techniques which measure out-of-plane effects. Principal curvatures of a plate surface can be measured with a spherometer, which is based upon the principle of measuring the vertical height of three points which lie as a straight line. Such a method involves touching the surface which can cause extraneous deformations but, more importantly, the instrument is very tedious to use and interpret.

Another method of measuring bending effects is to build a plate scanning device, such as the one (Fok 1980) illustrated in Fig. 5.4.6. When the vertical probes are moved up or down they bend a cantilever to which is attached a pair of strain gauges. These can be connected to a data logger which is programmed to interpret the readings as vertical displacements of the probe. The bank of probes is driven by a small electric motor across the surface of the deformed plate. In the instrument illustrated the motor was stopped automatically at each of the equispaced holes in the side plates by a switch. While the device was stopped the strain reading of each probe was converted to a deflection and stored in the data logger in matrix form. Fok (1980) then used this matrix as the data for a computer program which solved the Marguerre equations by the finite difference technique (Section 5.3.4). Thus he used the actual initial deflections whereas most other workers appear to have assumed trigonometric shapes. The deformation pattern under load was calculated and the results plotted by the computer as contour lines of deflection, stress, bending moments, etc. (Fig. 5.4.7) or as an isometric view with vertical displacements magnified (Fig. 5.4.8). This instrument did touch the plate surface but the forces were very small and did not lead to significant errors.

A powerful technique is the Moire fringe method. This is based upon an interference-like pattern developing between a fine grid of parallel

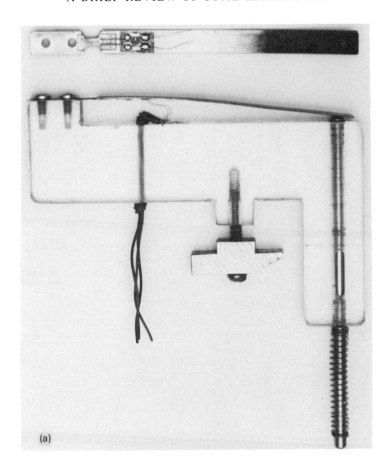

Fig. 5.4.6. Plate scanning device used to measure out-of-plane deflections at grid points used in the finite difference analysis. (a) Single probe; (b) assembled plate scanner with ten probes. (Fok 1980.)

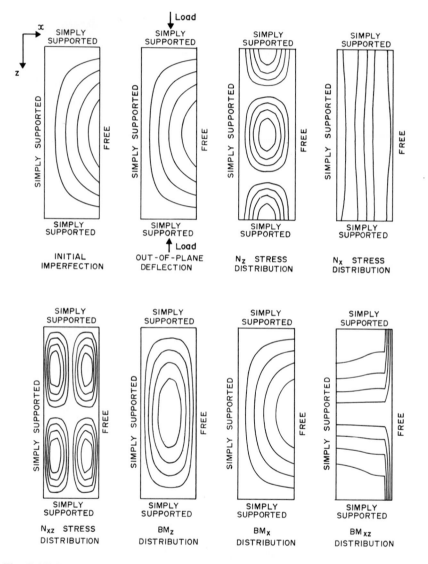

Fig. 5.4.7. Information which can be computed from data obtained by using a plate-scanning device (Fig. 5.4.6) (Fok 1980).

lines on a flat sheet of glass placed in front of the test plate and their shadows on the plate (Fig. 5.4.9). The 'interference' lines result when the lighted space between two adjacent shadow lines is blocked by a line on the grid-plate so that the whole of that small region appears to be dark. When this space is not blocked by a line on the grid the camera 'sees' a light region. These 'interference lines' are contours or lines of constant

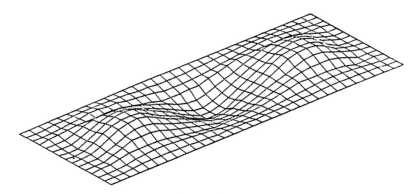

Fig. 5.4.8. Isometric view of plate deflections (magnified) obtained by using the plate-scanner shown in Fig. 5.4.6 and a computer. (Fok 1980.)

deformation of the test plate. A typical pitch of the lines which are photographically applied to the glass sheet is about 0.5 mm. The vertical distance between contours can be calculated from the geometry of the set-up but it is more convenient to use calibration wedges of known taper. Figure 5.4.10 shows two typical fringe patterns of buckled plates. The method is easy to use, the displacement of every point in the plate is obtained, and the development of the deflection pattern can be monitored

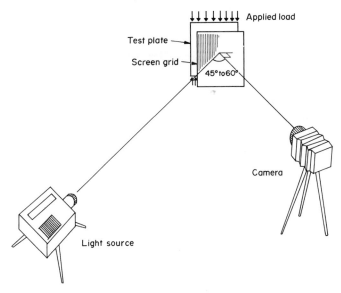

Fig. 5.4.9. Experimental set-up for measuring contours of out-of-plane deflection of a plate by using the Moire fringe method.

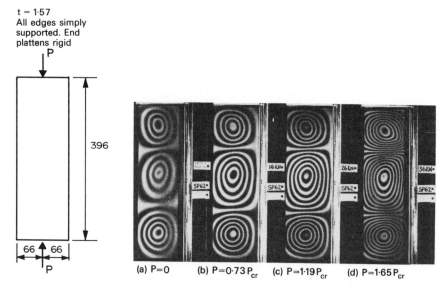

Fig. 5.4.10. Moire fringe patterns of a simply supported plate with initial dishing. (Fok 1980).

by photographing the fringe pattern at every load increment. The fringe pattern enables the curvatures and twist to be calculated at each point of the plate and from these the bending and twisting moments can be evaluated. One disadvantage of the method is that the pattern must be interpreted by hand measurement but this could be overcome by using a light-sensitive recording device. Another disadvantage is that while the technique is convenient to use under laboratory conditions it is not suitable for use on site.

One important problem in testing thin-walled structures in compression is that of defining the in-plane boundary conditions along the loaded edges. In most cases the shear stress is made zero, the edge is held straight and the longitudinal movements are uniform. This condition is relatively easy to achieve by using massive end-loading plattens. However, for some cases the end-load should be uniformly distributed. Hoff, Boley, and Coan (1948) developed flexible end-loading strips which could be adjusted during a test and thereby a uniformly distributed end-load could be imparted. Another feature of these tests was the use of orthotropically reinforced fibreglass for the test panels which enabled the theory of anisotropic plates to be checked by experiment.

One of the first successful attempts to experiment with plates in the large-deflection region was by Kaiser (1936) who had analysed a simply supported plate by using large-deflection theory and the finite difference method. The analysis was compared with some carefully conducted experiments using water pressure as the loading medium and dial gauges as the deflection-measuring devices. The membrane and bending stresses were obtained by using Huggenberger extensometers and excellent agreement with theory was obtained. This was a remarkable series of experiments and even with modern measuring equipment it is difficult to improve greatly on Kaiser's accuracy.

A series of model experiments to investigate the elastic buckling of stiffened plates was conducted by Tulk and Walker (1976). They used Araldite liquid and cast it between sheets of glass to obtain very precise flat plates to which the stiffeners were glued. Local imperfections were introduced by heating locally the finished model while it carried a small load. The loading rig could apply loads through the neutral plane of the cross-section or through a plane eccentric but parallel to it and this technique was used to simulate global imperfections. Because of the low elastic modulus of Araldite and the small size of the models the bench-top testing rig was very rigid relative to the model and the buckling phenomenon could be easily controlled by the mechanically-screwed loading plattens. The length of the models was chosen so that the Euler buckling load P_E was less than that at which local buckling occurred P_{cr}. After each test a short section of each panel was machined from it until $P_{cr} < P_E$ and in this way Euler and local buckling and their interaction could be studied. A subsequent paper by Thompson, Tulk, and Walker (1974) used these results to verify their theoretical studies of the sensitivity of stiffened plates to initial imperfections and the effects of the interaction of buckling modes.

Some tests of stiffened plates were conducted by Stein (1959) to check his theoretical analysis of plates in the post-buckling region (Section 5.3.5). Even though the experiments were very carefully conducted the results did not confirm that the third-order buckling analysis was more accurate than the second-order buckling analysis. Both curves pass fairly closely through the centre of the scattered experimental points, suggesting that second-order buckling analysis is sufficiently accurate even for most research purposes.

Fok's (1980) experiments on the elastic behaviour of imperfect plates have already been mentioned (Section 5.3.4). He used the plate-scanning device shown in Fig. 5.4.7 to measure initial deflections (Fig. 5.4.9) which were deliberately introduced into a plate by gently heating its surface. These measured initial deflections were then used as data for calculating by the finite difference method the elastic response of the plate at different load levels. Explicit functions (Section 5.3.5) were then calculated so that a continuous load-deflection curve could be drawn and used for comparison with experimental results. The full lines in Fig. 5.3.19 were obtained in this manner.

The purpose of this review has been to bring readers to the point where research on the theory of elastic behaviour is currently being carried out and to show what techniques are being used to increase the field of knowledge. The intention has not been to make a designer's catalogue of standard buckling cases (available books having already been mentioned) nor is it to review every paper written. The above selection is of some papers of historical interest and other more recent papers which are typical of the work being carried out currently. It is seen that, at present, interest centres around the effects of imperfections and the problems of analysing assembled structures rather than isolated plates. It seems that for theoretical analysis the finite difference method and the Runge–Kutta method (which is similar in many ways) have proved to be more fruitful than the finite element method. However, this may not always be true in the future.

6

BEHAVIOUR AND LOAD-CAPACITY
AT COLLAPSE

6.1. INTRODUCTION

The purpose of this chapter is to discuss the behaviour of thin-walled steel structures in the post-yield region. When a mild steel structure carries increasing loads eventually it will start to yield and from this point onwards the elastic theory described in previous chapters cannot be used to predict the behaviour of the structure. However, as explained in Chapter 1, its behaviour may be studied first by using elastic theory until one of the fibres starts to yield and then the final process of collapse can be analysed by applying rigid-plastic theory to the plastic mechanism which develops at this stage. Between the elastic range and the final collapse state at which a plastic mechanism is well-developed lies the so-called elasto-plastic region. In this region all existing theories that are based upon deterministic techniques are unfortunately too complicated for use in practice, although the region has been researched very well indeed. It is unfortunate also because the elasto-plastic region embraces the point at which the maximum load is attained. This state of affairs has led to the current situation in which there are very many theories and associated models of failure but nearly all of them seem to be based upon one or more empirical rules. This must leave a designer with nagging doubts about the accuracy of the theories and their range of validity. If he is conservative he can rely upon the more accurate elastic theories to predict the load at which yielding commences. This load is then taken as the maximum or failure load of the structure. This approach is always safe and indeed it is sometimes used as a criterion of failure. The Perry-Robertson formula (see Robertson 1928) for the design of columns is a case in point. Although this approach is safe it sometimes leads to very uneconomical designs because some structures fail at a much higher load than that at which yielding commences.

It is also desirable to have some appreciation of the 'suddenness' of collapse before one can say whether this criterion is realistic or too conservative. Those structures which buckle suddenly after the maximum load is reached are said to be 'brittle'. Although there is no question of the material failing in a brittle manner it is simply that the overall geometry of the structure leads to a sharp peak in the load–deflection curve and the unloading curve is very steep. Other structures behave in a more 'ductile' manner which is desirable in a structure because it will then give more warning before it fails. Also, ductile components of a structure allow it to develop alternative load paths if the designer has had the foresight to provide them.

These important matters can be studied by developing the rigid-plastic collapse curve (Fig. 1.2.6). There are considerable differences between these curves as they apply to structural frameworks and those derived for thin-walled structures. This is because usually the former structural type develops only a so-called *global plastic mechanism* while the latter often develops a combination of global and so-called *local plastic mechanisms* (Fig. 6.1.1).

Structural members with thin walls tend to undergo elastic local buckling as described in Chapters 4 and 5. This involves a deformation of the cross-section and after the elasto-plastic region is reached a local buckle becomes a local plastic mechanism. In contrast to this, members with thicker walls tend to buckle globally, e.g. as an Euler column, without local deformation of the cross-section, and the plastic hinges which form during collapse are simpler. Here they are referred to as *simple plastic hinges.* Local plastic mechanisms influence the structure while it is collapsing like a simple plastic hinge in the sense that their presence allows the structure to develop a global mechanism but it will be shown later that local plastic mechanisms are much weaker than simple plastic hinges and they increase the brittleness of the structure. In Section 6.2 the process of collapse of some thin-walled steel structures is first described and then its theoretical basis is developed by using rigid-plastic theory. This enables the rigid-plastic collapse curve to be sketched and from its shape and location the important matter of ductility can be quantified.

Following this, attention is focussed upon the maximum load itself. In Section 6.3 a brief review of some attempts to evaluate this load for some

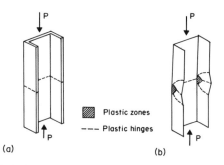

Fig. 6.1.1. (a) Global plastic mechanism, and (b) local plastic mechanism.

common thin-walled steel structures is given. From this it will be seen that many theoretical models have been developed. Some are easy to use but perhaps lack precision and some are accurate when applied to a narrow range of structural types but it is fair to say that currently there is no universally accepted formula for the designer. In the whole of the elasto-plastic region of behaviour there are many gaps in the current state of knowledge and further experimental and theoretical research is required to clarify many problems.

Perhaps the greatest difficulties stem from the fact that the number of parameters is very large. One of the simplest stiffened plates, for example, has rectangular stiffeners. Figure 6.1.2 shows the parameters which influence the behaviour of this type of stiffened panel. If a deterministic approach is to be adopted all of these parameters must appear in the theory. If an ambitious experimentalist decided to investigate the influence of each of these parameters and conducted experiments on specimens with only two values for each parameter he would need to test 16 384 specimens (=1 each day for 45 years), and then he could start on other shapes.

The approach which has generally been adopted is to check the theoretical models described in Section 6.3 against one or more series of tests. In this section some test programmes are also reviewed and used as a yardstick to check the accuracy of the theoretical models.

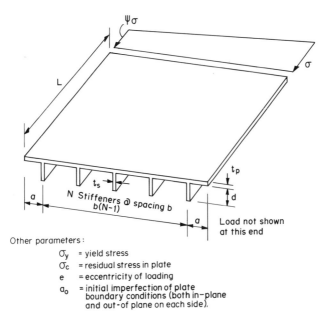

Fig. 6.1.2. Parameters which define the properties and behaviour of a simple stiffened plate.

6.2. RIGID-PLASTIC THEORY APPLIED TO THIN-WALLED STRUCTURES

The ideas behind rigid-plastic theory have already been introduced in Section 1.2 where they were applied to a pin-ended mild steel strut whose cross-section was rectangular. It was seen (Figs. 1.2.1 and 1.2.4) that by assuming that the stress–strain curve of mild steel is a step function with a step height between tension and compression yielding of $2\sigma_y$ the curve of load-carrying capacity could be derived in a fairly simple manner by applying an equation of equilibrium. This theory is an extension of the rigid-plastic theory of simple beams and frameworks (Horne 1971), the only additional effect taken into consideration being the axial stress which reduces the moment capacity of the strut (Figs. 1.2.4 and 1.2.5).

An exact rigid-plastic theory will satisfy all of the following conditions.

(a) Equilibrium—each part of the structure and the structure as a whole is in equilibrium with the applied loads and the reactions at the supports.

(b) Mechanism—sufficient (in number) plastic hinges are developed so that the whole or part of the structure can deflect as a mechanism.

(c) Yield—at no point in the structure can the bending moment exceed the plastic moment capacity of the cross-section.

However, except in the case of the simplest of structures it is not easy to satisfy simultaneously all three conditions, and it has been found expedient to resort to approximate methods which satisfy only two of the conditions. Thus the *upper bound method* satisfies the equilibrium and mechanism conditions and it will always give a calculated failure load which is either equal to or in excess of the actual. Although it is not a conservative design tool, nevertheless, it is a very useful one for engineers. In practice one can use the upper bound method and, with some care, get quite close to the right solution. The yield line method (Wood and Jones 1967) is an example of the upper bound approach. The *lower bound method* satisfies the equilibrium and yield conditions and gives a calculated failure load which is either equal to or less than the actual. Thus it is a conservative (safe) design tool. The Hillberborg (1956) strip method is an example of the lower bound approach.

In this book an upper bound approach is adopted but the problem inherent in it is largely avoided by using special techniques. The reason why the upper bound approach overestimates the failure load of a structure is that, in general, the designer (or analyst) will always choose a plastic mechanism which is not quite as 'good' as that which the structure itself will adopt.

As a simple example, the built-in beam shown in Fig. 6.2.1(a) will obviously develop hinges at A, B and C but it may be assumed (stupidly)

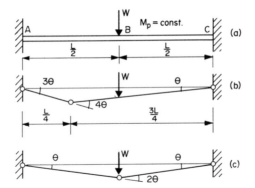

Fig. 6.2.1. A built-in beam is analysed by the upper bound method by choosing the locations of the hinges. (b) and (c) are alternative choices.

that the hinges occur at the locations shown in Fig. 6.2.1(b) and the mechanism condition is still satisfied. To satisfy equilibrium the 'principle of virtual displacements' can be used, i.e.

work done by the load = energy absorbed at the hinges

$$W(L\theta/2) = M_p(3\theta + 4\theta + \theta)$$

i.e. at collapse

$$W = 16M_p/L. \tag{6.2.1}$$

A similar analysis of the mechanism shown in Fig. 6.2.1(c) gives the collapse load as

$$W = 8M_p/L \tag{6.2.2}$$

which is less than the collapse load calculated in eqn (6.2.1). The structure will obviously choose the hinge locations shown in Fig. 6.2.1(c) in preference to those shown in Fig. 6.2.1(b).

A more systematic way to arrive at this conclusion would be to assume that the intermediate hinge is located a distance αL to the right of point A. It is easily shown that at failure

$$W = 4M_p/\alpha L. \tag{6.2.3}$$

Thus the beam will maximize α (i.e. $\alpha = 0.5$) so that the failure load has its minimum value given by eqn (6.2.2). In this section this kind of approach is adopted to try to avoid the disadvantage of the upper bound method. Unfortunately it is not always possible to obtain the optimum locations of the hinges in closed form and it is usually necessary to resort to numerical search techniques.

In the following analyses certain assumptions are made. Firstly it is assumed that strain-hardening can be neglected. Strain-hardening has the effect of spreading plastic hinges instead of allowing them to develop along a line as is assumed in the theory. This is similar to displacing the hinges a small distance away from their optimum location. It is probably true to say that in most cases this does not have a large effect provided the spreading is not too large compared to the unyielded lengths of plate adjacent to the hinges. Furthermore, interest usually centres on the behaviour of a plate structure in the region of collapse and a little beyond it. In this region the deflections and strains are small and strain-hardening effects are probably not too significant. However, strain-hardening can have a large effect upon certain mechanisms which require whole areas of plate (as compared to only hinge lines) to yield. This problem will be discussed later. Another assumption which is made is that the regions between the plastic hinges remain flat. Observations in the laboratory show that this assumption is reasonable once the plastic mechanism is well-developed. For these reasons it can be anticipated that there will be discrepancies between calculated results based upon rigid-plastic theory and experimental results. The main purpose of the analysis which follows is to gain some understanding of the way thin-walled structures behave in the vicinity of the maximum loads and to determine whether a structure is brittle or ductile.

6.2.1. The moment-capacity of plastic hinges

In Section 1.2 it was shown that the moment-capacity M'_p of a plastic hinge in rectangular strut is given by eqns (1.2.14) and (1.2.16), that is,

$$M'_p = M_p \left[1 - \frac{P^2}{P_y^2} \right] = \frac{\sigma_y bt^2}{4} \left[1 - \left(\frac{P}{\sigma_y bt} \right)^2 \right]. \tag{6.2.4}$$

In such a case the hinge is oriented at right angles to the direction of thrust P. However, thin-walled structures crumple with hinges lying in all directions when they collapse. Figures 6.2.2 and 6.1.1(b) show, for example, two ways in which a simple stiffened plate and two columns develop local plastic mechanisms. It is seen that not all of the hinges lie at right angles to the direction of thrust so it is first necessary to derive an expression for the moment-capacity of a plastic hinge inclined at an angle β to that direction.

In Fig. 6.2.3 a strip of flat plate is considered. For convenience it is assumed that there is a diagonal strip of material AB which has a yield stress σ_y while the remainder of the plate is infinitely rigid. As in Fig. 1.2.4 it is again assumed that a central core of material (EHGF) of depth

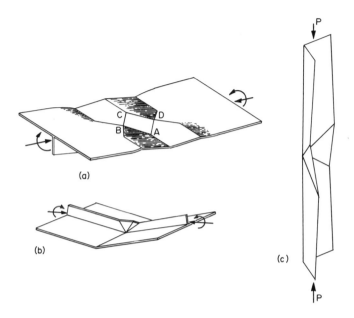

Fig. 6.2.2. Local plastic mechanisms in a stiffened plate and an angle column.

t_1 carries the axial load P and hence

$$P = \sigma_y b t_1. \tag{6.2.5}$$

Across AB there is a moment M_p''' carried by the remaining areas of the cross-section, i.e. ABEF and GHCD, and a twisting moment. M_p''' is evaluated from the stresses acting on these two areas, thus,

$$M_p''' = \sigma_y \frac{(t+t_1)}{2} \frac{(t-t_1)}{2} b \sec \beta$$

$$= M_p' \sec \beta \tag{6.2.6}$$

where M_p' is the moment-capacity if the plastic hinge is at right angles to the direction of thrust (see eqn (6.2.4)). For rotational equilibrium of the element ABJK the vector diagram shows that

$$M_p'' = M_p''' \sec \beta \tag{6.2.7}$$

whence

$$M_p'' = M_p' \sec^2 \beta \tag{6.2.8}$$

This result has been verified experimentally (Murray 1973a) by testing

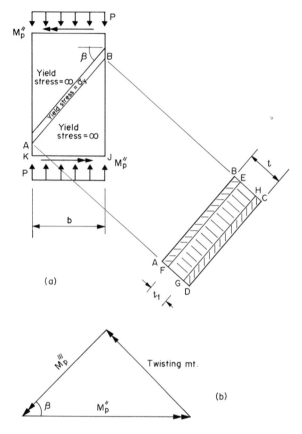

Fig. 6.2.3. The moment-capacity M_p'' of a plastic hinge inclined at an angle β to the direction of thrust is $M_p' \sec^2 \beta$,

a series of struts of the type illustrated in Fig. 6.2.4(a). The strut consisted of an inner plate which was strengthened by outer plates in such a manner that the failure mechanism shown in Fig. 6.2.4(b) developed. The load–deflection relationship of this mechanism is obtained by considering moment equilibrium of, say, the bottom half of the strut (Fig. 6.2.4(c))

$$P\Delta = M_p' + M_p'' = M_p' \, (1 + \sec^2 \beta) \tag{6.2.9}$$

where eqn (6.2.8) has been introduced. The validity of eqn (6.2.8) was checked by testing two struts one with $\beta = 30°$ and the other with $\beta = 60°$ and comparing their deflections Δ. For a given value of P the ratio of the central deflections were measured at 2.1 when the mechanism was well developed (Fig. 6.2.4(d)). The theoretical value is 2.14.

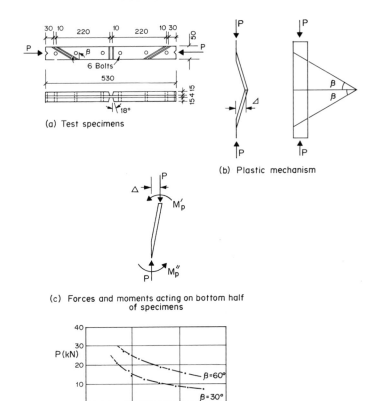

(a) Test specimens

(b) Plastic mechanism

(c) Forces and moments acting on bottom half
of specimens

d) Load – deflection curves of two specimens

Fig. 6.2.4. Experiment conducted to verify eqn (6.2.8). (Murray 1973a.)

6.2.2. Analysis of local plastic mechanisms

By conducting many laboratory tests on thin-walled structures one sees at first a confusing variety of shapes and sizes of local plastic mechanisms. However, a study (Murray and Khoo 1981) reveals that even the most complicated mechanism is only the sum of a number of simpler basic mechanisms which fit together so that their deflections are compatible with that of the whole mechanism. There are two major classes of plastic mechanism, the so-called *true mechanism* and the so-called *quasi-mechanism*. A true mechanism is one which is developed from the original thin-walled structure by folding the individual plates along the plastic hinge lines. A simple but not infallible test as to whether or not a mechanism is a true mechanism is to see whether it can be constructed

from cardboard and deflect freely. The mechanism of the column illustrated in Fig. 6.2.2(c) is an example of a true mechanism. A quasi-
mechanism is one in which some regions of the individual plates of the
structure must deform by yielding in their plane in order to allow the
plastic mechanism to deflect. If a cardboard model of such a mechanism is
constructed it will be necessary to cut out these regions before the model
will deflect freely. The plate mechanism of the stiffened plate illustrated
in Fig. 6.2.2(a) is an example of a quasi-mechanism. The region ABCD is
a square in its undeformed state and it becomes a rhombus when the
mechanism deforms. There are also tension and compression yield zones
in the lower and upper part of the stiffener, respectively, and the common
edge of these two zones acts like a real hinge or the pivot point of the
mechanism.

One of the commonest basic mechanisms is that seen in the stiffener of
Fig. 6.2.2(b) and it can be analysed as follows. Figure 6.2.5 shows the
mechanism in its deformed position. A strip of width dg and height g
above O intersects the hinges at A, B, and C. At C the hinge is normal to
the elemental force dP whereas at A the hinge is inclined to dP.

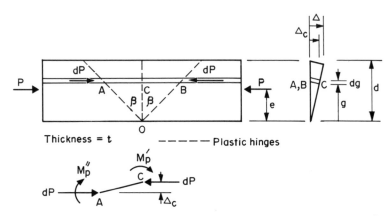

Fig. 6.2.5. Analysis of plastic mechanism formed by three concurrent plastic hinges (Type 3
basic mechanism).

Hence, from eqns (6.2.8) and (6.2.4)

$$M_p'' = M_p' \sec^2 \beta = \frac{1}{4} \sigma_y t^2 \left[1 - \left(\frac{dP}{\sigma_y t\, dg} \right)^2 \right] dg \sec^2 \beta. \qquad (6.2.10)$$

For rotational equilibrium of the element AC and about an axis lying in
the plane of the plate

$$dP\, \Delta_c = M_p'' + M_p' = \kappa_1 M_p' \qquad (6.2.11)$$

where

$$\kappa_1 = 1 + \sec^2 \beta. \tag{6.2.12}$$

From equations (6.2.10) and (6.2.11)

$$dP = \sigma_y t \left\{ \sqrt{\left[\left(\frac{2\Delta_c}{\kappa_1 t} \right)^2 + 1 \right]} - \frac{2\Delta_c}{\kappa_1 t} \right\} dg. \tag{6.2.13}$$

But, by simple proportion

$$\Delta_c = g\Delta/d \tag{6.2.14}$$

and when this is substituted into eqn (6.2.13), which is then integrated, the following expression for the total load P is obtained.

$$P = \frac{\sigma_y t d}{2} \left\{ \sqrt{\left[\left(\frac{2\Delta}{\kappa_1 t} \right)^2 + 1 \right]} - \frac{2\Delta}{\kappa_1 t} + \frac{\kappa_1 t}{2\Delta} \ln \left(\sqrt{\left[\left(\frac{2\Delta}{\kappa_1 t} \right)^2 + 1 \right]} + \frac{2\Delta}{\kappa_1 t} \right) \right\} \tag{6.2.15}$$

The moment of dP $(=g\,dP)$ can be integrated to obtain the moment of P about an axis normal to the plane of the plate through the point O.

$$Pe = M = \frac{\sigma_y t^3 d^2 \kappa_1^2}{12\Delta^2} \left\{ \left[\left(\frac{2\Delta}{\kappa_1 t} \right)^2 + 1 \right]^{3/2} - 1 - \left(\frac{2\Delta}{\kappa_1 t} \right)^3 \right\} \tag{6.2.16}$$

where e $(=M/P)$ is the eccentricity of the force P with respect to the point O.

These expressions for P and M appear as if they would become infinite when Δ is zero. A check on this by using L'Hospital's rule reveals that when Δ is zero P is at the squash load $(=\sigma_y t\, d)$ and M is equal to the full plastic moment of the cross-section $(=\sigma_y t\, d^2/4)$, which is the result to be expected.

Table 6.2.1 gives the load-deflection formulae for eight true basic mechanisms and it also gives the location of the resultant force P. It will be noticed that type 5 and type 8 mechanisms do rely upon some twisting and bending out-of-plane of the end panels but plates are much more flexible for these kinds of deformation so they fall within the above definition of a true mechanism.

The mechanism shown as type 8 is quite commonly observed and has been nicknamed the 'flip-disc'. Following observations in the laboratory its governing equation is derived by assuming that the shape of the plastic hinges is parabolic. Figure 6.2.6 shows one half of the disc; the equation for the location of the plastic hinge is

$$z = a(1 - 4x^2/b^2). \tag{6.2.17}$$

Table 6.2.1. True basic mechanisms

$$\kappa_1 = 1 + \sec^2\beta; \quad \kappa_2 = \sec^2\beta; \quad \kappa_3 = 2\sec^2\beta$$

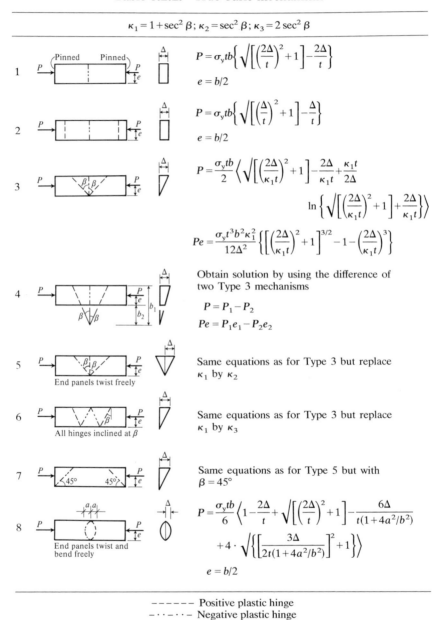

1. $$P = \sigma_y tb \left\{ \sqrt{\left[\left(\frac{2\Delta}{t}\right)^2 + 1\right]} - \frac{2\Delta}{t} \right\}$$
    $$e = b/2$$

2. $$P = \sigma_y tb \left\{ \sqrt{\left[\left(\frac{\Delta}{t}\right)^2 + 1\right]} - \frac{\Delta}{t} \right\}$$
    $$e = b/2$$

3. $$P = \frac{\sigma_y tb}{2} \left\langle \sqrt{\left[\left(\frac{2\Delta}{\kappa_1 t}\right)^2 + 1\right]} - \frac{2\Delta}{\kappa_1 t} + \frac{\kappa_1 t}{2\Delta} \ln\left\{ \sqrt{\left[\left(\frac{2\Delta}{\kappa_1 t}\right)^2 + 1\right]} + \frac{2\Delta}{\kappa_1 t} \right\} \right\rangle$$
    $$Pe = \frac{\sigma_y t^3 b^2 \kappa_1^2}{12\Delta^2} \left\{ \left[\left(\frac{2\Delta}{\kappa_1 t}\right)^2 + 1\right]^{3/2} - 1 - \left(\frac{2\Delta}{\kappa_1 t}\right)^3 \right\}$$

4. Obtain solution by using the difference of two Type 3 mechanisms
    $$P = P_1 - P_2$$
    $$Pe = P_1 e_1 - P_2 e_2$$

5. End panels twist freely

    Same equations as for Type 3 but replace κ_1 by κ_2

6. All hinges inclined at β

    Same equations as for Type 3 but replace κ_1 by κ_3

7. Same equations as for Type 5 but with $\beta = 45°$

8. End panels twist and bend freely
    $$P = \frac{\sigma_y tb}{6} \left\langle 1 - \frac{2\Delta}{t} + \sqrt{\left[\left(\frac{2\Delta}{t}\right)^2 + 1\right]} - \frac{6\Delta}{t(1+4a^2/b^2)} + 4 \cdot \sqrt{\left\{ \left[\frac{3\Delta}{2t(1+4a^2/b^2)}\right]^2 + 1 \right\}} \right\rangle$$
    $$e = b/2$$

------- Positive plastic hinge
−·−·−· − Negative plastic hinge

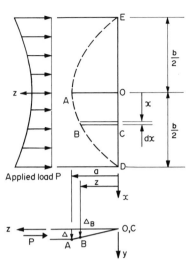

Fig. 6.2.6. Analysis of Type 8 basic mechanism in Table 6.2.1.

Because there is zero moment along the whole of DE the element BC has the following equation of equilibrium.

$$dP\,\Delta_B = M_P'' = M_P' \sec^2 \beta. \tag{6.2.18}$$

But

$$\sec^2 \beta = 1 + \left(\frac{dz}{dx}\right)^2 = 1 + \frac{64a^2x^2}{b^4} \tag{6.2.19}$$

and

$$\Delta_B = \Delta\,\frac{BC}{AO} = \Delta z/a = \Delta\left(1 - \frac{4x^2}{b^2}\right). \tag{6.2.20}$$

By solving for dP from eqns (6.2.10) and (6.2.18)

$$dP = \sigma_y t \left\{ \sqrt{\left[\left(\frac{2\Delta_B}{t\sec^2\beta}\right)^2 + 1\right]} - \frac{2\Delta_B}{t\sec^2\beta} \right\}$$

$$= \sigma_y t\left(\sqrt{\left\{\left[\frac{2\Delta b^2(b^2 - 4x^2)}{t(b^4 + 64a^2x^2)}\right]^2 + 1\right\}} - \frac{2\Delta b^2(b^2 - 4x^2)}{t(b^4 + 64a^2x^2)}\right) dx. \tag{6.2.21}$$

The integration of this elemental load cannot be carried out in closed form so an approximate value for P is found by using Simpson's Rule. The integrand is evaluated at $x = 0$, $b/4$ and $b/2$ and Simpson's Rule is applied in the usual way, whence,

$$P = \frac{\sigma_y tb}{6} \left\langle 1 - \frac{2\Delta}{t} + \sqrt{\left[\left(\frac{2\Delta}{t}\right)^2 + 1\right]} - \frac{6\Delta}{t(1 + 4a^2/b^2)} \right.$$
$$\left. + 4\sqrt{\left\{\left[\frac{3\Delta}{2t(1 + 4a^2/b^2)}\right]^2 + 1\right\}} \right\rangle \quad (6.2.22)$$

A study of this function reveals that a/b can have almost any value up to 0.3 before there is any significant increase in P. In other words the strength of the flip-disc is insensitive to its shape provided a/b is less than 0.3 and therefore, although one would expect the disc to adopt that shape which minimizes P, one should not be surprised if the shape has any a/b value up to 0.3. This is indeed observed in the laboratory. Because of the insensitivity to a/b it can be set to zero, whence the governing equation for a flip-disc is approximately

$$P = \frac{\sigma_y tb}{6} \left\{ 1 - \frac{8\Delta}{t} + \sqrt{\left[\left(\frac{2\Delta}{t}\right)^2 + 1\right]} + 4\sqrt{\left[\left(\frac{3\Delta}{2t}\right)^2 + 1\right]} \right\}. \quad (6.2.23)$$

It is often observed that the load–deflection relationships of plastic mechanisms in plated structures are, to a degree, insensitive to the shape of the mechanism. In laboratory tests it is difficult to reproduce exactly the same mechanism in two nominally identical specimens. Initial imperfections influence the location of the point of first yield and thereafter the mechanism is 'locked-in' to that location. The subsequent growth of the plastic hinges is influenced by many factors and the mechanism does not often develop that shape which minimizes the load. However the analysis of plastic mechanisms gives a guide to the lower limit of the actual behaviour of a large number of nominally identical specimens in a test series.

Attention is now turned to quasi-mechanisms. Certain regions of a mechanism of this kind must undergo large in-plane deformations by yielding of every element within that region. There appear to be only three types of region, that is, those that yield in direct tension, those that yield in direct compression, and those that yield in pure shear. The plate mechanism in Fig. 6.2.2(a) exhibits all three of these types of region. They are respectively the lower part of the stiffener, the upper part of the stiffener, and the square region ABCD. The last of these can also be thought of as carrying a compressive stress equal to half the yield stress parallel to the diagonal BD and a tensile stress of similar magnitude parallel to CA. While this is true when the plastic mechanism is well-developed it has been shown (Murray 1973b,c) that the tensile stress, $\sigma_y/2$, parallel to CA cannot develop until the deflections at the centre of the plate are large. For small deflections the region ABCD is in a state of pure compression yield and so is its contiguous region at the top of the stiffener. Therefore it has been necessary to develop both small- and large-deflection theories for some mechanisms.

A complete plastic mechanism consists of an assembly of basic mechanisms and fully plastic regions. They are combined so that the basic mechanisms are compatible with one another and satisfy equilibrium. Unfortunately it is usually not simply a matter of adding the basic mechanisms shown in Table 6.2.1 as the following example illustrates.

Figure 6.2.7 shows one longitudinal strip from a stiffened plate with its axial load P located e_1 above the upper surface of the plate. When the mechanism has deflected vertically through δ at D the stiffener has

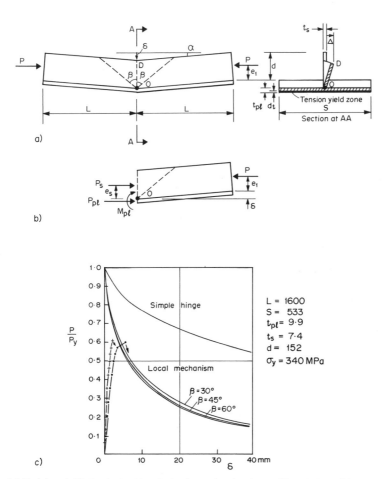

Fig. 6.2.7. (a) and (b) Analysis of local plastic mechanism in a stiffened plate; (c) test results of two stiffened plates compared with theories. (Murray 1973c.)

deflected horizontally through Δ also at D. It is desired to derive a relationship between P and δ. This appears to be a true mechanism but it is not merely a matter of adding a type 1 to a type 3 mechanism from Table 6.2.1. Obviously when δ is very large and the forces P rise above P_s in the stiffener the force P_{pl} must become tensile. The procedure is to assume that the depth of the tensile yield zone is d_t and use the equations of equilibrium and compatibility. Thus, for longitudinal and rotational equilibrium about O (Fig. 6.2.7(b))

$$P = P_{pl} + P_s \tag{6.2.24}$$

$$P(\delta + e_1) = M_{pl} - P_{pl}\frac{t_{pl}}{2} + P_s e_s. \tag{6.2.25}$$

Because of the sideways movement Δ at D the entire upper edge of the stiffener shortens by $\Delta^2/d \tan \beta$ and therefore

$$\alpha = \Delta^2/(2\,d^2 \tan \beta). \tag{6.2.26}$$

Hence

$$\delta + e_1 = \Delta^2 L/(2\,d^2 \tan \beta) + e_1. \tag{6.2.27}$$

The force and moment in the plate are

$$P_{pl} = \sigma_y S(t_{pl} - 2d_t); \qquad M_{pl} = \sigma_y S\, d_t(t_{pl} - d_t). \tag{6.2.28}$$

From eqns (6.2.24) and (6.2.28)

$$d_t = 0.5t_{pl} - (P - P_s)/(2\sigma_y S) \tag{6.2.29}$$

or

$$d_t = \bar{A}P + \bar{B} \tag{6.2.30}$$

where

$$\bar{A} = -\frac{1}{2\sigma_y S} \quad \text{and} \quad \bar{B} = 0.5t_{pl} + P_s/(2\sigma_y S). \tag{6.2.31}$$

On substituting from eqn (6.2.28) into eqn (6.2.25)

$$P(\delta + e_1) = \bar{C}d_t^2 + \bar{D}d_t + \bar{F} \tag{6.2.32}$$

where

$$\bar{C} = -\sigma_y S; \qquad \bar{D} = 2\sigma_y t_{pl} S; \qquad \bar{F} = P_s e_s - \sigma_y t_{pl}^2 S/2. \tag{6.2.33}$$

By substituting eqn (6.2.30) into eqn (6.2.32) a quadratic equation in P is obtained. Its solution is

$$P = \frac{\delta + e_1 - \bar{A}\bar{D} - \bar{B} - \sqrt{\{[\delta + e_1 - \bar{A}\bar{D} - \bar{B}]^2 - 4\bar{A}^2\bar{C}(\bar{C}\bar{B}^2 + \bar{D}\bar{B} + \bar{F})\}}}{2\bar{A}^2\bar{C}} \tag{6.2.34}$$

This equation is used by first assuming a value of Δ, substituting it into eqn (6.2.27) to find $\delta + e_1$ and then into the equations for type 3 mechanisms in Table 6.2.1 to find P_s and $P_s e_s$. Finally, P is evaluated from eqn (6.2.34) and the process is repeated for a new value of Δ. In this way a plastic collapse curve is obtained.

In evaluating P it is necessary to assume a value for the angle β which defines the shape of the mechanism in the stiffener. Again it is easy to use a computer to search for the value of β which minimises P.

Several full-scale tests on panels of this kind have been reported by Horne and Narayanan (1976a, b), by Horne, Montague, and Narayanan (1976a, b), and by Murray (1973b, c, 1975). In the latter series the stiffeners were bulb-flats and the additional effect of the bulb at the outer edge of the stiffener was investigated. Figure 6.2.7(c) shows two typical test results which are compared with theoretical curves. A search for the value of β which minimises P is seen to suggest that β is almost 60° but that it is fairly insensitive to changes in β of 15° on either side of this value. In the laboratory β was found to be close to 60°. The investigation showed that the size of the bulb is an important factor in determining both the ductility and strength of the panel. A small increase in bulb dimensions results in a more favourable cross-section and a considerable increase in the strength and ductility of the panel. Because of the large size of these specimens it was not possible to obtain the plastic collapse curve experimentally. To do this requires a very stiff testing machine. Results show that stiffened plates of this kind exhibit very 'brittle' behaviour. The uppermost curve is that which is obtained if it is assumed that the panel develops a simple plastic hinge at mid-span instead of a local mechanism. It is seen that such an assumption gives misleading results and the conclusion is that collapse analyses must be based upon local mechanisms. A further analysis of the effect of a small eccentricity of the end load was made and it was found that it had a very strong influence in reducing the load-carrying capacity of this panel. The effect of an eccentricity of 6.6 mm was to lower the intercept of the collapse curve with the load axis from $(P/P_y =)$ 1 to 0.7. Thus it can be concluded that this type of panel is sensitive to initial imperfections.

One way of comparing the ductility of structures is to calculate an index by measuring the deflections δ_1 and δ_u at 90% of the failure load, where δ_1 is the deflection as the structure approaches its maximum load and δ_u is that as the structure collapses. They can usually be closely estimated from the elastic and rigid-plastic curves, respectively. The ductility index of the structure is then defined at $(\delta_u - \delta_1)/\delta_1$. For the stiffened plate analysed here the ductility index is about 1.4 which is very small. Another feature of these panel tests was that the central deflection at maximum load was very small (less than 5 mm in a panel which was 3.2 m long). Collapse therefore occurred with very little visible warning.

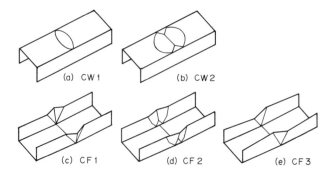

Fig. 6.2.8. Five plastic mechanisms observed in laboratory tests on channel-columns. (Khoo 1979.)

It is seen that the analysis described enables many important properties of this cross-section to be investigated and evaluated.

The plastic mechanism shown in Fig. 6.2.7(a) is also found to occur in some thin-walled channel sections used as columns. In order to investigate the plastic collapse curve more thoroughly Khoo (1979) tested many channels of different cross-sections as compression members and observed five mechanisms (Fig. 6.2.8). Figure 6.2.9(a) shows the results of two tests, confirming the theoretical method described above. The four other mechanisms were also analysed and generally it was found that the above theory gives good agreement with test results.

In the case of quasi-mechanisms it has been found that, if the lower yield stress is used in their analysis for the regions which are fully plastic, the theoretical plastic collapse curve always lies below the experimental points. When the mechanism is well-developed these regions must deform by large amounts and the steel is stressed well into the strain-hardening range. In these regions the ultimate strength of the steel was therefore used and better agreement was obtained.

Fig. 6.2.9(b) shows the load–deflection curves and the plastic mechanism of four thin-walled steel I-columns and again good agreement has been obtained.

Sherbourne and Korol (1971, 1972) and Korol and Sherbourne (1972) used an analysis similar to that described above. Their analyses which equates the work done by the applied load in compressing the mechanism to the energy absorbed by the plastic hinges, also allows for the reduction in plastic moment with axial load. Their theoretical work was confirmed by experiments. Another interesting application of a simpler version of the foregoing theory is that due to Rawlings and Shapland (1975) who studied the collapse patterns of thin box-columns used as energy absorbers. Such devices can be used, for example, at the bottom of lift-wells

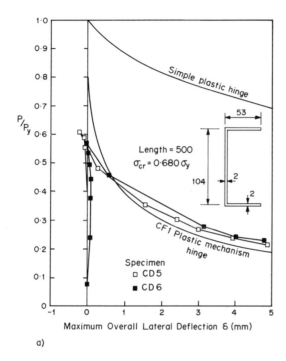

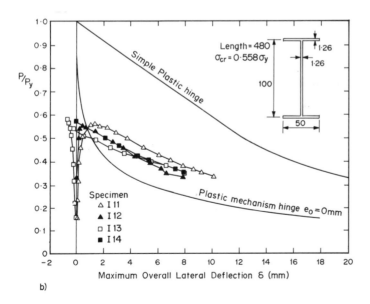

Fig. 6.2.9. Comparisons of tests results and rigid-plastic analyses of (a) two channel-columns and (b) four I-columns. (Khoo 1979.)

and come into action in the event of a mishap. Their theory ignores the reduction in plastic moment with axial load but this is reasonable for the large deflections involved in these devices.

6.3. THEORIES OF FAILURE AND TEST RESULTS OF THIN-WALLED STRUCTURES

In the previous section the behaviour of some thin-walled structures during the collapse process is described. Rigid-plastic theory enables this to be done when the collapse process has reached a fairly advanced stage but it cannot be used directly to predict failure loads, i.e. the maximum loads, because at these loads the structure is always in the elasto-plastic region. Elasto-plastic theory is too complicated to use except for research purposes so special techniques for estimating failure loads have to be devised. It is the aim of this section to outline a few of these techniques. In the following sections some methods used to estimate failure loads of five classes of thin-walled structure, that is, isolated plates, stiffened plates, thin plate girders, box columns, and box girders, are described. In the plates and columns the primary loading is considered to be in-plane compression in the longitudinal direction (the z-direction) while there may also be present minor in-plane shear stresses and relatively small transverse forces acting normal to the plane of the plate, i.e. in the y-direction. In the case of the girders the main loads are the shear forces, such as occur at the supports and which have a tendency to cause buckling of the web, and the bending moments.

6.3.1. Theories of failure and test results of isolated plates and stiffened plate panels

The determination of the failure stress of thin-walled structures under compressive loads is a topic which has attracted considerable attention over the past hundred years. T. Box (1883) appears to have been the first to propose a formula for the failure stress σ_m of a simply supported mild-steel panel with uniform in-plane stress in one direction only, that is,

$$\sigma_m = 1236(b/t)^{-\frac{1}{2}} \text{ MPa}. \tag{6.3.1}$$

Over the last 50 years there has been a proliferation of formulae which enable designers to estimate failure stresses but nearly all of these formulae to a lesser or greater extent include empirical rules. While it is obviously desirable for designers to have at their disposal simple rules such as eqn (6.3.1) the rules should be based upon sound theoretical concepts with a back-up of careful experiments. Failure of a mild-steel panel is a complicated elasto-plastic process which depends upon the geometry of the panel, its initial imperfections, the yield stress, the boundary conditions (both in-plane and out-of-plane), the magnitude and distribution of residual stresses due to welding and rolling, and so on. It is self-evident that a simple formula such as eqn (6.3.1) cannot account for all of these factors.

An 'exact' elasto-plastic analysis of a thin-walled structure up to and beyond σ_m is complicated even with the aid of present-day computing techniques. There have so far been few attempts to follow the strain history of even a simple imperfect plate into the elasto-plastic range. Such an analysis is expensive in computer time and it is unlikely that designers can use these techniques directly as design tools. Nor is it likely that design tables could be produced to cover the whole range of problems likely to confront designers. However, 'exact' theories, although limited in scope, are important because they provide a yardstick for judging other theories. There are many methods of analysis which are usually easier to use than 'exact methods and they cover a wider range of problems. However, they are often approximate, over-conservative and—their worst feature—it is not always clear what their short-comings and range of validity are.

Because of this uncertainty it seems that the most satisfactory way to introduce the topic of theories of failure of thin-walled structures is to start with a review of those papers which are based upon more rational analyses and to describe some experimental results which confirm their findings. By doing this it is hoped that the reader will gain some feel for the way actual structures behave before some of the more empirical failure theories are reviewed. The simplest way of assessing theories of failure of structures is to compare theoretically predicted failure loads with those observed during tests in a laboratory under carefully controlled conditions. If a theory is exact and the experiments were without error the failure loads would always be predicted correctly. However, in most cases neither of these conditions prevails because firstly, most theories incorporate empirical approximations at some point in their development and, secondly, experimental results can only be measured within finite limits of accuracy. To minimize the influence of experimental errors various theories can be applied to the same set of test specimens, which are both large in number and which have been carefully tested.

Suppose that from a set of N specimens the nth specimen has a ratio of its theoretically predicted failure load to its measured failure load of r_n. A failure theory can be said to be reliable if it satisfies *all* of the following conditions.

(a) The mean values of r_n should be close to 1.
(b) The standard deviation ρ of the set of r_n values should be very small.

(c) So that the theory can be used safely for design there should be only a few values of r_n greater than 1 and these values should exceed 1 by a small margin which is less than the experimental error.

In assessing the failure theories of stiffened plates described below the mean and standard deviations of r_n are quoted when they are applied to a reliable set of test results published by Dowling *et al.* (1973) and Horne and Narayanan (1976a). The former investigators tested stiffened plates which were part of a series of box-girders while the latter tested isolated stiffened plates.

The first theory investigated in this way was developed between 1964 and 1976 by a group at Cambridge University, England, under the general direction of J. B. Dwight. They have minimized the amount of empiricism by resorting to an elasto-plastic, large-deflection Rayleigh–Ritz analysis of isolated and stiffened plates. It will be seen that their method has a rational basis and that they have developed a simplified design procedure which appears to yield accurate results.

Survey of Cambridge University investigations

The first aim of these investigations was to derive curves of the axial stress at failure (σ_m) of isolated and stiffened plates as functions of b/t and L/r. The general form of the functional relationship between σ_m and b/t is shown in Fig. 6.3.1 where it is seen that actual plates generally fail at a lower stress than do ideal plates. This is because of the presence of imperfections in the form of initial dishing and residual stresses due to welding. Moreover from Fig. 6.3.1 it is evident that plates with higher b/t ratios buckle elastically long before they reach their maximum axial stress. The second aim was to develop a simple procedure for the design and analysis of simple stiffened plate panels.

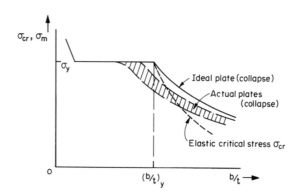

Fig. 6.3.1. Graphs showing relationship of yield stress, collapse curves, and elastic critical stress for an isolated plate.

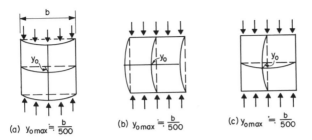

Fig. 6.3.2. (a) Longitudinal, (b) transverse, and (c) spherical initial dishing.

The investigations were carried out on plates with b/t less than 80 and included the effects of initial dishing and residual stresses. Shrinkage of hot weld metal relative to that of the parent plate causes both of these effects to arise. The magnitude of these parameters can be controlled to some extent by the fabricator and the Cambridge investigations were aimed at obtaining representative values to use as data for their subsequent theoretical analyses.

Measurements indicated three types of initial dishing (Fig. 6.3.2) with typical maximum values shown. In the theoretical analyses typical values of 0, $b/2000$, and $b/200$ were used for the two cases of cylindrical dishing and 0, $b/1000$ for spherical dishing.

Residual stresses due to welding were first studied by means of a simple model consisting of a single run of weld laid down on a flat plate. This type of weld causes a trapezoidal zone of tension yield stresses (Fig. 6.3.3)

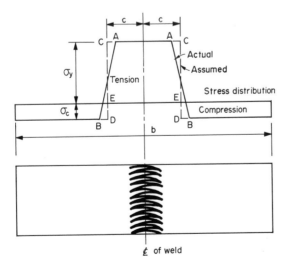

Fig. 6.3.3. Distributions of residual stresses due to a single pass weld.

which must be balanced by compressive stresses in the regions of the plate remote from the weld. Dwight and Moxham (1969) replaced the trapezoid AABB by a rectangle CCDD and called the force in the tension block CCEE the shrinkage force F_s. Theory suggested that:

$$F_s = H(Q/v) \, (\text{kN})$$

where Q = Current × voltage = power released at the arc (watts)
$\quad v$ = velocity of welding (mm sec^{-1})
and $\quad H$ = constant determined by test data (dimensionless).

The width c of the tension block and the compressive stress σ_c are easily derived from longitudinal equilibrium.

$$c = H(Q/v)/2t\sigma_y$$

$$\sigma_c = \frac{2c\sigma_y}{b - 2c} \tag{6.3.2}$$

Kamtekar (1975) used a finite difference method to obtain theoretical values for H but agreement with experimental results was improved by introducing a factor p for the efficiency of the welding process, thus

$$H = 0.18p \tag{6.3.3}$$

where $p = 0.75$ for manual metal arc and metal inert-gas welding

$\qquad = 0.85$ for submerged arc welding.

Tests carried out by White (1977a) showed that, contrary to the relationship shown in eqn (6.3.2), σ_c is independent of the yield stress σ_y. He introduced the concept of a tendon force F which is resisted by the whole of the plate $(b \times t)$ and not just the compression area $[(b - 2c) \times t]$. Tests showed that when this approach is used the scatter of results is considerably reduced. He proposed the following formulae for F and σ_c

$$F = 0.2p(Q/v) \tag{6.3.4}$$

$$\sigma_c = F/bt. \tag{6.3.5}$$

He also suggested that when preheat $(T_i > 100° \text{C})$ is used a modified tendon force F' is given by

$$F' = (1.1 - T_i/1000)F. \tag{6.3.6}$$

White (1977b, d) then investigated multipass weldments using eqn (6.3.4) as the basis with $p = 0.8$. He showed that each weld pass sets up a zone of influence whose area A_0 is equal to the effective yield tension area for that pass on its own. For the nth pass

$$A_{0n} = F_n/(\sigma_y + \sigma_c) \tag{6.3.7}$$

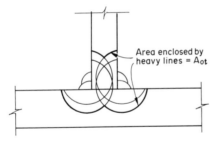

Fig. 6.3.4. The zone of influence of a multi-pass weld is obtained by constructing radii from the centroids of each pass.

where F_n is the tendon force calculated for the nth pass considered in isolation in accordance with eqn (6.3.4) allowing for preheat where necessary. These zones of influence overlap in the manner indicated for a T-fillet weld in Fig. 6.3.4. Other multipass weldments are treated in a similar manner. The full tendon force F_t is then given by

$$F_t = A_{0t}(\sigma_y + \sigma_c). \tag{6.3.8}$$

This equation is based upon the observed phenomenon that the influence of a previous weld pass can be partially or wholly washed out by subsequent passes. The computational procedure is to guess a value for σ_c, calculate A_{0n} for each pass, and so determine A_{0t} graphically. F_t is calculated from eqn (6.3.8) and this force must then be equilibrated by a new value of σ_c. Although the procedure is iterative one cycle is usually sufficient to evaluate σ_c. The method is generally applicable to multipass welds in which the passes are close to one another or spaced some distance apart. Flame-cut edges used in the preparation of the profile of the plates prior to welding can be considered as equivalent to a weld pass. The tendon force due to a flame cut profile (Fig. 6.3.5) is given by the equation (h' is measured in mm)

$$F_{fc} = 30 + 3h' (\text{kN}) \tag{6.3.9}$$

When the above theory is applied to a welded T-section there is one important additional factor to consider. For T-sections which are firmly tacked before welding the tendon force F_t is calculated according to the method described above for a plate. F_t is then applied to the full profile of the T-section as an eccentrically applied force in the axial direction. It

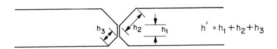

Fig. 6.3.5. Definition of h' in eqn (6.3.9).

causes bending about the neutral axis and this gives rise to a uniform compressive stress σ_c in the plate and a linearly varying stress in the web. The latter stress may become tensile at the tip of the web and in some cases it is therefore beneficial from the point of view of delaying stiffener buckling. This picture is modified considerably when the T-section is not firmly tacked before welding because slip can then occur between the two plate elements due to the differential rates of heating. Slip causes mismatch and an additional bending moment whose sense depends upon whether the web is heated more than the plate or less than it. In the former case the additional bending moment will reduce or even reverse the tensile stress at the tip of the web and, with normal geometries, this is the more likely event. White (1977b) recommends that slip should not be relied upon to produce beneficial effects and suggests the following procedures.

(a) Heavy tacking before welding.
(b) Low heat input to first pass.
(c) Intermittent welding for the first pass (with the possibility of a continuous run on the other side).
(d) Use of a wide outstand; however this may be counter-productive for strength.

Thus the method developed at Cambridge University for calculating residual stresses has been described. It is based upon sound theoretical concepts but the answers it gives should not be thought of as being any more than approximate. Residual stresses can easily be modified by heat treatment, flame gouging, and even by the deflections which occur during handling and transportation (a phenomenon known as 'shake-out').

Isolated plates have been analysed theoretically in the elasto-plastic range by Ractliffe (1966), Graves-Smith (1967) and Moxham (1971a, b) and it is Moxham's work which is described here. The plates he considered were simply supported at each loaded end, one buckle long, and with longitudinal edges which were unloaded and free to pull in. The loaded edges were straight and remained parallel during loading and various combinations of initial dishing and residual stresses were considered. The plate was divided into 1620 volumes or finite-elements (18 in both the longitudinal and transverse directions and 5 through the plate thickness). The aspect ratio of the plate a/b was taken as 0.875 to conform with the observation that although a plate buckles elastically into square panels the subsequent plastic buckles have a shorter length. The assumed shape of initial dishing was sinusoidal and after each increment of loading the Fourier coefficients of the deflected form of the plate were adjusted to minimize the strain energy. In the analysis of the plates near the buckling loads, it was found that some elements undergo a reversal in

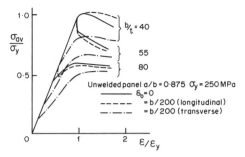

Fig. 6.3.6. Graphs of axial-stress–end-shortening for initially dished unwelded panels. (Moxham 1971a.)

the direction of strain and these strain reversals are catered for by using the Prandtl-Reuss flow rule and the von Mises equations of plasticity.

Some typical results of Moxham's analysis are presented in Figs. 6.3.6 and 6.3.7 which show the influence of welding, b/t ratio, and initial deflection on the graphs of average axial-stress–axial-strain. The important points to notice are that transverse imperfections have a greater influence in reducing the failure load than those in the longitudinal direction and that sudden collapse occurs when b/t is in the middle range approximating to 55. The elastic critical stress σ_{cr} of a mild steel square panel ($k = 4$ in eqn (5.3.24)) is equal to the yield stress of 250 MPa when $b/t = 54$. Thus, sudden collapse is most likely to occur in panels which are designed on the basis of trying to achieve simultaneous overall yielding and local buckling. The benefits of efficient design are eroded by the presence of this undesirable property. More recently Little (1980) has published axial-stress–axial-strain curves for 960 plates of different geometry and initial stress and initial deformation patterns.

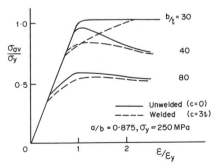

Fig. 6.3.7. Graphs of axial-stress–end-shortening for welded and unwelded plates with spherical dishing $\delta/b = 10^{-3}$. (Moxham 1971a.)

The effect of residual stresses due to welding was also studied theoretically by Moxham by assuming that the plate had initially a strip of material at the tensile yield stress along each side of width $3t$ and the remainder of the plate had a balancing compressive stress σ_c. The graphs of average axial-stress–axial-strain have a similar shape to those shown in Fig. 6.3.6 for transverse distortion. However when b/t is 30, which is a typical value for a stocky plate, the graph is as shown in Fig. 6.3.7, indicating that there is a considerable loss of stiffness when the average axial stress σ_{av} causes general yielding of the plate, i.e. when

$$\sigma_{av} = \sigma_y - \sigma_c = \sigma_y(b-4c)/(b-2c) \qquad (6.3.10)$$

where (again) $2c$ is the width of the tension yield zone at the weld. When $b/t = 30$ this reduction in stiffness occurs when $\sigma_{av} = 0.75\sigma_y$. It should be noted that if such a panel were built into a structure the surrounding panels would have to pick up the additional load arising from this loss of stiffness. This will, of course, occur only when the panel is loaded beyond $0.75\sigma_y$ for the first time, after which it will obtain a new pattern of residual stresses. This problem has its most important influence when the deflections are considered because then it will be necessary for designers to allow for this loss of stiffness. Also during erection this effect can occur and may result in components being permanently distorted beyond the tolerances specified by the designer.

Moxham (1971b) verified his theoretical analyses described above by some carefully conducted experiments on 144 isolated plates with built-in or pinned sides and generally having an aspect ratio of 4. The end load was applied by means of a screw-driven wedge which allowed the uniform end-shortening to be controlled and the unloading curves of the plates to be obtained. Further tests on box-columns were also carried out under strain-controlled conditions. A comparison of experimental and theoretical results is shown in Fig. 6.3.8. Dwight (1971) suggested the following empirical expressions for the maximum axial stress in steel panels with spherical initial dishing ($y_0 = b/1000$).

For unwelded panels:

when $b/t < 0.85(b/t)_y$ $\sigma_m = \sigma_y$

$$b/t > 0.85(b/t)_y \qquad \sigma_m = \frac{0.85(b/t)_y}{(b/t)}\sigma_y. \qquad (6.3.11(a))$$

For welded panels:

$$\text{for all } b/t \ \ \sigma_m = \frac{0.85(b/t)_y}{(b/t)}\sigma_y - \frac{(b/t)_y}{(b/t)}\sigma_c$$

$$= \frac{(b/t)_y}{(b/t)}\frac{0.85b - 3.7c}{b-2c}\sigma_y \qquad (6.3.11(b))$$

(but $\sigma_m \not> \sigma_y$)

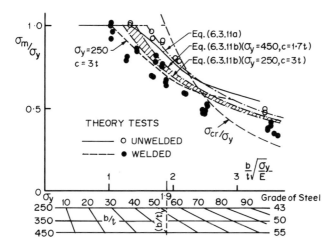

Fig. 6.3.8. Comparison of theoretical, experimental, and empirical failure stress for isolated simply supported plates. (Dwight 1971.)

where $(b/t)_y$ is the value of (b/t) at which the critical stress of a simply supported square panel is equal to its yield stress.

Figure 6.3.8 shows a comparison of some experimental and theoretical results with the plots of eqns (6.3.11). Agreement is good over the whole range of b/t values, suggesting that Fig. 6.3.8 can be used for the design of isolated pinned plates which are loaded in compression. Furthermore the good agreement suggests that Moxham's results and the Rayleigh–Ritz method he developed can be used for predicting the failure loads of other types of plates.

This idea was used by Little and Dwight to obtain design curves which enable the failure loads of stiffened plates to be predicted. Little (1976a) divided a panel stiffened with bulb flats into short segments in the longitudinal direction and then Moxham's results were used to derive M–ϕ–P graphs of the relationship between the applied bending moment M, the curvature ϕ of a segment, and the applied axial load P. Two levels of residual stress corresponding to very light and very heavy welds were assumed. A typical set of M–ϕ–P graphs is shown in Fig. 6.3.9. By using a trial procedure, which assembles the short segments into the complete panel, it was possible to study the influence of the primary variables b/t, L/r, and the residual compressive stress σ_c in the plate upon the axial stress at the point of collapse. Thus Little and Dwight were assuming that the simple supports along the sides of a stiffened panel have no influence upon its collapse behaviour. In other words the stiffened panel was assumed to behave as a wide column. This assumption has been justified by observations in the laboratory and it is now widely accepted for panels with about four or more stiffeners.

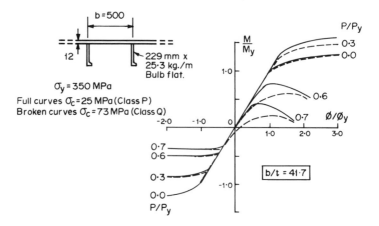

Fig. 6.3.9. Theoretical moment–curvature–axial-load $(M-\phi-P)$ relationships for cross-section shown.

Figure 6.3.10 shows how the primary variables, that is, b/t, L/r, and σ_c, affect the failure stress of a family of pin-ended plates each with $229\ \text{mm} \times 25.3\ \text{kg m}^{-1}$ bulb-flat stiffeners spaced at 500 mm and with a yield stress of 350 MPa. At the higher values of L/r the graphs approach the Euler column curve. It is also seen that heavy welding does not have a particularly strong influence in reducing the failure stress except when b/t is in the vicinity of $(b/t)_y$ [46 for the panels in Fig. 6.3.10] and L/r is less than 100. The maximum percentage reduction in failure stress is only about 12%. The graphs show the stronger influence of both b/t and L/r in reducing the failure stress.

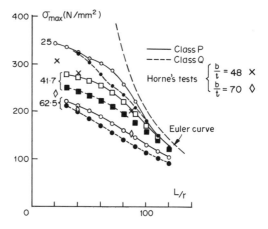

Fig. 6.3.10. Theoretical failure stress of stiffened plate shown in Fig. 6.3.9. (Horne's test results on similar panels are shown for comparison.)

The effects obtained by varying two secondary variables, y_p/r and A_s/bt (y_p the distance between the centroid of the full cross-section of the panel and that of the plate alone and A_s is the area of the stiffener) were studied but they were found to have only a marginal effect upon the failure stress. Little has calculated a table of failure stresses for pin-ended stiffened-plate columns and Table 6.3.1 is a shortened version of it. It is important to note that this table has been derived on the assumption that failure is triggered by local failure of the plate and not by failure of the stiffener.

If the table is used for design purposes a panel with axial compressive stresses should also be checked for stiffener failure. It is generally accepted that for open stiffeners failure may be taken as the point at which the outer edge of a bulb-flat or rectangular stiffener reaches the yield stress. The panel is treated as a wide column with a suitable overall imperfection of the form $\Delta_0 \sin \pi z/L$ in a direction which favours compressive stresses at the free edge of the stiffener, where Δ_0 is the initial

Table 6.3.1. Theoretical failure stresses of pin-ended stiffened-plate columns (Little 1976)

					b/t ratio				
	Class		25			41.7			62.5	
	of					Bulb-flat type				
L/r	weld	A	B	C	A	B	C	A	B	C
30	P	338	335	333	—	278	278	—	221	227
	Q	338	335	331	—	249	251	—	211	217
40	P	329	324	320	273	272	272	209	212	216
	Q	328	322	317	242	242	242	195	198	203
50	P	315	311	309	266	264	263	200	201	204
	Q	312	303	296	235	233	232	183	184	187
60	P	306	302	299	257	254	251	189	189	190
	Q	289	278	270	225	223	221	170	170	171
80	P	262	257	252	224	218	214	163	160	159
	Q	238	234	230	199	194	190	142	140	139
100	P	175	179	182	173	160	166	133	130	128
	Q	174	179	176	161	167	153	116	113	112
120	P	123	125	127	126	126	124	107	104	102
	Q	122	125	127	123	120	118	94	91	90

Note: Bulb flat type A is $178 \text{ mm} \times 16.9 \text{ kg m}^{-1}$
Bulb flat type B is $229 \text{ mm} \times 25.3 \text{ kg m}^{-1}$
Bulb flat type C is $280 \text{ mm} \times 35.3 \text{ kg m}^{-1}$
Class P weld has residual compressive stress in plate of 25 MPa
Class Q weld has residual compressive stress in plate of 73 MPa
$b = 500$ mm, $\sigma_y = 350$ MPa and $y_0 = 200/L$ throughout

Failure is assumed to be triggered by local failure of plate, i.e. stiffener failure must also be checked.

deflection at the centre of the panel relative to a straight line joining the end. If Δ_0 is magnified according to the hyperbolic law (Fig. 1.2.3) the average stress at failure σ_m is obtained by solving the following quadratic equation which equates the stress at the free edge of the stiffener to the yield stress.

$$\sigma_y = \sigma_m + \frac{\sigma_m A \Delta_0 y_s}{\left(1 - \frac{\sigma_m}{\sigma_E}\right) I} \qquad (6.3.12)$$

where σ_E is the stress at which Euler buckling occurs
y_s is the distance from the neutral axis of the cross section to the free edge of the stiffener
A is the area of cross-section of the panel
I is the second moment of area of the panel about the neutral axis.

The solution of this equation is the well-known Perry–Robertson formula commonly used in codes of practice for the design of columns, viz.,

$$\sigma_m = \tfrac{1}{2}[\sigma_y + \sigma_E(1 + \eta)] - \tfrac{1}{2}[\{\sigma_y + \sigma_E(1 + \eta)\}^2 - 4\sigma_y \sigma_E]^{\frac{1}{2}} \qquad (6.3.13)$$

where

$$\eta = A \Delta_0 y_s / I \qquad (6.3.14)$$

which is a measure of the overall imperfection of the panel. Representative values of Δ_0 are specified in codes of practice (see Chapter 7).

Returning to panels which fail by plate collapse, Dwight and Little (1976) proposed a design method which is more general than Table 6.3.1. Two basic plate strength curves are presented for lightly and heavily welded profiles in Fig. 6.3.11(a) and (b), respectively. Fig. 6.3.11(c) shows how these curves are used and it is seen that the nomogram on the left of each graph enables the effects of shear to be added. The curves have the same form as the Perry–Robertson formula (eqn 6.3.13)) but allowances are made on an empirical basis for initial stresses and imperfections. The simple design procedure of Dwight and Little (1976) is set out in Table 6.3.2.

In the preliminary section the stiffener size is chosen so that they are sufficiently stocky to avoid stiffener buckling (Fig. 6.2.2(b)). The stresses applied to the panel are obtained from a global analysis of the structure and σ_3 and τ_1 (Fig. 6.3.12) are representative stresses at the middle plane of the plate. The failure stresses σ_{pm} of the individual plates are calculated from Fig. 6.3.11(b) in order to evaluate σ_{mav} which is an averaged value of the compression strength of the cross-section. Figure 6.3.13(d) is a column curve which allows for the reduction in strength of the panel

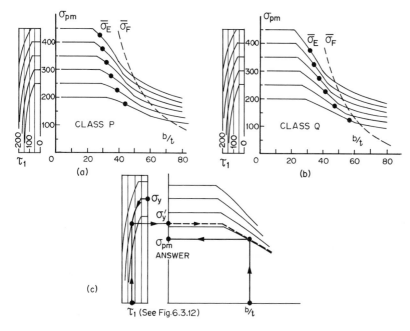

Fig. 6.3.11. Plate strength curves: (a) lightly welded ($\sigma_c = 25$ MPa); (b) heavily welded ($\sigma_c = 73$ MPa); (c) how to use curves (a) and (b).

due to its slenderness (L/r ratio). In other words σ_{mav} is treated as though it were the effective yield stress of an equivalent column, and the average stress for column failure (σ_{col}) is derived by following the curve which cuts the stress axis of Fig. 6.3.13(d) at σ_{mav} to the appropriate L/r value.

In section 2 of Table 6.3.2 it is seen that four checks are specified to complete the simple procedure. This design can be refined by following the steps outlined in section 3 of the table.

In their paper Dwight and Little (1976) compare failure stresses predicted by their method with those observed by Dowling *et al.* (1973) and by Horne and Narayanan (1976a, b) in twenty-five laboratory tests. By applying the method of Dwight and Little described in Table 6.3.2 with the three refinements an average value of r_n (= ratio of predicted failure load to measured failure load) of 0.958 was obtained with a standard deviation of 0.055. The corresponding figures obtained by applying eqn (6.3.1) due to Box (1883) to the same set of results are 1.002 and 0.237. Obviously both methods satisfy the first criterion of a reliable theory but only the Dwight and Little method satisfies the second and third criterion. An examination of Dwight and Little's results reveals that the method is safe except for two cases in which the loads were deliberately applied with a large eccentricity. Their method of analysis is

Table 6.3.2. Simplified design method for the design of axially loaded plates stiffened with flats or bulb-flats. (Dwight and Little 1976.)

---

1. Preliminary:
   (a) Stiffeners
       (No. $= N - 1$)

   > Flats: use $d/t_2 \not> 0.35\sqrt{(E/\sigma_y)}$
   >
   > Bulb flats: use heaviest section

   (b) Applied stresses

   > Calculate applied stresses $\sigma_1, \sigma_2, \sigma_3, \tau_0$ (see Fig. 6.3.12)
   > Calculate $\tau_1 =$ greater of $\frac{2}{3}\tau_0$ and $(1 - 4/N)\,\tau_0$

   (c) Characteristic stresses of panel
       Calculate failure stresses of individual plates of cross-section from Fig. 6.3.11(b), i.e. σ_{pm}.
       Hence calculate

   $$\sigma_{m\,av} = \frac{\text{Sum of failure loads of individual plates}}{\text{Total area of plates}} = \frac{\sum \sigma_{pm} A}{\sum A}.$$

   Hence use Fig. 6.3.13(d) to find σ_{col} ($=$ average stress for column failure).
   Follow curve which cuts the stress-axis at $\sigma_{m\,av}$ to appropriate L/r value and read off σ_{col} horizontally from this point.

2. Checks:

Check I

$$\text{(local failure) } \sigma_1 \not> \sigma_{m\,av}$$

Check II

$$\text{(column failure) } \sigma_3 \not> \sigma_{col}$$

Check III

   (stiffener failure): check compression yield at the free edge of stiffener when initial deflection favours this direction of failure (use eqn (6.3.13)).

Check IV

   (serviceability check): check that no plate element bulges at working load.

3. Refinement:

   (a) *First refinement—plate elements*

   For lightly welded plating (e.g. small multi-pass welds) use Class P diagram (Fig. 6.3.11(a)).

   (b) *Second refinement—column curves for check II*

   Instead of using Fig. 6.3.13(d) in Check II define two curves of σ_{col} against L/r as follows ($y_p =$ distance of centroid of stiffener–plate unit to mid-thickness of plate)

   Curve 1: Refer to Fig. 6.3.13(c) (or Fig. 6.3.13(d) if $y_p > 0.7\,r$) Select the curve which intersects the stress-axis at $\sigma_{m\,av}$. This is called Curve 1.
   Curve 2: Refer to Fig. 6.3.13(a) (or Fig. 6.3.13(b) if $y_p > 0.7\,r$) Select the curve which intersects the stress-axis at the least of σ_{pm}, $\bar{\sigma}_E$, and $\bar{\sigma}_F$ where each

Table 6.3.2.—*contd.*

of these quantities is found from Fig. 6.3.11(a) or (b). This is called Curve 2. $\bar{\sigma}_E$ is read directly from Fig. 6.3.11(a) or (b) as a function of σ'_y only and $\bar{\sigma}_F$ as a function of b/t only.
For the given L/r values now select the σ_{col} value which is the higher of the two values given by Curves 1 and 2. Check III must be carried out, i.e.
Stress at centroid of stiffener-plate unit $\ngtr \sigma_{col}$ from Fig. 6.3.13(c)

(c) *Third refinement—effective column length*

It is permitted to use L_{eff} instead of L in Checks II and III where L_{eff} is the lesser of K_1L and $2wK_1\sqrt{(r/t)}$, K_1 is obtained from Fig. 6.3.14 and w = width of stiffened plates (Fig. 6.3.12).

---

incorporated into the British Standard BS2573 (British Standards Institution 1977) for the design of cranes.

Summary of Manchester University investigations on isolated stiffened plates

The investigations carried out at Manchester University by Horne, Montague, and Narayanan (1974, 1975, 1976) consisted of laboratory tests on thirty-one full-scale panels and a theoretical study of their collapse loads. The experiments were designed to study the influence on the collapse load of many parameters such as end fixity (pinned or fixed), weld detail (continuous or intermittent), the ratios L/r and b/t, imperfections deliberately introduced into the plate and the stiffeners, and the shape and size of the stiffener profile (flats and bulb-flats were investigated).

One additional reason for this test program was to evaluate the Merrison rules (1973) which had just been developed for the design of

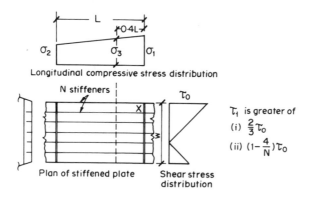

Fig. 6.3.12. Stresses acting on stiffened plate.

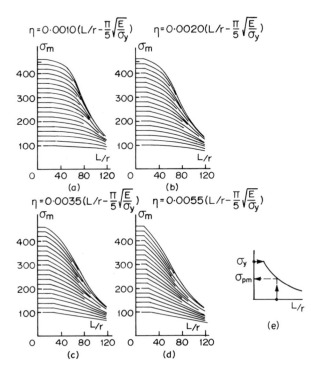

Fig. 6.3.13. Column strength curves. Initial imperfection increases from (a) to (d).

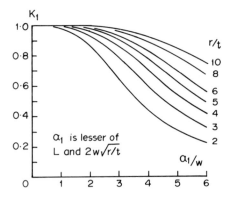

Fig. 6.3.14. Effective column length factor K_1 used in Section 3(c) of Table 6.3.2.

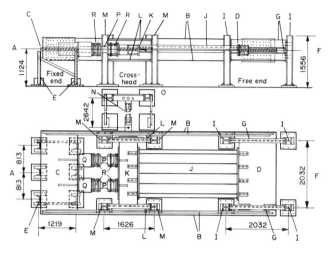

A	Anchor end	K	Yoke free to move longitudinally
B	Tension straps		but not laterally
C	Heavy cross-plate	L	Guides for K
D	Cross-plate at free end	M	Supports for L
E	Braced frame	N	Restraint against horizontal
F	Free end		lateral movement for K
G	Guide beams to allow	O	Rigid support for N
	elongation of straps	P	Load cells
I	Support frame for G	Q	Jack (2 @ 5000 kN, motor or manual)
J	Test specimen	R	Cylindrical end bearings of jacks

Fig. 6.3.15. Rig used by Horne and Narayanan (1976a) for testing full-scale stiffened plates in compression.

stiffened plates in box-girder bridges. The Merrison rules are discussed more fully in Chapter 7. A special test rig (Fig. 6.3.15) capable of applying axial loads of 10 000 kN to panels up to 1.8 m wide × 3.3 m long was manufactured. The rig was designed as a very stiff testing machine in order to obtain the collapse curve but it was only in the case of the panels which failed by plate buckling that this was possible. When the stiffeners buckled first, collapse was too sudden for the rig to record the collapse curve. The test results lead to the following conclusions (Horne and Narayanan 1976a):

(a) High residual stresses in plate panels due to single-pass continuous welds do not have any large effect on the collapse loads of stiffened panels, although there appears to be some benefit from the lower residual stresses induced by intermittent welding for panels of high slenderness ($L/r \simeq 90$).

(b) In panels of low slenderness, some separation of stiffeners and plate occurred in intermittently welded panels and this appears to have some adverse effect on the carrying capacity.

(c) The measured residual stresses due to plate–stiffener welds were of the order 1.5–3 times the values calculated by Merrison rules.

(d) Post-buckling behaviour showed a much more rapid unloading characteristic for stiffener failure than plate failure.

(e) For panels collapsing by plate failure, intermittent welding induced a more rapid unloading characteristic than continuous welding.

(f) An increase of plate imperfections from less than Merrison tolerance to three times the prescribed tolerance caused a small drop in the collapse load.

(g) Local (torsional) imperfections in the stiffeners of up to three times the Merrison tolerance produced decreases in the collapse load of the order of 8%.

(h) Overall panel imperfections (column imperfections) to five times the Merrison tolerance produced decreases in collapse load of about 25%.

The theoretical study resulted in two methods for predicting the failure loads of stiffened panels (Horne and Narayanan 1975, 1976b). The methods are developed by first considering an isolated rectangular plate with pinned sides and longitudinal in-plane loading which is confined to the elastic range. Both of these methods use the well-known concept of an effective width. As seen in Section 5.3.3 the region of the plate adjacent to the stiffeners will carry greater membrane stresses, and therefore proportionately more of the axial load than the central strip of plate after local buckling has occurred. Therefore, it is argued, a central strip of the plate can be thought of as completely ineffective, all of the load being carried by an effective width of plate which is contiguous with the stiffener and has a uniform membrane stress [Fig. 6.3.16(a)] equal to the edge stress σ_e in the actual plate. The effective width b_e is obtained by considering longitudinal equilibrium

$$\sigma_{av} b t = \sigma_e b_e t \qquad (6.3.15)$$

whence

$$\frac{b_e}{b} = \frac{\sigma_{av}}{\sigma_e} = K_{bs} \qquad (6.3.16)$$

where K_{bs} is called the *secant effective width factor*. The reason for this name is seen from Fig. 6.3.16(b) which was derived as Fig. 5.3.16(b) in Chapter 5. At a general point S on the curve where the average stress is σ_{av} the edges of the plate are straight so the strain at the edge is $\varepsilon_e = \sigma_e/E$ which is the same as the apparent longitudinal strain of the whole plate

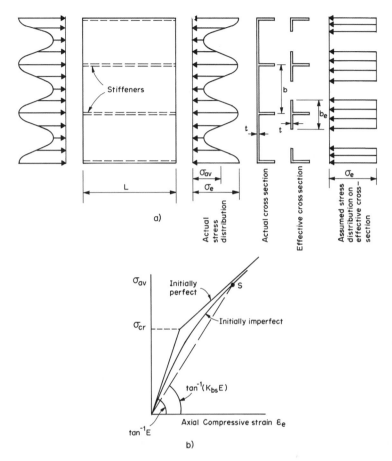

Fig. 6.3.16. (a) The stress distribution in a buckled stiffened plate is replaced by an assumed stress distribution acting on the effective cross-section; (b) the secant effective width factor K_{bs} is a measure of the average stiffness of a buckled plate.

(=axial shortening/L). Hence from eqn (6.3.16)

$$\frac{\sigma_{av}}{\varepsilon_e} = K_{bs}E. \qquad (6.3.17)$$

Thus K_{bs} is not only an index of the effective width of the plate and the stress distribution but it is also an indicator of its in-plane stiffness averaged up to the applied load ($=\sigma_{av}bt$).

An expression for K_{bs} can be derived in the following way. An initially imperfect square plate with boundary conditions of case (b) (Section 5.3.3) is considered in its loaded position. From eqn (5.3.46) with

$\beta = \lambda = \pi/b$, $K_1 = K_2 = 0$, the compressive stress in the plate is

$$\sigma_{av} = \frac{w_0 \beta^2 E}{2} \left[\frac{2t^2}{3(1-\nu^2)(w_0+y_0)} + \frac{w_0+2y_0}{8} \right]. \tag{6.3.18}$$

Also from the first of eqns (5.3.50) the compressive stress at the edge of the plate is

$$\sigma_e = E \frac{w_0(w_0+2y_0)\beta^2}{8} + \sigma_{av}. \tag{6.3.19}$$

It is convenient to introduce the magnification factor m and the variable C defined as follows

$$m = (w_0 + y_0)/y_0 \tag{6.3.20}$$

$$C = \frac{3(1-\nu^2)}{16} \left(\frac{y_0}{t} \right)^2. \tag{6.3.21}$$

By substituting from the last four equations into eqn (6.3.16) the following expression for K_{bs} is obtained.

$$K_{bs} = \frac{1+2m(m+1)C}{1+4m(m+1)C} \tag{6.3.22}$$

This expression was first derived by Horne and Narayanan (1976b). It enables the effective width of a square plate with case (b) boundary conditions to be calculated in terms of the initial central deflection y_0 and an arbitrary central deflection $(w_0+y_0) = my_0$.

The reader will gain some idea of the influence of m and C upon the effective width of a plate from Table 6.3.3. It is seen that the effective width of a given plate decreases as both the initial imperfection and the load increase. It is also seen that K_{bs} is never less than 0.5 which could incidentally be used as a safe design value.

Table 6.3.3. Typical K_{bs} values from equation (6.3.22).

y_0/t	$m = 1$	$m = 2$	$m = 4$
0.5	0.87	0.75	0.61
1.0	0.71	0.60	0.53
2.0	0.58	0.53	0.51

A stiffened plate which has many longitudinal stiffeners behaves like a wide column and the influence of the restraint along the longitudinal edges can be ignored. Such a panel buckles as an Euler column with the transverse stiffeners acting as the pinned ends. When the direction of buckling is towards the stiffeners the plate will carry higher compressive stresses than the stiffeners. Furthermore the plating mid-way between the transverse stiffeners (i.e. at mid-span) will carry higher average compressive stresses than the plating near the ends (i.e. adjacent to the transverse stiffeners). Table 6.3.3 shows that the effective width of the plate must therefore be a maximum at the ends and a minimum at mid-span. In the first method of Horne and Narayanan (1976b) this variation of effective width in the longitudinal direction is taken into account. They consider one element, i.e. a stiffener and its associated plate, as being replaced by an equivalent column whose cross-section has the same stiffener, but the plate has a reduced thickness t' and width b' (Fig. 6.3.17). Furthermore the weld at the web-plate junction and the heat-affected zone give rise to a longitudinal tensile force which interacts with the initial imperfection. After a thorough analysis of a general cross-section the whole column is divided into segments of equal length in a similar manner to the technique adopted by Little (1976a, b). The properties of the segments and the continuity conditions are assembled in a computer and the applied axial load is incremented up to and beyond the maximum (failure) load. The necessary steps for this fairly long and involved procedure are outlined in their paper (1976b).

Results of this theoretical analysis confirmed the experimental findings. Of special interest are the following conclusions.

(a) Residual stresses in the plate of up to 40 MPa have a negligible effect upon the strength of a panel.

(b) When imperfections are limited to tolerances defined by the Merrison rules (1973) (see Chapter 7) the reduction in strength is less than 16%. This appears to be the most importance influence upon the strength of a panel after that of the b/t and L/r ratios.

As described in an earlier part of this section both of these conclusions were confirmed by their experimental work. The first method of Horne and Narayanan (1976b) has been applied to eighteen panels in their own

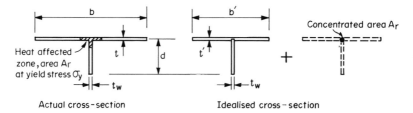

Fig. 6.3.17. In the analysis of Horne and Narayanan (1976b) the actual cross-section is replaced by the idealized cross-section shown.

series of tests described earlier and it is found that the mean value of r_n (=predicted failure load/measured failure load) is 0.964 and the standard deviation is 0.062. Thus, from the point of view of accuracy the method just described is almost identical with that of the simplified method put forward by Dwight and Little ($r_{nav} = 0.958$, standard deviation $= 0.055$) but the amount of computation required to determine the failure load of a given panel is considerably more for Horne and Narayanan's first method.

In their second method Horne and Narayanan (1975) assume that the effective width of the plate between two adjacent stiffeners is constant throughout the length of the panel and equal to the value attained at mid-length (where K_{bs} has its minimum value). Furthermore it is assumed that failure occurs when the stress at the edge of the effective plate at mid-length has reached the yield stress σ_y. From eqns (6.3.16) and (6.3.19) with $\sigma_e = \sigma_y$ the secant effective width factor than has the value

$$K_{bs} = 1 - \frac{(m^2 - 1)\pi^2}{8} \left(\frac{E}{\sigma_y}\right)\left(\frac{y_0}{t}\right)^2\left(\frac{t}{b}\right)^2. \tag{6.3.23}$$

The parameter m can be eliminated from eqns (6.3.22) and (6.3.23) by using numerical methods and it is seen that K_{bs} is a function of y_0/t and $\frac{b}{t}\sqrt{\left(\frac{\sigma_y}{E}\right)}$. (For convenience, the effect of the yield stress is allowed for by replacing the last parameter by $\frac{b}{t}\sqrt{\left(\frac{\sigma_y}{245}\right)}$.)

Figure 6.3.18 shows this functional relationship.

The following effects have been ignored in the foregoing analysis.

(a) In the actual plate yielding will have penetrated into the plate for some distance.

(b) Transverse stresses required to keep the edges straight have been ignored.

(c) At higher b/t values yielding may commence first at the centre of the plate due to bending (see Chapter 5).

(d) Residual stresses due to welding will influence the behaviour of the plate.

Horne and Narayanan (1975) recommend that y_0 should be replaced by the following empirical expression to account for these effects.

$$y_0 = (\bar{y}_0 + \tfrac{1}{2}\bar{y}_{0m}) \frac{b}{30t} \sqrt{\frac{\sigma_y}{245}} \tag{6.3.24}$$

where $\bar{y}_0$ is the measured imperfection

$\bar{y}_{0m}$ is the plate tolerance recommended by the Merrison rules, that

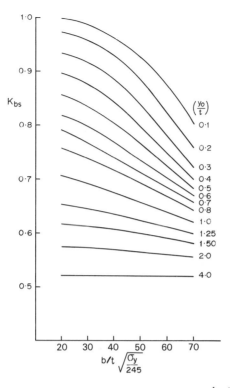

Fig. 6.3.18. K_{bs} at failure plotted as a function of $\dfrac{y_0}{t}$ and $\dfrac{b}{t}\sqrt{\left(\dfrac{\sigma_y}{245}\right)}$.

is,

$$\frac{b}{30t}\left(1+\frac{b}{5000}\right) \text{ mm for } t<25 \text{ mm}$$

$$\frac{b}{750}\left(1+\frac{b}{5000}\right) \text{ mm for } t\geqslant 25 \text{ mm}$$

(6.3.25)

When a single unit of a wide stiffened panel (i.e. the stiffener and its associated plate) is considered the plate is assumed to buckle when its critical stress is reached. At this stress the effective cross-section changes so there is a change in the location of the neutral axis (Fig. 6.3.19). Thus the effective column can be thought of as a new column and the axial load is now applied with an eccentricity e which is the shift in the neutral axis. If the original column has some initial overall imperfection Δ_0 the effective column can be assumed to be a pin-ended column with an

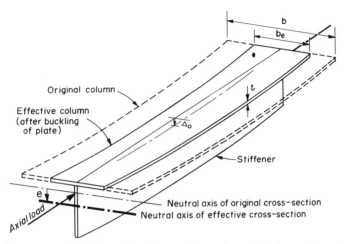

Fig. 6.3.19. After local buckling of the plate a stiffener-plate unit behaves like a pin-ended column with end-eccentricity e and initial deflection Δ_0.

equivalent initial overall imperfection Δ_0' at mid-height where

$$\Delta_0' = \Delta_0 + 1.2e. \tag{6.3.26}$$

As stated earlier, failure of this pin-ended column is assumed to occur when the membrane stress in the plate reaches the yield stress σ_y so the expression for the maximum stress is obtained from an analysis similar to that used to derive the Perry–Robertson formula (eqn (6.3.13)). However, the variables in that equation must be interpreted in the context of the effective column (Fig. 6.3.19) which is now buckling in the direction which results in higher compressive stresses in the plate than at the free edge of the stiffener. After studying their test results Horne and Narayanan found that the best agreement with this theory was obtained when

σ_m is replaced by $\sigma_m' =$ mean axial stress at which effective column fails

σ_E is replaced by $\sigma_E' =$ Euler buckling stress of the effective column

y_s is replaced by $y_p' =$ distance of centroid of effective column from the middle fibre of the plate

A is replaced by $A' =$ area of effective column $(= A_s + K_{bs}bt)$

I is replaced by $I' =$ second moment of area of effective column

Δ_0 is replaced by $\Delta_0' =$ equivalent imperfection of effective column (see eqn (6.3.26))

r is replaced by $r' =$ radius of gyration of the effective column and

η is replaced by $\eta' = 0$ for $L < L_0 = \dfrac{0.2\pi I'}{A'}\sqrt{\dfrac{E}{\sigma_y}}$

$$\eta' = \frac{A'y_p'\Delta_0'(L - L_0)}{LI'} \quad \text{for} \quad L > L_0. \tag{6.3.27}$$

Figure 6.3.20 shows graphs prepared by Horne and Narayanan (1975) of σ'_m/σ_y against L/r' for various values of the parameter

$$\alpha = \frac{A' y'_p \Delta'_0}{LI'} \tag{6.3.28}$$

for three grades of steel and the plate imperfections recommended by the Merrison rules (1973). A given stiffened panel is analysed by this method in the following way.

(1) The plate imperfection y_0 is calculated from eqn (6.3.24).
(2) K_{bs} is read off Fig. 6.3.18.
(3) The properties of the effective section and the parameter α are calculated.

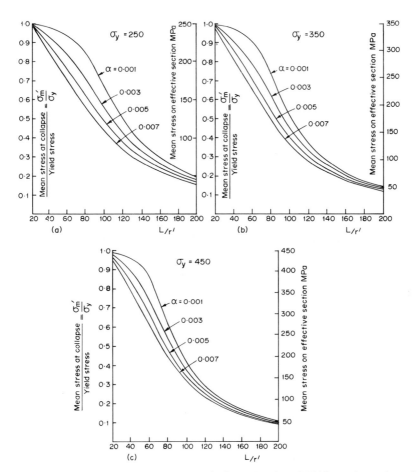

6.3.20. Mean stress on effective cross-section (σ'_m) as a function of L/r' for various values of α (see eqn (6.3.28)).

(4) η' is calculated from eqn (6.3.27) and σ'_m is calculated from eqn (6.3.13), or alternatively Fig. 6.3.20 is used.

(5) The failure load of one stiffener-associated plate unit of the panel is

$$P = \sigma'_m(A_S + K_{bs}bt). \tag{6.3.29}$$

This simplified method was tested by Horne and Narayanan by using it to predict the failure loads of thirty-one panels tested by themselves. The average value of r_n (=ratio of predicted to measured failure load for the nth panel) was 0.971 and the standard deviation was 0.070 (compare 0.958 and 0.055 for Dwight and Little's simplified method (1976) and 0.964 and 0.062 for Horne and Narayanan's (1976b) more accurate (first) method). The simplified method of Horne and Narayanan (1975) is as easy to use as that due to Dwight and Little and it has a comparable accuracy.

Commentary on effective yield stress method compared to effective width method and a review of other methods

In Dwight and Little's simplified method (1976) the actual yield stress of the steel is replaced by an effective yield stress σ_{mav} when the interaction between local and global buckling is considered (see step 1(c) in Table 6.3.2). In both methods of Horne and Narayanan some of the plating between the stiffeners is declared ineffective after it has buckled and a so-called effective width is used in the calculation of the failure load. Dwight and Little claim that the effective yield stress method is more reliable because the effective width method assumes that local buckling of the plate always occurs whereas it is well-known that for high L/r values local buckling is of minor importance only. However, the accuracy tests applied above to panels with proportions likely to be used in practice (b/t values between 24 and 70 and L/r values between 38 and 93) show that the two approaches give results with about the same accuracy. The effective width method has one important advantage, that is, that it is well-established and designers are familiar with the concept.

An effective width method was also developed by Murray (1975) and it also appears to give reliable results when checked against test results on full-scale stiffened plates tested at Monash University. This method is very similar to the simplified method of Horne and Narayanan (1975) but allowance is made for the fact that the shift in the neutral axis (Fig. 6.3.19) occurs when the effective column already carries a stress which is equal to that causing the plate to buckle. Therefore, it is argued, the end eccentricity should be reduced according to the well-known hyperbolic relationship (Fig. 1.2.3) of an isolated strut.

Another effective width method for the design of stiffened plates used as the compression flanges of box girders has also been presented by

Chatterjee and Dowling (1977). Although it is based upon several empirical rules it appears to give accurate results when applied to a wide variety of stiffened panels.

There also a few methods which are entirely empirical but often they give very good estimates of failure load. Their best advantage is their ease of application. A typical stiffened panel has three characteristic stresses which depend upon the geometry of the panel and the properties of the steel.

The first of these is the yield stress σ_y of the steel and it is obvious that the average stress at failure σ_m cannot exceed σ_y. The second of these stresses is the stress σ_E $(= \pi^2 E/(L/r)^2)$ at which the panel will buckle as an Euler column, assuming that yielding and local buckling do not occur first. It is also obvious that σ_m cannot exceed σ_E. The third characteristic stress is that at which local buckling of the plate occurs $(\sigma_{cr} = \pi^2 E/[3(1 - \nu^2)(b/t)^2])$. This assumes that the stiffener is stocky and does not contribute to local buckling. If it does, σ_{cr} can be evaluated by using the finite strip method (see Chapter 4) or by using the tables of Murray and Thierauf (1981). Because of the rising shape of the elastic post-buckling curve of a plate [Fig. 5.1.1(b)] it is seen that σ_m can exceed σ_{cr} especially when b/t is large. Therefore σ_{cr} should be treated differently from σ_y and σ_E.

Allen (1975) proposed the following empirical relationship for *each* plate element of the cross-section

$$\sigma_m^{-n} = \sigma_{cr}^{-n} + \sigma_E^{-n} + \sigma_y^{-n} \tag{6.3.30}$$

and suggested that $n = 2$. It is easily proved that this rule does not allow σ_m to exceed σ_{cr} so it cannot give good results for the higher b/t ratios when post-buckling strength can be relied upon.

Herzog (1976) proposed a similar formula which was an attempt to surmount this difficulty. For the whole cross-section

$$\sigma_m^{-2} = k\sigma_{cr}^{-2} + \sigma_E^{-2} + \sigma_y^{-2} \tag{6.3.31}$$

and he proposed that $k = 0.22$. de George (1979) adjusted the value of k to fit his own test results from trough-stiffened panels and found that $k = 0.60$ gave a more accurate formula.

Summary of Braunschweig University investigations on isolated stiffened plates

A comparative test program on thirty-nine stiffened plates of various forms has been carried out at Braunschweig University by Barbré *et al.* (1976). Many of the tests were performed to study the performance of certain construction details associated with stiffened plates, for example, splices (used to joint two plate panels) and intermittent welding, but most of the panels were plain. From these studies Barbré et al. (1976) developed an effective-width method for determining the failure stress of a stiffened panel when it fails towards the stiffener and it is loaded with axial stress only. They used the following empirical relationship to evaluate the effective width.

$$b_e = 0.9b(\sigma_{cr}/\sigma_y)^{\frac{1}{3}} \quad \text{(but } b_e \not> b) \tag{6.3.32}$$

where σ_{cr} is the stress required to cause local buckling of the plate–stiffener combination. As in the methods described earlier in this section the panels were analysed as if they were Euler columns with the effective cross-section.

The panels were designed by the rules of the German stability code DIN 4114. This code places restrictions upon the slenderness of stiffeners and if these rules are followed the full cross-section of the panel may be used when it is being analysed as a column failing towards the plate. This is because these rules guarantee that local buckling of the stiffener will not occur. When these restrictions are violated it is mandatory to carry out a full stability analysis of the cross-section. This can be done by using the effective depth of the stiffener or by using the tables prepared by Murray and Thierauf (1981).

Barbré et al. (1976) applied this method to thirty-three plain panels (i.e. without splices). They followed the rules set out in DIN 4114 for the initial imperfection Δ_0 at the centre of the equivalent column, that is,

$$\Delta_0 = r(0.05 + 0.002L/r) \tag{6.3.33}$$

where r is the radius of gyration. The eccentricity at the ends was taken as the shift in the neutral axis as the plate or stiffener buckles locally and the original cross-section is replaced by the effective cross-section. Failure was defined as first yield of the extreme fibre of the plate or stiffener according to the requirements of DIN 4114 Ri 7.9. When this procedure was applied to the thirty-three plain panels tested in the program the mean value of r_n (=ratio of predicted to measured average axial stress at failure) was 0.993 and the standard deviation 0.077. There is a slight but not serious tendency for this method to overestimate the failure stress.

Also at the Braunschweig University, Schmidt (1975) reviewed the 'state-of-the-art' of research in the field of stiffened plates and, in particular, discussed the problem of the interaction of local and global buckling in the elasto-plastic range. He proposed a simple interaction diagram (Fig. 6.3.21) which enables designers to read off the failure stress σ_m of a stiffened plate. The graph of the left of this diagram is that which relates the failure stress of an isolated square plate with the same width b

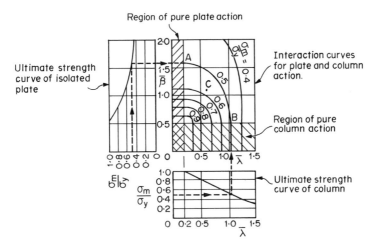

Fig. 6.3.21. Interaction diagram of plate and column action proposed by Schmidt (1975).

and thickness t as the plate being designed to the parameter $\bar{\beta}$ where

$$\bar{\beta}^2 = \left(\frac{b}{t}\right)^2 \frac{3(1-\nu^2)}{\pi^2 E} \sigma_y = \sigma_y/\sigma_{cr\ plate}. \qquad (6.3.34)$$

It is seen that $\bar{\beta}$ depends only upon the dimensions of the cross-section and the yield stress. The diagram below the interaction diagram is the ultimate strength curve of a column which relates σ_m to $\bar{\lambda}$ where

$$\bar{\lambda}^2 = \frac{L^2 \sigma_y}{r^2 \pi^2 E} = \sigma_y/\sigma_E. \qquad (6.3.35)$$

It is seen that $\bar{\lambda}$ is a function of the slenderness ratio L/r of the panel, and of the yield stress. The interaction diagram which links the plate and column failure curves is formed empirically by drawing ellipses shown in the upper right hand region. The dotted lines show how the end points A and B of the ellipse for $\sigma_m/\sigma_y = 0.5$ are established in this region. All of the interaction curves in this region are drawn in this way. In the other regions, which are shaded, either pure plate or pure column behaviour occurs so straight lines are drawn across them. It is only in the upper right-hand region that interaction between column and plate behaviour occurs.

This diagram is used to predict the failure stress of a given panel in the following way. $\bar{\lambda}$ and $\bar{\beta}$ are evaluated and the point with these values as coordinates is located in the interaction diagram. The ratio σ_m/σ_y for the panel is then obtained by interpolating between ellipses and hence the failure stress σ_m is found. For example, a panel whose dimensions are such that $\bar{\beta} = 1.10$ and $\bar{\lambda} = 0.60$ is represented on the interaction diagram by the point C. Hence for this panel $\sigma_m = 0.58\,\sigma_y$.

This kind of diagram currently appears in DIN 4114 (see Chapter 7). Schmidt (1975) tested this method with fifty-eight experimental results from Lehigh, Braunschweig, Monash, Cambridge and Nagoya Universities. The average value of r_n (=predicted failure stress/actual failure stress) was 0.845 and its standard deviation was 0.107. Although this mean value is lower and the standard deviation is greater than for other methods described earlier it should be remembered that the specimens came from many different sources and there is no uniformity in their quality of manufacture and other properties. However, the method is easy to use in a design office.

The behaviour of stiffened plates under combined axial and bending loads

Dowling (1971) drew attention to the fact that at mid-span of a box girder bridge the upper deck will generally carry a compressive axial load due to the overall bending of the girder and at the same time the upper deck must carry the wheel loading from the traffic, so a stiffened panel in the upper deck will be subjected to combined axial and bending loads. It is usual for the transverse stiffeners to be sufficiently stiff so that they will cause the stiffened plate to behave like a wide column whose length is equal to the spacing of the transverse stiffeners. In the design of these panels it is necessary to consider two loading conditions as follows:

Case (a):

Axial stress σ due to overall bending of the box girder and simultaneous wheel loads acting vertically downwards. This loading causes sagging of the panel.

Case (b):

Axial stress σ due to overall bending of the box girder and simultaneous wheel loads acting vertically downwards on the adjacent panels only. This loading causes hogging of the panel.

It is necessary to consider both of these conditions because some open stiffeners are relatively weak against case (b) loading. In the design of ships' hulls case (a) loading arises from the overall bending of the ship as a beam resting on two adjacent waves and the lateral pressure of the water.

These problems were studied experimentally at Monash University first for plates stiffened with bulb flats by Michelutti (1976) and then for plates stiffened with trough stiffeners by de George (1979). The two series were reviewed by de George *et al.* (1979).

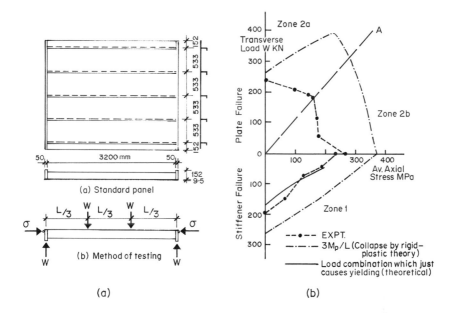

Fig. 6.3.22. (a) Standard panel; (b) interaction diagram for collapse. (Michelutti 1976.)

In the first series of tests fourteen nominally identical panels (Fig. 6.3.22(a)) were tested with an average axial stress σ and two transverse loads W located at the third points of the span. The axial stress σ was applied to its maximum value first so as to represent the stresses due to the overall bending of the box girder. The transverse loads W were then increased until collapse, which occurred without warning when the average axial stress σ was more than $\sigma_y/4$. This is because once failure commences the axial load drives the specimen to collapse and it has no capacity to support W.

Figure 6.3.22(b) shows the interaction curves of failure and it is seen that there are three distinct zones. In zone 1 the tip of the bulb flat carries additional compressive stresses because the direction of W is upwards. Measurements and calculations show that in this zone failure occurs very soon after the tip of the bulb flat reaches the yield stress. The observed mode of failure is a lateral buckling of the tip of the bulb flat. Thus the criterion of failure which designers can adopt for this zone is the commencement of yielding at the tip of the stiffener. The panel is treated as a pin-ended column with a mid-span initial imperfection Δ_0' which is the sum of the actual initial imperfection Δ_0 and that arising from the loads W, that is,

$$\Delta_0' = 0.0356 WL^3/EI + \Delta_0 \qquad (6.3.36)$$

where I is the second moment of area of the cross-section of the panel. As the axial stress σ increases Δ_0' is magnified according to the hyperbolic factor. The value of W at failure is therefore found by solving the equation

$$\sigma_y = \sigma\left[1 + \frac{Ay_s}{I}(0.0356WL^3/EI + \Delta_0)/(1 - \sigma/\sigma_E)\right] + \frac{WLy_s}{3I} \qquad (6.3.37)$$

where y_s is the distance between the centroid of the cross-section and the tip of the stiffener and σ_E is the Euler stress of the panel. This equation was used to determine the full line in the lower part of Fig. 6.3.22(b).

In the upper part of Fig. 6.3.22(b) there is an alarming decrease in load capacity when σ is approximately equal to $0.44\sigma_y$. The reason for this decrease is that there is a change in the collapse mechanism as σ increases beyond zone 2a. In zone 2a the axial stress is relatively small and because of its greater distance from the neutral axis the first fibre to yield is that at the tip of the stiffener. Thus a region of tensile yield penetrates deeply into the stiffener before yielding commences in the plate. There is relatively little evidence of elastic buckling of the plate and it is only for those specimens carrying higher σ values that a plastic mechanism eventually forms in the plate. In contrast to this in zone 2b the specimen exhibits first elastic buckling in the plate and then it starts to yield while the stiffener is still completely elastic. A small increase in W beyond this state causes a local plastic mechanism to form in the plate as the specimen collapses suddenly. It has been shown in Section 6.2 that the plastic collapse curves for local mechanisms lie well below those for simple plastic mechanisms. Therefore, the load capacity of the specimens in zone 2b is considerably less than it is for those in zone 2a. The chain-dotted line is these zones in Fig. 6.3.22(b) is the theoretical curve for failure assuming that only simple plastic hinges form in the specimen. It again shows that simple plastic theory cannot be used for these panels. Although this investigation has been limited to one particular panel and one particular loading pattern, certain general conclusions can be drawn.

The following is a summary of the findings of this investigation.

(a) Transverse loading in both directions must be considered.

(b) Only zones 1 and 2a can be used in practice.

(c) The dividing line OA between zones 2a and 2b is located by equating the tensile stress at the tip of the stiffener to the compressive stress in the plate. For the panels tested the equation of OA is easily shown to be

$$W = 6I\sigma/L(y_s - y_p) \qquad (6.3.38)$$

where y_s and y_p are the distances from the centroid to the tip of the stiffener and the middle surface of the plate, respectively.

(d) In zone 1 the design criterion for collapse is the load combination which first causes yielding at the tip of the stiffener.

The tests carried out by de George (1979) were performed on a set of nominally identical one-sixth scale models (Fig. 6.3.23(a)) of an actual bridge deck and again it is not possible to draw general conclusions from such a restricted test series. The resulting interaction curve of this profile (often called a battledeck profile) is shown in Fig. 6.3.23(b) and it is seen that compared to the panels tested by Michelutti this battledeck panel is much less susceptible to local buckling. The experimental results all lie fairly close to the theoretical curves obtained by assuming a failure mechanism with simple hinges and by using rigid-plastic theory. The reasons for this improvement are, firstly, that stiffeners with closed profiles tend to be more stable than those with open profiles of similar

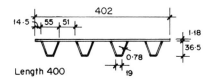

STANDARD SERIES 4 $\frac{1}{6}^{\text{th}}$ SCALE PANEL

(a)

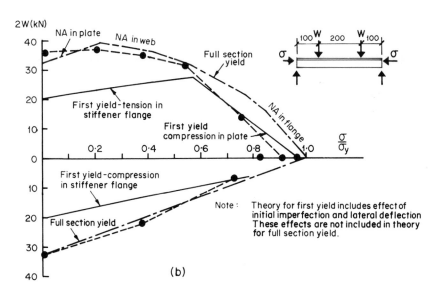

(b)

Fig. 6.3.23. (a) Standard panel; (b) interaction diagram for collapse. (de George 1979.)

dimensions and, secondly, the slightly higher b/t ratios in the first series ($b/t = 56$ compared with 47 for the plate and 52 for the web of the trough stiffener) encourage local buckling.

Another series of tests on scale models of stiffened plates representing bridge decks has been conducted by Chan *et al.* (1977). Simultaneous axial and wheel loading were applied to all of the panels which had b/t ratios of 25 and L/r ratios of 70 and 60. With such a low b/t ratio local buckling was not a major consideration and the models behaved like elasto-plastic columns which could be analysed by using expressions similar to eqn (6.3.37).

The kindred problem of ship's plating which is subjected to simultaneous longitudinal compression and transverse water pressure was studied by Smith (1975). He conducted full scale tests on twelve panels stiffened with tee-bars. The b/t ratios varied between 34 and 97 and L/r ratios of 21 to 68; the latter ratio being small enough to suppress purely elastic Euler buckling. Various modes of elasto-plastic buckling are reported; they include local buckling of the plate, local torsional buckling of the stiffeners, and column buckling of the panels between the transverse stiffeners. A theoretical analysis was carried out by means of an elasto-plastic finite element method. A typical plate-stiffener unit was divided into elements which were each rectangular parallelopipeds. After each load increment each element was checked to see whether it had yielded and, if that was the case, its stiffness was made equal to zero in the next load increment. This method allowed the effects of initial imperfections and residual stresses to be studied, the failure load to be calculated, and the growth of the plastic zones to be traced. The method was used to carry out parametric studies on two typical ship's panels and the following is a brief summary of the findings.

(a) Failure which is towards the plating occurs at a lower load than when it is towards the stiffener. (This confirms the shape of the interaction diagram shown in Fig. 6.3.22(b)).

(b) Compressive residual stresses at the free edge of a stiffener can greatly reduce the failure load because it leads to premature stiffener buckling.

(c) The incremental finite element method described above gives results which agree with experimental observations. The 'use of tangent moduli† in conjunction with elastic buckling analysis cannot provide a reliable estimate of collapse load unless 'structural' tangent moduli are

---

† The tangent modulus E_t is the slope of the effective stress–effective strain curve of a column. It is often used when the curve is non-linear, especially when a column is in the elasto-plastic region. For such cases Young's modulus E appearing in stability formulae is replaced by E_t which is of course less than E.

established referring to the appropriate stiffener geometry and allowing correctly for initial deformations and residual stresses'.

The last comment is interesting because Engesser's tangent modulus is used extensively for *design* codes in Europe (see Chapter 7). Smith's comment is about the buckling *analysis* of a given structure. Design codes recommend E_t values which are supposed to allow for the effects referred to by Smith so as to leave adequate margins of safety on collapse.

6.3.2. Theories of failure and test results of plate girders

Plate girders (Fig. 6.3.24) are an important class of thin-walled structure because they are used so extensively for bridges, crane runways, structural building frameworks, and so on. They are popular with designers because they can be tailor-made to suit any situation whereas rolled steel joists are only rolled in a limited number of profiles, particularly when the size of the profile is very large. Especially in the case of the larger sizes buckling may occur and designers must know how to prevent it from happening at working loads.

When a plate girder carries only small loads the engineers' theory of bending may be used to determine how the internal forces, that is, the transverse shear force V and the bending moment M, are carried by the web and flanges. It is well known that in practical girders the web carries 95% or more of the shear and that, although the flanges resist most of the bending moment, a considerable part may be carried by the web. When the applied loads are increased the combination of stresses acting on the web may initiate yielding, buckling, or both yielding and buckling. When the applied loads are further increased the web must carry an even greater shear force because the flanges cannot carry shear force. In order to do this the web must reduce its share of the bending moment. Thus the flanges have to carry not only the increase in the bending moment from the applied loads but that which the web sheds in order to carry the increase in shear force. This phenomenon is known as *load shedding*. Horne (1980) described and demonstrated the principles of load shedding

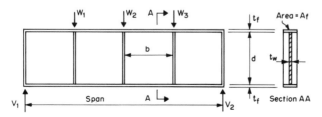

Fig. 6.3.24. A typical plate girder.

by analysing stocky plate girders. However, most practical plate girders have thin web plates which tend to buckle before they yield so that mechanisms of failure can be many and varied. Figure 6.3.25 shows some of the modes of buckling which can occur as well as some other ways in which a plate girder can fail without buckling. This section is concerned with the analysis of these failure modes and the design methods which have been developed to ensure against them. It is assumed that sufficient bracing is provided to prevent lateral buckling of the girder.

The failure modes of plate girders result from the applied transverse shear forces, bending moments, and a combination of these loadings. In the girder illustrated in Fig. 6.3.24 the web is reinforced by vertical stiffeners whose function is to restrict the growth of a shear buckling pattern similar to that illustrated in Figs. 5.3.1 and 5.3.12. Tests indicate that a stiffened web may still have a tendency to buckle even when vertical stiffeners are incorporated into the design. This is because the combination of diagonal tension and diagonal compression can cause

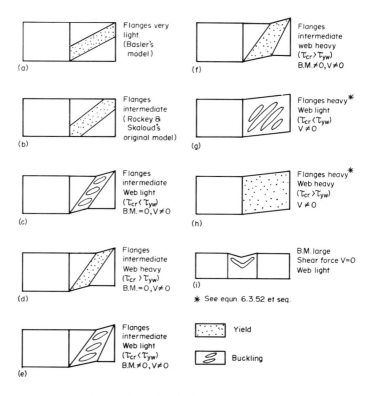

Fig. 6.3.25. Modes of failure of a plate girder.

closely spaced buckles to form across a diagonal. In deep girders not only are the vertical stiffeners usually placed so that their spacing is less than the depth of the girder but sometimes horizontal stiffeners are also added to the compression side in order to reduce the size of the plate panels and thereby increase their buckling stress.

Basler (1961) carried out shear tests on plate girders with vertical stiffeners and developed a theory of failure which was based upon a theoretical model (Fig. 6.3.25(a)) with a diagonal band of tension yield, and which ignored the contribution of the flanges to the strength of the girder. Observations showed that these girders could carry a much larger load than that required to initiate elastic buckling of the web. In the elasto-plastic region it was seen that the buckled web panel behaved like a truss with a diagonal tension member and that failure occurred when the panel became a plastic mechanism. At this point the panel carried part of the applied shear force by the truss action just described and part by the web plate with in-plane shear stresses at their elastic critical value τ_{cr}. The truss-like action is now known as *tension field action*.

Basler's work was extended by Chern and Ostapenko (1969) who investigated plate girders with unequal flanges and who later (Ostapenko and Chern 1971) developed a more sophisticated theory which dealt with both bending and shear loadings and longitudinal stiffening. Whereas Basler (1961) had assumed that the flanges were too flexible to influence the behaviour of the webs, Rockey and Skaloud (1968, 1972) found that the collapse mechanism involved plastic hinges in the flanges and that they often had a strong influence upon the behaviour of the panel. When the flanges are very light the collapse mechanism approximates to that assumed by Basler [Fig. 6.3.25(a)] but if they are heavy their plastic hinges form at the four corners of the panel [Fig. 6.3.25(g) and (h)] which fails like a Vierendeel girder. For intermediate flanges the plastic hinges are located as shown in Fig. 6.3.25(c)–(f). In their earlier papers Rockey and Skaloud (1968, 1972) assumed that for the case when the shear force acts alone (i.e. $M = 0$) the plastic mechanism had the form shown in Fig. 6.3.25(b) where it is seen that the tension field is assumed to be parallel to the diagonal of the panel. This restriction was relaxed in later papers by Rockey, Evans, and Porter (1978) and Evans, Porter, and Rockey (1978) and they developed a design method which also allowed for the combined loading of shear force and bending moment. Their method can cater for all of the collapse modes illustrated in Fig. 6.3.25 and when it was tested against the measured collapse loads of eighty-eight girders reported by various investigators it was found that the average value of r_n (=predicted collapse load/measured collapse load) was 0.997 with a standard deviation of 0.064. It is this method of analysis and design which is described in the following pages.

The method of analysis is introduced here by considering the most general of the collapse mechanisms illustrated in Fig. 6.3.25, that is, that shown in Fig. 6.3.25(e). It will be seen later that the other cases can be treated as a special case of this mechanism. The panel which fails is subjected to a bending moment M, and a shear force V. Figure 6.3.26 shows that when V acts alone ($M = 0$) XY and WZ are parallel. This is because of the antisymmetry of the panel but when $M \neq 0$ no antisymmetry exists and XY and WZ are not parallel. It should be appreciated that a minimum kinematic requirement (for the panel to deform as a mechanism) is that the whole region XYZW should yield but this does not mean that yielding is confined to this region.

In the analysis of the mechanism shown in Fig. 6.3.26 it is considered that there are three stages in the load path up to collapse.

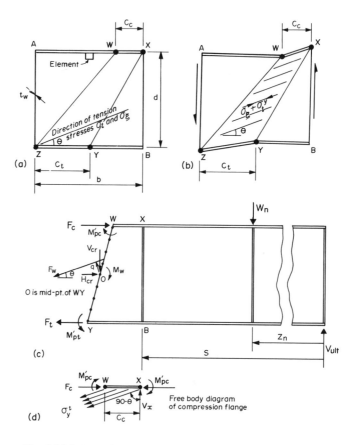

Fig. 6.3.26. Analysis of panel in plate girder shown in Fig. 6.3.25(e).

Stage 1: *unbuckled behaviour*

The web plate remains perfectly flat until it reaches its critical stress. If the bending moment is zero there will be principal stresses (one tensile and one compressive) at 45° and 135° until the shear stress τ reaches its critical value τ_{cr}. For a rectangular plate

$$\tau_{cr} = k \left[\frac{\pi^2 E}{12(1 - v^2)} \right] \left(\frac{t_w}{d} \right)^2 \tag{6.3.39}$$

where

$$k = 5.35 + 4(d/b)^2, \text{ valid if } b/d > 1; \; k = 5.35 \, (d/b)^2 + 4, \text{ valid if } b/d < 1. \tag{6.3.40}$$

If the shear force V is zero and M acts alone the critical bending stress is (Fig. 5.1.4)

$$\sigma_{crb} = 23.9 \left[\frac{\pi^2 E}{12(1 - v^2)} \right] \left(\frac{t_w}{d} \right)^2 \tag{6.3.41}$$

When V and M act simultaneously it is possible to determine their critical values by using the finite difference method described in Section 5.3.2. However, fairly accurate values are obtained by using the following equation

$$\left(\frac{\sigma_{mb}}{\sigma_{crb}} \right)^2 + \left(\frac{\tau_m}{\tau_{cr}} \right)^2 = 1 \tag{6.3.42}$$

where

σ_{mb} is the critical value of the bending stress at the edge midway between the ends A and X of the panel when V and M both act;

τ_m is the critical value of τ when V and M both act;

σ_{crb} is the critical value of the bending stress with $V = 0$, $M \neq 0$ (see eqn (6.3.41));

τ_{cr} is the critical value of the shear stress with $V \neq 0$ and $M = 0$ (see eqn (6.3.39)).

Stage 2: *post-buckled behaviour*

Once the critical stress is reached the web starts to buckle and it cannot carry any increase in compressive stress. Any additional load has to be supported by tension field action. It is assumed that the shear and bending stresses remain at their critical values τ_m and σ_{mb} and that there are *additional* membrane stresses σ_t which are inclined at an angle θ to the horizontal and which carry any increases in the applied load.

Stage 3: *ultimate load behaviour*

When the material in the region XYZW reaches yield the panel becomes a plastic mechanism and it cannot sustain any further increase in load. In this ultimate condition the additional membrane stress σ_t reaches its maximum value σ_t^y. This mechanism (Fig. 6.3.26) is analysed in the following paragraphs in order to evaluate the load-carrying capacity of the panel.

A small rectangular element located at the edge (Fig. 6.3.26(a)) is first considered at the start of Stage 2. The stresses on the faces of a similar element obtained by a rotation through angle θ are found by using Mohr's circle

$$\sigma_\eta = -\sigma_{mb} \sin^2\theta - 2\tau_m \sin\theta \cos\theta$$

$$\sigma_\xi = -\sigma_{mb} \cos^2\theta + 2\tau_m \sin\theta \cos\theta \qquad (6.3.43)$$

$$\tau_{\eta\xi} = -\sigma_{mb} \sin\theta \cos\theta + \tau_m (\sin^2\theta - \cos^2\theta)$$

The failure condition (Stage 3) (Fig. 6.3.26(b)) is reached by adding σ_t^y to σ_ξ and introducing a yield criterion. In two-dimensional problems it is common to use the von Mises-Hencky criterion (Hill 1967), which was first discovered by Clerk Maxwell in 1856 (van Iterson 1947)

$$\sigma_y^2 = \sigma_1^2 + \sigma_2^2 - \sigma_1\sigma_2 + 3\tau^2 \qquad (6.3.44)$$

where σ_y is the yield stress,

  σ_1, σ_2 are the direct stresses acting on two orthogonal planes, and

  τ is the shear stress acting on the same two planes.

Hence, when the yield stress of the web is σ_{yw}.

$$\sigma_{yw}^2 = (\sigma_\xi + \sigma_t^y)^2 + \sigma_\eta^2 - \sigma_\eta(\sigma_\xi + \sigma_t^y) + 3\tau_{\eta\xi}^2 \qquad (6.3.45)$$

A quadratic equation can be obtained from eqns (6.3.43) and (6.3.45) and solved for σ_t^y. Thus

$$\sigma_t^y = -\tfrac{1}{2}A + \tfrac{1}{2}[A^2 - 4(\sigma_{mb}^2 + 3\tau_m^2 - \sigma_{yw}^2)]^{\frac{1}{2}} \qquad (6.3.46)$$

with

$$A = 3\tau_m \sin 2\theta + \sigma_{mb} \sin^2\theta - 2\sigma_{mb} \cos^2\theta. \qquad (6.3.47)$$

These equations have been derived for a point on the boundary but they can be used to evaluate σ_t^y at any point in the web, and in particular, at equispaced points along the line WY (Fig. 6.3.26(c)). Thus the resultant force F_W, which arises from the tension field, can be evaluated and located.

When the average stresses in the compression and tension flanges are σ_{cf} and σ_{tf} and the yield stress of the flanges is σ_{yf}, the reduced plastic moments of the flanges are (see (eqn (6.2.4))

$$M'_{pc} = M_{pc}\left[1 - \left(\frac{\sigma_{cf}}{\sigma_{yf}}\right)^2\right] \quad \text{for the compression flange}$$

$$M'_{pt} = M_{pt}\left[1 - \left(\frac{\sigma_{tf}}{\sigma_{yf}}\right)^2\right] \quad \text{for the tension flange.}$$

(6.3.48)

The plastic hinges in the flanges may now be located by considering equilibrium of the forces acting on the flanges. During a virtual displacement ϕ of the mechanism in the compression flange [Fig. 6.3.26(d)] the work done by the shear force at X is equal to the energy absorbed by the plastic hinges at X and W and the work done *against* the tension field stress σ_t^y. It is convenient to take an average value for σ_t^y, that is, σ_{tc}^y at the midpoint of WX.

$$V_X c_c \phi = t_w \sigma_{tc}^y (\sin^2\theta)\frac{c_c^2}{2}\phi + 2M'_{pc}\phi. \tag{6.3.49}$$

The minimum value of V_X is obtained by differentiation with respect to c_c, whence

$$c_c = \frac{2}{\sin\theta}\sqrt{\left(\frac{M'_{pc}}{\sigma_{tc}^y t_w}\right)}. \tag{6.3.50}$$

Similarly in the tension flange

$$c_t = \frac{2}{\sin\theta}\sqrt{\left(\frac{M'_{pt}}{\sigma_{tt}^y t_w}\right)}. \tag{6.3.51}$$

These equations define the location of the plastic hinges in the flanges but there is a restriction that c_c and c_t must be less than b. Therefore from eqn (6.3.50)

$$M'_{pc} < \frac{t_w b^2 \sin^2\theta}{4}\sigma_{tc}^y \tag{6.3.52}$$

where σ_{tc}^y is evaluated from eqns (6.3.46) and (6.3.47) at the midpoint of WX. When M'_{pc} is greater than this critical value the plastic hinges are located at X and A and the panel deforms like a Vierendeel girder. Flanges which satisfy the criterion of eqn (6.3.52) are said to be light while those which do not are said to be heavy. In the latter case the tension field occupies the whole of the web and $c_c = b$ as shown in Fig. 6.3.25(g) and (h). Similar conclusions are obtained when the tension flange is considered.

The average axial stress in the compression flange between W and X is obtained by considering horizontal equilibrium of the section WX (Fig.

6.3.26(d))

$$\sigma_{cf} = \frac{F_c - \frac{1}{2}(\sigma^y_{tc} \sin \theta \cos \theta + \tau_m)c_c t_w}{A_{cf}} \tag{6.3.53}$$

where F_c is the force in the compression flange at W
and A_{cf} is the cross-sectional area of the compression flange.

The corresponding formula for the tension flange is

$$\sigma_{tf} = \frac{F_t + \frac{1}{2}((\sigma^y_{tt} \sin \theta \cos \theta + \tau_m)c_t t_w}{A_{tf}} \tag{6.3.54}$$

The forces F_c and F_t are found by considering equilibrium of the girder to the right of the line WY (Fig. 6.3.26(c)).
Resolving vertically and noting that $V_{cr} = \tau_m t_w d$

$$V_{ult} = F_w \sin \theta + \tau_m t_w d + \sum_n W_n \tag{6.3.55}$$

Resolving horizontally and noting that $H_{cr} = \tau_m t_w (b - c_c - c_t)$

$$F_c - F_t = F_w \cos \theta - \tau_m t_w (b - c_c - c_t). \tag{6.3.56}$$

Moments about O give rise to the following equation

$$F_c + F_t = \frac{2}{d} \left[V_{ult} \left\{ S + \frac{b + c_c - c_t}{2} \right\} + M'_{pt} - M'_{pc} + F_w q - M_w - \sum_n W_n z_n \right]. \tag{6.3.57}$$

where W_1 to W_n are external loads applied to this part of the beam and M_w is the bending moment in the web at the end of stage 2, i.e. $M_w = \dfrac{\sigma_{mb} b d^2}{6}$. From these expressions the flange forces are

$$F_c = \frac{V_{ult}}{2d} (d \cot \theta + 2S + b + c_c - c_t)$$

$$+ \frac{1}{d} \left(M'_{pt} - M'_{pc} + F_w q - M_w - \sum_n W_n z_n \right)$$

$$- \frac{1}{2} \tau_m t_w (d \cot \theta + b - c_c - c_t) \tag{6.3.58}$$

$$F_t = \frac{V_{ult}}{2d} (d \cot \theta + 2S + b + c_c - c_t)$$

$$+ \frac{1}{d} \left(M'_{pt} - M'_{pc} - F_w q - M_w - \sum_n W_n z_n \right)$$

$$+ \frac{1}{2} \tau_m t_w (d \cot \theta + b - c_c - c_t). \tag{6.3.59}$$

An iterative procedure was adopted by Evans, Porter, and Rockey (1978) for solving eqns (6.3.55)–(6.3.59). After adopting a value for θ the process was started by assuming that σ_{cf} and σ_{tf} were both zero. This enabled c_c and c_t to be evaluated through eqns (6.3.48)–(6.3.51) and subsequently approximate values of F_c and F_t would be found from eqns (6.3.58) and (6.3.59). Thus better estimates of σ_{cf} and σ_{tf} could then be made from eqns (6.3.53) and (6.3.54) for the next cycle of iteration.

So far the analysis has been concerned with the general case illustrated in Fig. 6.3.25(e) where both V and M act and the plastic hinges W and Y are located between the corners of the panel. It has been pointed out that when the flanges have a reduced plastic moment exceeding that given in eqn (6.3.52) the plastic hinges occur at the corners. In these cases the above analysis is valid but c_c and c_t are both equal to b. There are other constraints which must be incorporated into the theory and it will be seen that they not only affect the failure load but also the mode of collapse. These constraints are now discussed with reference to the various failure modes shown in Fig. 6.3.25.

Mode (g)

In this girder the web plate is light and the flanges have M_p' values in excess of the critical value (eqn (6.3.52)). The web plate buckles elastically under the combination of applied bending and shear stresses before it yields under the same combination of applied stresses. In order to ensure that this is so the combination of stresses (σ_b, τ) to cause yielding can be found from eqn (6.3.44) and it should be greater than that found from eqn (6.3.42).

Mode (h)

In this girder both the web plate and the flanges are heavy so the web yields before it buckles elastically under a combination of shear and bending stresses. The test for whether a panel fails in this mode or in mode (g) is described above. If it is found that mode (h) (i.e. yielding) governs, the combination of stresses given by eqn (6.3.44) should be used and σ_t^y is zero.

Mode (c)

In this girder the applied bending moment is zero and therefore $c_c = c_t$. Also the flanges are light, so that $c_c = c_t < b$, and the web is light so that it buckles elastically before it yields. From eqn (6.3.44) it is seen that a flat plate carrying pure shear yields when the shear stress reaches $\sigma_y/\sqrt{3}$. Hence for this mode of failure to occur in the web it is necessary that

$$\tau_{yw} = \sigma_{yw}/\sqrt{3} > \tau_{cr} \qquad (6.3.60)$$

where τ_{cr} is given by eqn (6.3.39). In applying the theory to this case $\tau_m = \tau_{cr}$ and $\sigma_{mb} = 0$.

Mode (d)

This girder is similar to that which develops a mode (c) failure mechanism but it has a heavier web. In this case

$$\tau_{yw} = \sigma_{yw}/\sqrt{3} < \tau_{cr} \tag{6.3.61}$$

so extensive yielding of the web occurs rather than elastic buckles. In applying the foregoing theory to this case $\tau_m = \tau_{yw}$ and $\sigma_{mb} = 0$.

Mode (f)

This girder is similar to that which develops a mode (e) failure mechanism but it has a heavier web. For the same reasons as those described for mode (d) $\tau_m = \tau_{yw} = \sigma_{yw}/\sqrt{3}$ when the foregoing theory is used.

Mode (i)

In this case the bending moment is large and the web is light. Failure occurs as a result of the inwards failure of the compression flange and a simultaneous crumpling of the thin web plate in the region of the flange buckle. The foregoing theory does not apply to this case and it is generally accepted that the following empirical equation due to Cooper (1965, 1971) can be used when $M_{ult} \not> M_p$

$$\frac{M_{ult}}{M_y} = 1 - 0.0005 \frac{A_w}{A_f} \left[\frac{d}{t} - 5.7 \sqrt{\left(\frac{E}{\sigma_{yf}}\right)} \right] \tag{6.3.62}$$

where M_y is the bending moment to cause yielding in the extreme fibres assuming that the web is fully effective.

M_p is the plastic moment of the girder assuming a fully effective web.

σ_{yf} is the yield stress in the flange.

A_w, A_f are the cross-sectional areas of the web and each of the flanges, respectively.

Mode (a)

This is Basler's model and it can be analysed by the foregoing theory by assuming $M'_{pc} = M'_{pt} = 0$.

The theory described above was applied by Evans, Porter, and Rockey (1976) to a large number of designs of plate girders with the aid of a computer. As a result of these parametric studies a simplified design procedure was developed and this is described in the following paragraphs.

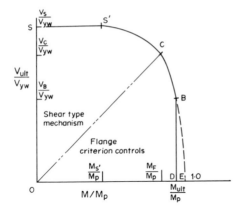

Fig. 6.3.27. Interaction diagram for failure of a plate girder. (Rockey, Evans, and Porter 1978.)

A panel of given geometry will fail in different modes depending upon the ratio of applied shear force to bending moment. The parametric studies referred to above showed that the interaction diagram for failure has the form shown in Fig. 6.3.27. In the simplified design method the points S, S′, C, B and D are located according to simple formulae and the curves connecting these points are either straight lines or parabolae. The denominator V_{yw} is the shear force required to make the web fully plastic, i.e.

$$V_{yw} = \tau_{yw} t_w d = \sigma_{yw} t_w d / \sqrt{3}. \qquad (6.3.63)$$

In the interaction diagram the ordinate OS is determined by considering vertical equilibrium of PXBY (Fig. 6.3.28) with, for pure shear

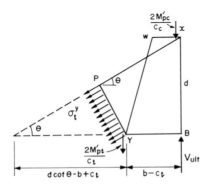

Fig. 6.3.28. Analysis of part of web with pure shear loading.

loading, $c_c = c_t = c$, $\sigma_{tt}^y = \sigma_{tc}^y = \sigma_t^y$. Thus

$$V_{ult} = V_S = (PY)\sigma_t^y t_w \sin\theta + \tau_{cr} t_w d + 4M_p'/c. \tag{6.3.64}$$

From the geometry of PXBY and eqns (6.3.50) and (6.3.51) and after dividing by V_{yw} (eqn (6.3.63))

$$\frac{V_S}{V_{yw}} = \frac{\tau_{cr}}{\tau_{yw}} + \frac{\sqrt{3}\sigma_t^y}{\sigma_{yw}}\sin^2\theta\left(\cot\theta - \frac{b}{d}\right) + 4\sqrt{3}\sin\theta\sqrt{\frac{\sigma_t^y}{\sigma_{yw}}}\sqrt{\frac{M_{pf}}{d^2 t_w \sigma_{yw}}}. \tag{6.3.65}$$

The control on eqn (6.3.65) is again that $c \not> b$. From eqn (6.3.50)

$$\frac{2}{\sin\theta}\sqrt{\left(\frac{M_{pf}}{\sigma_t^y t_w}\right)} \not> b. \tag{6.3.66}$$

With $\theta = 45°$ and by using the relationship $\sigma_{yw} = \sqrt{3}\tau_{yw}$

$$M_{pf} \not> \frac{b^2 t_w \sigma_{yw}}{8}\left[-\frac{\sqrt{3}\tau_{cr}}{2\tau_{yw}} + \left(1 - \frac{\tau_{cr}^2}{4\tau_{yw}^2}\right)^{\frac{1}{2}}\right]. \tag{6.3.67}$$

When this inequality is not satisfied it is necessary to analyse the panel as a Vierendeel girder with an in-fill panel. The ultimate shear force is obtained by using eqn (6.3.55) where F_w is evaluated across the diagonal AB (Fig. 6.3.26), i.e.

$$F_w = \sigma_t^y d \sin(\theta + \theta_d)/\sin\theta_d \tag{6.3.68}$$

where θ_d is the inclination of the diagonal of the panel to the flanges and σ_t^y is given by eqn (6.3.46) with $\sigma_{mb} = 0$.

The parametric studies carried out by Evans, Porter and Rockey (1976) showed that V_S/V_{yw} has a stationary value when $\theta \simeq 2/3\theta_d$. Their 1978 paper presents various design aids which enable designers to evaluate V_S/V_{yw} quickly.

The coordinates of the point C are given by the following empirical formulae derived by parametric studies

$$\frac{M_C}{M_p} = \frac{M_F}{M_p} \tag{6.3.69}$$

$$\frac{V_C}{V_{yw}} = \frac{\tau_{cr}}{\tau_{yw}} + \left[\frac{\sigma_t^y}{\sigma_{yw}}\sin\frac{4\theta_d}{3}\right]\left[0.554 + \frac{36.8M_{pf}}{M_F}\right]\left[2 - \left(\frac{b}{d}\right)^{\frac{1}{8}}\right] \tag{6.3.70}$$

or

$$\frac{V_C}{V_{yw}} = \frac{V_S}{V_{yw}} \quad \text{if} \quad V_C > V_S$$

where M_{pf} = plastic moment of resistance of each flange ($=\frac{1}{4}\sigma_{yf}b_f t_f^2$)
M_F = plastic moment of resistance of flanges acting alone ($=\sigma_{yf}A_f(d + t_f)$).

The point S' in Fig. 6.3.27 has the coordinates $(M_S/M_p, V_S/V_{yw})$ where

$$M_{S'} = V_S b \quad \text{but} \quad \not> 0.5 M_F. \tag{6.3.71}$$

A straight line joins S and S′ and a parabola with its crown at S′ is fitted between S′ and C.

The coordinates of point B which corresponds to the inward collapse of the flange when the panel carries a large bending moment are $(M_{ult}/M_p, V_b/V_{yw})$ where M_{ult} is given by eqn (6.3.62). The curve CBE is also a parabola with its crown at E so the shear load acting with M_{ult} at B is

$$V_B = V_c \sqrt{\left(\frac{M_p - M_{ult}}{M_{pw}} \right)} \tag{6.3.72}$$

where M_{pw} is the plastic resistance of the web alone $(= \sigma_{yw} t_w d^2/4)$. As stated earlier the interaction diagram is found to predict the collapse loads of plastic girders with a high degree of accuracy. For a given design of panel it is easy to establish the points S, S′, C, B, and D of the interaction diagram and then to use it to determine the combination of V and M which will cause it to collapse.

A general study of the behaviour of thin plates with initial imperfections and combined direct, bending, and shear in-plane loading has been carried out by Harding, Hobbs, and Neal (1977a, 1977b and 1979) at Imperial College, London. They developed a finite difference method which allowed plasticity to develop and solved the governing equations by dynamic relaxation. The theory treated the panel as a system of layers similar to that developed by Moxham (1971a) and used the von Mises–Hencky yield criterion and the Prandtl–Reuss flow rule to enable plastic effects to be studied. Strain hardening was also allowed for as well as initial imperfections (dishing) and initial stresses. Several types of deflection and strain conditions and in-plane stress distributions along the boundaries were assumed. Thus the following factors which influence the failure load of a panel were studied.

(a) Edge restraint: if the edges are free to move in the plane of the plate tension field action cannot develop. Two cases were considered, that is, completely free and completely restrained boundaries.

(b) Out-of-plane initial imperfections: a standard initial imperfection at the centre of each plate whose value is $y_0 = 0.145 b \sqrt{(\sigma_y/E)}$ was used. This is the value adopted by the new British Standard BS 5400 (Bridge Code) (1978–1981).

(c) Shape of initial imperfection: in most cases the initial imperfections were a single half sine wave but in some panels with a higher aspect ratio a double half sine wave was assumed.

(d) Residual stresses: standard distributions of initial stresses with peak values of $0.1\sigma_y$ and $0.3\sigma_y$ were used.

(e) b/t: b/t values of 30, 60, 120 and 180 were used.

(f) Shape of edge displacements and stress distributions: the shapes assumed were a constant value, a simple triangle with the maximum value at one edge and the usual linear distribution of bending stresses. Shear was usually introduced as a constant displacement.

(g) Horizontal stiffeners: in some panels horizontal stiffeners were introduced to study their influence upon maximum load

The results of their analyses were plotted as curves which show the interactive effects of longitudinal direct stress and shear stress on failure of the panels. When there is no buckling of a web panel, failure is determined by the stress combination which causes yielding. When the transverse stress σ_x is zero the von Mises–Hencky criterion (eqn (6.3.44)) can be rewritten in the form $(\sigma_z/\sigma_{yw})^2 + (\tau/\tau_{yw})^2 = 1$. This equation defines a circle which forms the outer boundary of most of the curves derived by Harding and Hobbs (1979), but there are some exceptions. Figure 6.3.29 shows two typical sets of results which are interpreted in the following way. Suppose, for example, that the compressive stress σ_z and the shear stress τ are applied in a proportional manner so that $(\sigma_z/\sigma_{yw}) = (\tau/\tau_{yw})$. A line at 45° from the origin intersects the curves at points which define failure. These particular sets of curves show that as the slenderness (b/t) of the plates increase there is a reduction in the failure stress which is more marked when the panels are unrestrained. It is also seen that initial imperfection y_0 has only a limited influence upon that failure stress. It is also seen from Fig. 6.3.29 that if the web panel is restrained and if it is allowed to shed some of its bending stresses to the flanges, it can develop tension field action and develop a much higher shear stress τ. Finally these curves show that a web panel at the point of bending failure has a considerable reserve of strength to carry shear stresses as well without any decrease in its bending capacity.

Further studies on the effects of the parameters listed previously lead to the following conclusions. Many of these conclusions have now been incorporated into BS 5400 (1978–1981).

(a) Edge restraint: this generally causes an upwards and outwards bowing of the curves in the compression part of the graph. When this happens there are gains to be made from the use of tension field action.

(b) Out-of-plane initial imperfections: in no cases did initial imperfections have a great effect upon the stresses at failure and hence upon failure load.

(c) Shape of initial imperfection: it was only in a few isolated zones of the graphs that the shape of the initial imperfection had some effect.

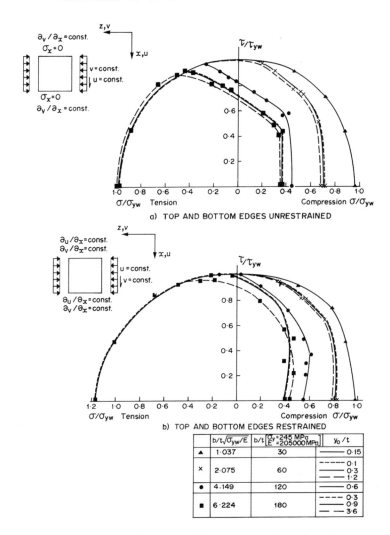

Fig. 6.3.29. Typical interaction diagrams of failure showing effects of combined shear and direct stress in imperfect web panels. (Harding and Hobbs 1979.)

(d) Residual stresses: residual stresses generally speaking have little effect upon the failure load of a restrained web panel but they can reduce failure loads in unrestrained panels by as much as 25%. The first of these conclusions, incidentally, agrees with the experimental findings of Horne and Narayanan (1976a) in their work on stiffened plates with uniform longitudinal in-plane stresses.

(e) b/t: this factor has more effect upon the compressive failure load of a plate than any other single parameter.

(f) Shape of edge displacements and stress distributions: these factors also have a strong influence upon the shape of the interaction diagrams. This appears to confirm that in the case of a practical web panel the designer must be careful in his choice of assumptions because there are many possible failure mechanisms (see Fig. 6.3.25). This is particularly true during load shedding when the distribution of boundary stresses on a web panel must be changing rapidly.

The theory developed by Rockey, Evans, and Porter has been adopted in the new BS 5400 Bridge Code as the design method for web panels without horizontal stiffeners. The findings of Harding, Hobbs, and Neal have been used in that code as the basis for the design method for web panels which are reinforced with horizontal stiffeners. These topics are therefore discussed further in Chapter 7.

6.3.3. Theories of failure and tests of thin-walled box columns

Thin-walled steel box columns are used in a wide variety of situations, from small cold-formed verandah posts up to the enormous towers in a suspension or cable-stayed bridge. For a given column length and cross-sectional area of plating used, the designer can either avoid local buckling by using low b/t ratios or avoid global buckling by using high b/t ratios. However, in the first case the global buckling load will be relatively small and in the second case the local buckling load will be relatively small. This section is concerned with the problem of the analysis and design of box columns.

As in the case of stiffened plates there are two essentially different philosophies, that is, the effective width and the effective stress philosophies. A typical effective width philosophy is that described by Schubert (1975), which is treated here first; an effective stress philosophy was developed by Little (1976b, 1979) and his theory and experiments are described later in this section.

Tests on box columns have been performed at Cambridge University between 1963 and 1972 by various workers. It was found that the importance of residual stresses due to welding in reducing the collapse load depended upon b/t and L/r ratios. A number of tests were also reported by Klöppel et al. (1969).

The effective width theory for box columns developed by Schubert (1975) uses Marguerre's equations (eqns (5.2.9) and (5.2.10)) as a starting point but he included a second Airy stress function Φ_0 in the first of these equations to allow for residual welding stresses. Equations (5.2.9) and

(5.2.10) are rewritten thus[†]

$$\nabla^4(\Phi - \Phi_0) + Et(y''\ddot{w} - 2y'\dot{w} + \ddot{y}w'' + w''\ddot{w} - \dot{w'}^2) = 0 \qquad (6.3.73)$$

$$D\nabla^4 w - [\ddot{\Phi}(y+w)'' - 2\dot{\Phi}'(y+w)' + \Phi''(y+w)\ddot{}] - Y = 0 \qquad (6.3.74)$$

Along each of the boundaries of the individual plates the conditions of continuity consider both the in-plane and out-of-plane effects. Equations (6.3.73) and (6.3.74) were solved by Schubert by assuming the following forms for the deflection of the flanges and web (Fig. 6.3.30) and the Airy

[†] Schubert's equations appear to differ from eqns (6.3.73) and (6.3.74) but this is only because his w is the distance from the x–z plane to the central plane of the plate. In this book that distance is $y + w$.

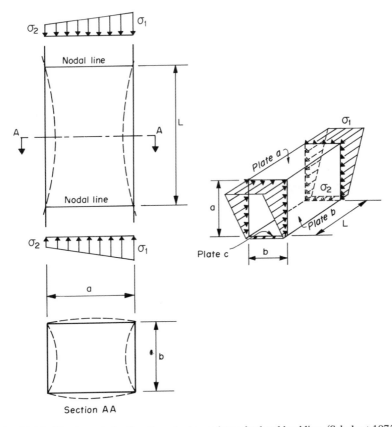

Fig. 6.3.30. Elastic analysis of section of a box column for local buckling. (Schubert 1975.)

stress functions Φ_a, Φ_b and Φ_c

$$w_a = \cos\frac{\pi z}{L}\left[(f_{a1}-f_{a10})\cos\frac{\pi x}{b}+(f_{a2}-f_{a20})\left(1+\cos\frac{2\pi x}{b}\right)\right]$$

$$w_b = \cos\frac{\pi z}{L}\left[(f_{b1}-f_{b10})\cos\frac{\pi x}{a}+(f_{b2}-f_{b20})\left(1+\cos\frac{2\pi x}{a}\right)\right.$$

$$\left.+(f_{b3}-f_{b30})\sin\frac{2\pi x}{a}\right]$$

$$w_c = \cos\frac{\pi z}{L}\left[(f_{c1}-f_{c10})\cos\frac{\pi x}{b}+(f_{c2}-f_{c20})\left(1+\cos\frac{2\pi x}{b}\right)\right]$$

$$\Phi_i = -\cos\frac{2\pi z}{L}\left[A_i\cosh\frac{2\pi x}{h_i}+B_i\frac{2\pi x}{h_i}\sinh\frac{2\pi x}{h_i}+C_i\sinh\frac{2\pi x}{h_i}\right.$$

$$\left.+D_i\frac{2\pi x}{h_i}\cosh\frac{2\pi x}{h_i}\right]-G_i\frac{x^2}{2}-H_i\frac{x^3}{6}+\Phi_{ip}+\Phi_{i0}$$

(6.3.75)

where $f_{a1}, \ldots, f_{c2}, A_i, \ldots, H_i$ are constants

$f_{a10}, \ldots, f_{c20}$, are Fourier coefficients of the initial shape

Φ_{ip} is a particular solution of equations (6.3.73) and (6.3.74), i.e. a solution which includes the transverse surface load Y acting on the ith plate.

Φ_{i0} is the Airy stress function for the distribution of initial stresses.

h_i is the width of the ith plate (i.e. $h_i = a$ or b).

The initial deflection of each plate was assumed to be equal to that specified in the German Code DIN 1079 and an empirical expression for the inital stress distribution was used in the analysis.

These expressions for the deflections and stress functions contain altogether twenty-five constants which are evaluated firstly by substituting into the boundary conditions and secondly by substituting into eqns (6.3.73) and (6.3.74). This leaves three independent parameters which Schubert evaluated by minimizing the total energy of the system with the aid of a computer.

One of the aims of Schubert's investigation was to decide on a satisfactory criterion of failure. Five criteria were selected as follows:

(a) The average stress in the critical plate reaches

$$\sigma_{cr}\left[=\frac{\pi^2 E}{3(1-\nu^2)}\left(\frac{t}{b}\right)^2\right]$$

with DIN 4114 reduction factors in the plastic region.

(b) The maximum membrane stress at the edge reaches the yield stress.

(c) The maximum stress (membrane plus bending) in the plate reaches the yield stress.

(d) The deformation at the centre of the critical plate reaches $0.01b$.

(e) The deformation at the centre of the critical plates reaches $0.0125b$.

The behaviour near collapse of a number of experimental models was studied and compared with that predicted by the foregoing theory. It is concluded that criteria (b) and (c) give reasonable predictions of the collapse load. This theory is concerned with the collapse of a box column due to local buckling. The effects of column length and of interaction between local and global buckling are not considered. Also the test specimens had L/r ratios of about 10 so global buckling did not occur. A theory of collapse which considers both local and global buckling was developed by Little (1976b) and it is described in the next pages.

The method developed by Little (1976b, 1979) for the determination of collapse load of a pin-ended box column has many similarities to his method for determining the collapse load of a stiffened plate (Section 6.3.1). In both investigations he uses 'effective stress' rather than 'effective width', the length of the member is divided into elements whose M–ϕ–P (moment–curvature–axial load) relationship is derived by considering the axial-stress–axial-strain curves of a simple plate developed by Moxham (1971a), and the investigations cover firstly, the more difficult theoretical analysis and, secondly, a simplified design technique. These methods of analysis and design are described below.

Little's method of analysis of a pin-ended box column is introduced by considering the behaviour of an isolated simply-supported plate. The maximum (failure) average stress σ_{pm} of a plate is a function of b/t, σ_y and the initial imperfections due to the residual welding stress σ_c in the plate (Fig. 6.3.3) and the initial dishing y_0 at the centre. Little found that when test results were plotted as in Fig. 6.3.1 a curve of the form given by the Perry–Robertson equation [eqn (6.3.13)] could be fitted to them. For a column this equation can be written as

$$(\sigma_y - \sigma_m)(\sigma_E - \sigma_m) = \eta \sigma_E \sigma_m \qquad (6.3.76)$$

where σ_y is the yield stress

σ_E is the Euler stress

σ_m is the average stress at failure

η is an imperfection factor $\left(= \dfrac{\Delta_0 A y_s}{I} = \dfrac{\Delta_0}{r} \dfrac{y_s}{r} \right)$.

For an isolated plates eqn (6.3.76) is interpreted as follows

σ_m: σ_m is replaced by the maximum average stress in the isolated plate at failure, σ_{pm}. This is the maximum average stresses reached in plate curves shown in Fig. 6.3.7.

σ_E: For low b/t, σ_E must be replaced by $\sigma_{cr} = \dfrac{\pi^2 E}{3(1-\nu^2)} \left(\dfrac{t}{b}\right)^2$ and for high b/t, σ_E must be replaced by the post-buckled strength σ_{pi} of a perfect plate. This is because σ_{pi} is the asymptote for all imperfect panels (see Fig. 6.3.8).

In the first place, for a given yield stress σ_y and residual stress σ_c, two values of b/t are established.

(a) $(b/t)_y$ is the value of b/t at which $\sigma_{cr} = \sigma_y$, i.e. $1.9\sqrt{(E/\sigma_y)}$ (Fig. 6.3.1) and

(b) $(b/t)_0$ is the extent of the yield plateau, where empirically,

$$(b/t)_0 = 0.65(b/t)_y - 0.3\alpha_p(b/t)_y^2 \tag{6.3.77}$$

where

$$\alpha_p = 0.0015 + 0.45\sqrt{(\sigma_c/E)}. \tag{6.3.78}$$

Thus, σ_E is replaced by σ_{cr} for $b/t < (b/t)_y$ or it is replaced by σ_{pi} for $b/t > (b/t)_y$, where empirically,

$$\sigma_{pi} = \sigma_y \left[\frac{(b/t)_y}{(b/t)} - \left\{ \frac{(b/t)_y}{(b/t)} \right\}^4 + \left\{ \frac{(b/t)_y}{(b/t)} \right\}^5 \right] \tag{6.3.79}$$

η: The effect on the failure load of the initial imperfections σ_c and y_0 were found to be almost independent of the yield stress so η is taken as

$$\eta = 0 \quad \text{for} \quad b/t < (b/t)_0$$

$$= \alpha_p \left[\frac{b}{t} - \left(\frac{b}{t} \right)_0 \right] \text{ for } \left(\frac{b}{t} \right)_0 < \frac{h}{t} < \left(\frac{b}{t} \right)_y$$

$$= \alpha_p \left[\left(\frac{b}{t} \right)_y - \left(\frac{b}{t} \right)_0 \right] \text{ for } \left(\frac{b}{t} \right)_y < \frac{b}{t} \tag{6.3.80}$$

Thus it is seen that η depends upon σ_c and to save designers the tedious task of calculations σ_c three types of construction have been identified as follows.

Class O: $\sigma_c = 5$ MPa, $\alpha_p = 0.0035$ (used for rivetted and unwelded box column)

Class P: $\sigma_c = 25$ MPa, $\alpha_p = 0.0065$ (used for lightly welded construction)

Class Q: $\sigma_c = 73$ MPa, $\alpha_p = 0.0100$ (used for heavily welded construction).

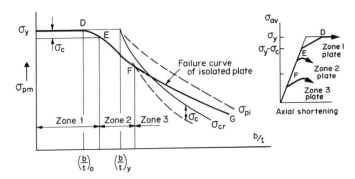

Fig. 6.3.31. Failure curves of isolated plates can be divided into three zones. Typical stress–strain curves are also shown.

In all cases the initial bowing of the plate is $y_0/b = 10^{-3}$.

Further studies by Little into the axial-stress–axial-strain curves of imperfect plates with different b/t ratios enabled him to identify three zones which are located by the points E and F shown in Fig. 6.3.31. Typical axial shortening curves for these three zones are shown in Fig. 6.3.7 which was developed by Moxham using an elasto-plastic Rayleigh–Ritz technique described in Section 6.3.1.

For a welded plate in zone 1 (low b/t) yielding commences when the applied stress σ_{av} reaches $(\sigma_y - \sigma_c)$ (see curve with $b/t = 30$ in Fig. 6.3.7). A welded plate with intermediate b/t (zone 2) has a stress–strain curve which follows the initial elastic line up to σ_{pm} (=maximum stress of the plate element). In zone 3 (high b/t) the curve deviates from the elastic line because of elastic buckling of the plate when $\sigma_{av} = \sigma_{cr} - \sigma_c$.

So far the discussion of Little's method has centred around isolated plate elements. These concepts are now applied to a square box column (Fig. 6.3.32) with the assumed pattern of residual stresses shown. Theoretical M–ϕ–P relationships similar to those shown in Fig. 6.3.9 were first computed for a column of length $0.875b$, this length being approximately equal to the individual plastic buckles at the point of failure observed in many column tests. In these calculations two extreme conditions for the webs were considered.

(a) The webs were assumed to remain flat, i.e. to deform in-plane as perfect elasto-plastic plates. This led to the so-called 'no-web buckling' method.

(b) The webs were assumed to buckle and in so doing to have a stress–strain curve which is the average for the flange plate carrying the highest compressive stress. This led to the so-called 'web buckling' method.

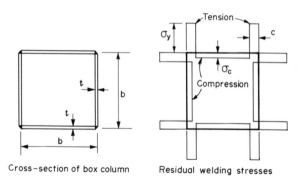

Cross-section of box column   Residual welding stresses

Fig. 6.3.32. Box column analysed by Little (1976b, 1979).

These M–ϕ–P curves were checked by experimental specimens with b/t ratios in the range 26 to 56 and it was found that for low L/r value the 'web buckling' method provided the correct column strength curve but that for high L/r the strength curve was asymptotic to the 'no-web buckling' curve. The tests also confirmed that for low b/t and L/r ratios residual stresses have little effect upon the strength of a box column ($\sigma_m \simeq \sigma_y$) but when $40 < L/r < 100$, class Q columns are significantly weaker than class P columns. This is because of the large decrease in stiffness of a class Q column when the applied axial stress reaches ($\sigma_y - \sigma_c$) (see Fig. 6.3.7).

Having developed the above methods of analysis for plates and box columns, Little then developed design curves for the box columns with pinned ends and zero end-eccentricity. These design curves take into account the interaction which exists between global and local buckling as well as the effects noted above in connection with the tests. The design curves are again based upon the Perry–Robertson equation for columns (eqn (6.3.76)), but the parameters are interpreted in the following way.

σ_E: σ_E is the Euler stress as before.

σ_m: σ_m is the maximum (failure) average stress in the box column.

η: For each column strength curve (Fig. 6.3.13) a plateau with $\sigma_m =$ constant for L/r values has been established. The limit of this plateau is $(L/r)_0$ where

$$(L/r)_0 = \frac{\pi}{5} \sqrt{\left(\frac{E}{\sigma_y}\right)} \tag{6.3.81}$$

For $L/r < (L/r)_0 \;\; \eta = 0$
$\quad L/r > (L/r)_0 \;\; \eta = \text{constant factor} \times [L/r - (L/r)_0]$

where the constant factor depends upon the level of imperfection. In the new European column-curves this factor is given the value of 0.0020, 0.0035 and 0.0055 and these are the values adopted by Little.

σ_y: In the case of a normal column σ_y is the intercept of the column's strength curve with the vertical axis, but for a box column the intercept, hereafter called Σ, will usually be different from σ_y. For high L/r values the criterion of failure adopted in this theory is the attainment of the average stress at which there is a sudden loss of stiffness of the critical plate only. This occurs at the 'knee' stress σ_k of the average-stress–average-strain curves shown in Fig. 6.3.7 for plates, where for

Zone 1 plating $\sigma_k = \sigma_y - \sigma_c$

Zone 2 plating $\sigma_k = \sigma_{pm}$ $\hspace{3cm}$ (6.3.82)

Zone 3 plating $\sigma_k = \sigma_{cr} - \sigma_c$

Thus for high L/r the column strength curve should be selected from Fig. 6.3.13 as that which intercepts the vertical axis at σ_k. This is curve 2 in Fig. 6.3.33 which shows a typical column design curve.

For low L/r the column is assumed to have a length of $0.875b$ and that it fails only when all of the elements reach their failure load. This is the criterion of failure. Hence the appropriate column strength curve selected from Fig. 6.3.13 intercepts the vertical axis at $\sigma_{m\,av}$ (=sum of failure loads of individual plate elements found from plate stresses of Fig. 6.3.7/area of cross section). This establishes curve 1 in Fig. 6.3.33. Thus two strength curves for box columns are established by replacing σ_y by Σ_1 and Σ_2 where

$$\Sigma_1 = \sigma_{m\,av}, \qquad \Sigma_2 = \sigma_y - \sigma_c \hspace{2cm} (6.3.83)$$

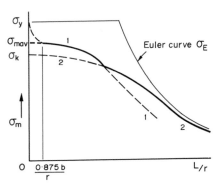

Fig. 6.3.33. Typical strength curve for interaction of local and global buckling of a pin-ended box column.

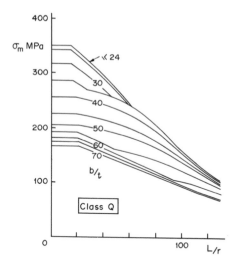

Fig. 6.3.34. Design curves for average stress at collapse of pin-ended box columns. (Little 1976b, 1979.)

and curve 1 is selected from Fig. 6.3.13(d) while curve 2 is selected from Fig. 6.3.13(b). In using these curves the higher one is always used as indicated by the thicker line in Fig. 6.3.33.

The steps in the analysis of a typical box column are as follows:

(a) Find σ_{pm} for the plate element (=peak stress in Fig. 6.3.7) and the zone in which it lies from the construction shown in Fig. 6.3.31.

(b) Select the appropriate values of Σ_1 and Σ_2 (eqn (6.3.83)) and the corresponding curves from Figs. 6.3.13(d) and (b) respectively.

Figure 6.3.34 shows a typical design chart which was established by following these two steps. The discontinuities in the slopes of the curves arise either from the plateau (eqn (6.3.77)) or from the use of separate curves (see eqn (6.3.83)) in the low and high L/r ratios. It is unfortunate that a comprehensive set of such charts is not yet available to designers. Also further experimental work should be carried out in order that the theoretical work can be applied with greater confidence.

6.3.4. Theories of failure and tests of thin-walled box-girders

A box girder is a thin-walled beam of hollow (closed) profile. Its behaviour under uniform and non-uniform torsion and before the onset of buckling has been described in Chapters 1–3. The behaviour of a box girder under both torsion and bending loads up to the point of collapse is described in Chapter 7. In this section one method for determining its collapse load due to bending loads is presented. The form of box girder which is considered here and the terminology used is illustrated in Fig. 6.3.35. It is seen that the upper flange is a plate stiffened in the longitudinal direction by stiffeners which are relatively closely spaced at distance b' apart. Transverse stiffeners which are widely spaced at distance L apart $(L \gg b')$ are used to prevent very long buckles from developing in the upper flange. The transverse stiffeners have to be sufficiently stiff to ensure that if the upper flange buckles the nodal lines, or lines of contraflexure, will coincide with the transverse stiffeners. The design of these stiffeners to achieve this is considered in Chapter 7.

When an increasing bending moment M is applied to a box girder, which has proportions within the range used in practice, eventually its compression flange will buckle in one or more of the following ways:

(a) In those flange panels which buckle upwards the stiffeners may buckle because their unsupported outer edges carry the highest compressive stresses. If the stiffeners are light and open they may buckle by a local torsional instability. This buckling mode usually results in sudden collapse and it should be avoided by designing with stocky stiffeners. If the stiffeners have closed profiles they may initiate collapse of the box girder by local plate buckling of the stiffener but this buckling mode

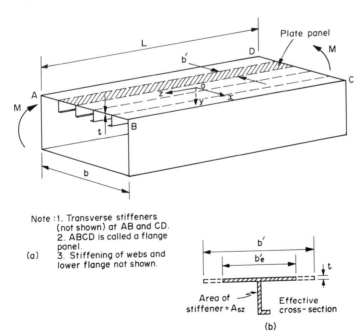

Note : 1. Transverse stiffeners (not shown) at AB and CD.
2. ABCD is called a flange panel.
(a)   3. Stiffening of webs and lower flange not shown.

Fig. 6.3.35. Diagrams to show terminology used in the collapse analysis of a box girder.

usually has some post-buckling strength. In this section it is assumed that the stiffeners are sufficiently stocky that they do not initiate collapse of the box girder.

(b) If the plate panels between the stiffeners have high b'/t ratios (>50 or so) they may buckle locally into roughly square regions ($b' \times b'$, see Fig. 5.1.3). For high b'/t ratios this form of buckling has some post-buckling strength but for lower b'/t ratios collapse occurs before the elastic buckles are fully developed. This aspect of plate behaviour was discussed in Section 6.3.1 (see Fig. 6.3.1).

(c) The flange panels may buckle into one or more buckles of width b and length L/n where n is the number of buckles between adjacent transverse stiffeners. If b and L are roughly equal and there are many stiffeners n will be 1 but if b/L is less than 0.5 it is possible for two or more buckles to develop between adjacent transverse stiffeners. For these latter cases designers have to find the value of n ($n = 1, 2, 3$, etc.) which minimizes the buckling load.

When a box girder is loaded in bending it is possible for modes (b) and (c) above to interact. Thus, suppose a flange panel (Fig. 6.3.35(a)) $b \times L$ begins to buckle downwards in mode (c). This will cause the plate panels to carry a compressive stress which is greater than the average stress in the cross-section and as a result the plate panels are induced to buckle in mode (b). This in turn leads to a reduction in effective stiffness of the flange panel (see Fig. 6.3.35(b)) and this causes it to buckle further.

One further phenomenon which has been observed in laboratory tests is the post-buckling stress distribution shown in Fig. 6.3.36. When the stiffeners are relatively weak there is a superposition of buckling modes (b) and (c), the smaller ripples in the curve being caused by mode (b) buckles and the large dropping dotted curve is caused by the global buckling of the flange panel (mode (c)). As is indicated in the diagram, stronger stiffeners tend to suppress mode (c). Figure 6.3.37 shows such a distribution of stress in the flange of a box girder recently observed by the author. Similar distributions have been published by Massonnet and Maquoi (1973). It is shown in Section 6.3.1 that the elastic behaviour of a stiffened flange panel is influenced by its initial dishing and that it is necessary to use large-deflection theory (Marguerre's equations) in order to derive the relationship between load and deflection. Maquoi and Massonnet (1971) developed such a method for determining the collapse load of a thin-walled box-girder whose compression flange consists of a stiffened plate. The main assumptions used in their analysis are as follows:

(a) The global behaviour of the box girder can be derived by 'smearing' the stiffeners over the surface of the compression flange which is then

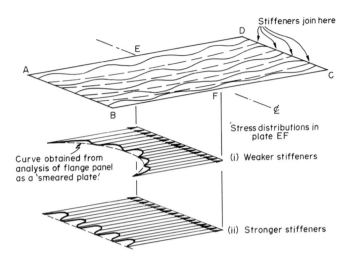

Stiffeners join here

Stress distributions in plate EF

(i) Weaker stiffeners

Curve obtained from analysis of flange panel as a 'smeared plate.'

(ii) Stronger stiffeners

Fig. 6.3.36. A flange panel buckling in modes (b) and (c) simultaneously.

replaced by an equivalent orthotropic plate which is called here the smeared plate or smeared flange. The way this smearing is done by the Guyon–Massonnet–Bares method (Bares and Massonnet 1968) is described below.

(b) Collapse is assumed to occur when the mean longitudinal membrane stress along an unloaded edge reached $\bar{\sigma}_y = 1.065\sigma_y$, i.e.

$$\frac{1}{L}\int_{-L/2}^{L/2} (\sigma_z)_{x=\pm b/2}\, dz = \bar{\sigma}_y. \tag{6.3.84}$$

The fictitious yield stress $\bar{\sigma}_y$ is used instead of σ_y because it was found that on average the following theory underestimated the failure load by 6.5%.

(c) The webs of the box girder are not significant in resisting bending moments.

(d) The flange panel has the same boundary conditions as those of the isolated plate which is designated as case (c) in Section 5.3.3.

(e) All of the material used in the construction is isotropic and has uniform properties. The anisotropy of the flange panel arises only from the geometry of the construction.

The governing equation of an initially imperfect flange panel can be derived in a similar manner to that used to obtain the Marguerre equations (eqns (5.2.9) and (5.2.10)). The full derivation of the following

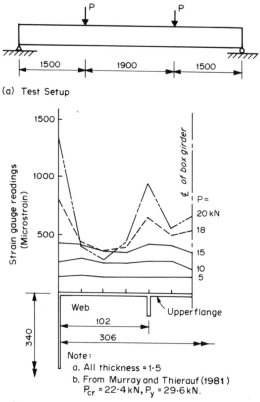

Fig. 6.3.37. Test results showing distribution of membrane stress in the upper flange of a box girder exhibiting combined local and overall flange buckling. (Murray, unpublished.)

equations is presented by Maquoi and Massonnet (1971).

$$\frac{\Phi''''}{B_z} + \frac{2\Phi''\ddot{}}{\bar{B}} + \frac{\Phi\cdots}{B_x} + (1 - \bar{\nu}^2)[y''\ddot{w} - 2y'\dot{}w'\dot{} + \ddot{y}w'' + w''\ddot{w} - w'\dot{}^2] = 0 \tag{6.3.85}$$

$$\bar{D}_z w\cdots + 2\bar{D}w''\ddot{} + \bar{D}_x w'''' - [\Phi\ddot{}(y + w)'' - 2\Phi'\dot{}(y + w)'\dot{} + \Phi''(y + w)\ddot{}]$$

$$- Y = 0 \tag{6.3.86}$$

where the section and material properties are defined as follows†

$$B_z = E(A_{sz} + b_z t)/b_z = E \times \text{smeared thickness in } z\text{-direction}$$

† This notation does not conform with that used by Maquoi and Massonnet. *B* and *D* have been interchanged to agree with the notation of Chapter 5 of this book.

$B_x = E(A_{sx} + b_x t)/b_x = E \times$ smeared thickness in x-direction

$$\bar{B} = \frac{1 - \nu}{1 - \dfrac{\nu B^2}{B_x B_z}} B$$

$$B = \frac{Et}{1 - \nu^2}$$

$$\bar{\nu} = \frac{B}{\sqrt{(B_x B_z)}} \nu$$

$$D = \frac{Et^3}{12(1 - \nu^2)} \qquad\qquad (6.3.87)$$

$D_x, D_z =$ flexural rigidities of the stiffened flange plate in
the x- and z-directions, respectively.

$$\bar{D} = D + \tfrac{1}{2}(D_{xz} + D_{zx}) + \frac{\nu}{1 - \bar{\nu}^2} e_x e_z B = \text{sum of Saint Venant}$$

torsional rigidities of the plate end stiffeners and an
allowance for the effect of the eccentricities e_x and e_z

$$\bar{D}_z = D_z - \frac{\bar{\nu}^2}{1 - \bar{\nu}^2} e_z^2 B_z$$

$$\bar{D}_x = D_x - \frac{\bar{\nu}^2}{1 - \bar{\nu}^2} e_x^2 B_x.$$

The boundary conditions are defined in Fig. 6.3.38 where the overall

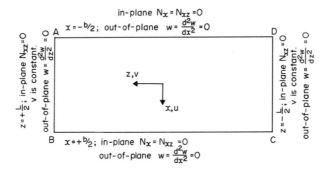

Fig. 6.3.38. Boundary conditions of flange panel assumed in Maquoi and Massonnet's (1971) theory of box girders.

shortening of the flange panel is given by the expression

$$2(v)_{z=L/2} = \int_{-L/2}^{L/2} \left[\frac{1}{1-\bar{\nu}^2} \left(\frac{\Phi''}{B_z} - \frac{\bar{\nu}}{\sqrt{(B_z B_x)}} \Phi^{\cdot\cdot} \right) \right.$$
$$\left. - \tfrac{1}{2}(y^{\cdot} + w^{\cdot})^2 + \tfrac{1}{2}y^{\cdot 2} \right] dz \quad (6.3.88)$$

which should be independent of x. The surface load Y is assumed to be zero.

The method of solution of eqns (6.3.85) and (6.3.86) follows the general pattern of that used to solve the Marguerre equations (see case (c), Section 5.3.3) but the assumed form for Φ used by Maquoi and Massonnet is slightly different, that is,

$$\Phi = A_1 \cos \frac{2\pi z}{L} + A_2 \cos \frac{2\pi x}{b} + p \frac{x^2}{2}$$
$$+ \cos \frac{2\pi z}{L} [A_3 \cosh \beta_1 x + A_4 \cosh \beta_2 x] \quad (6.3.89)$$

All of the boundary conditions, including eqn (6.3.88), are satisfied when

$$A_1 = -\frac{B_x(1-\bar{\nu}^2)L^2 w_0(w_0+2y_0)}{32b^2}$$

$$A_2 = -\frac{B_z(1-\bar{\nu}^2)b^2 w_0(w_0+2y_0)}{32L^2}$$

$$A_3 = \frac{j_2 \sinh \pi j_2 \, b/L}{j_1 \sinh \dfrac{\pi j_1 b}{L} \cosh \dfrac{\pi j_2 b}{L} - j_2 \sinh \dfrac{\pi j_2 b}{L} \cosh \dfrac{\pi j_1 b}{L}} A_1 = \mu_1 A_1 \qquad (6.3.90)$$

$$A_4 = \frac{j_1 \sinh \pi j_1 \, b/L}{j_1 \sinh \dfrac{\pi j_1 b}{L} \cosh \dfrac{\pi j_2 b}{L} - j_2 \sinh \dfrac{\pi j_2 b}{L} \cosh \dfrac{\pi j_1 b}{L}} A_1 = \mu_2 A_1$$

$$j_1 = [B_z/\bar{B}]^{\frac{1}{2}}\{1 + [1 - (\bar{B}^2/B_z B_x)]^{\frac{1}{2}}\}^{\frac{1}{2}}$$

$$j_2 = [B_z/\bar{B}]^{\frac{1}{2}}\{1 - \bar{B}^2/B_z B_x]^{\frac{1}{2}}\}^{\frac{1}{2}}$$

$w_0, y_0 =$ values of w and y at the origin.

When these values are substituted into eqn (6.3.89), which is then used to find the axial shortening of the flange panel from eqn (6.3.88),

$$2(v)_{z=L/2} = \frac{pL}{B_z(1-\bar{\nu}^2)} - \frac{\pi^2}{8L} w_0(w_0+2y_0). \qquad (6.3.91)$$

where p is the average force per unit width of plate. It is noted that this expression is independent of x so that the nodal lines at the ends of the flange panel remain straight which was a required boundary condition.

As in Chapter 5 it remains to determine the relationship between the axial load parameter p and the deflection w_0 at the centre of the flange panel and this is again achieved by using Galerkin's method with eqn (6.3.86). After much algebra,

$$
\begin{aligned}
(\sigma_z)_{av} = & -\frac{w_0}{w_0 + y_0} \frac{\pi^2}{L^2 t} (\bar{D}_z + 2\bar{D}L^2/b^2 + \bar{D}_x L^4/b^4) \\
& -\frac{\pi^2 B_x (1-\bar{\nu}^2) L^2}{16 b^4 t} \left(1 + \frac{b^4 B_z}{L^4 B_x}\right) w_0 (w_0 + 2y_0) \\
& -\frac{\pi^2 B_x (1-\bar{\nu}^2) L^2}{16 b^4 t} \left(\frac{\mu_1 L}{\pi j_1 b} \sinh \frac{\pi j_1 b}{L} + \frac{\mu_2 L}{\pi j_2 b} \sinh \frac{\pi j_2 b}{L}\right) \\
& \hspace{6cm} w_0 (w_0 + 2y_0) \quad (6.3.92)
\end{aligned}
$$

This equation is the load–deflection relationship of an orthotropic plate in the elastic range. When the failure criterion (eqn (6.3.84)) is satisfied p is defined as having reached its ultimate value p_f. With $\sigma_z = N_z/t$ given by eqn (5.2.4) the value of p_f is given by

$$
\frac{4\pi^2}{b^2} A_2 + p_f = -\bar{\sigma}_y t \left(1 + \frac{\bar{m} A_{sz}}{bt}\right) \tag{6.3.93}
$$

where

$$
\bar{m} = b/b' \tag{6.3.94}
$$

The value of A_2 (see eqn (6.3.90)) is a function of w_0 whose value at the point of failure, that is, w_{0f}, is not known. Therefore w_{0f} is found by substituting for A_2 from eqn (6.3.90) and for p_f from eqn (6.3.93) which gives the following equation for w_{0f}.

$$
\begin{aligned}
& \frac{w_{0f} \bar{\sigma}_{cr} (1 + \bar{m} A_{sz}/bt)}{w_{0f} + y_0} - \frac{\pi^2 B_x L^2 (1-\bar{\nu}^2)}{16 b^4 t} (1 + 3B_z b^4/B_x L^4) w_{0f}(w_{0f} + 2y_0) \\
& -\frac{\pi^2 B_x L^2 (1-\bar{\nu}^2)}{16 b^4 t} \left(\frac{\mu_1 L}{\pi b j_1} \sinh \frac{\pi j_1 b}{L} + \frac{\mu_2 L}{\pi b j_2} \sinh \frac{\pi j_2 b}{L}\right) w_{0f}(w_{0f} + 2y_0) \\
& \hspace{7cm} = -\bar{\sigma}_y \left(1 + \frac{\bar{m} A_{sz}}{bt}\right) \quad (6.3.95)
\end{aligned}
$$

where

$$
\bar{\sigma}_{cr} = -\frac{\pi^2}{\left[L^2 t \left(1 + \dfrac{\bar{m} A_{sz}}{bt}\right)\right]} \left(\bar{D}_z + \frac{2L^2}{b^2} \bar{D} + \frac{L^4}{b^4} \bar{D}_x\right). \tag{6.3.96}
$$

The effectiveness ρ_t of the smeared flange plate at the point of failure is defined as the ratio of the average stress at the loaded end to the fictitious yield stress. It is shown by a simple integration of N_z that the average stress is $p/t(1+\bar{m}A_{sz}/bt)$, and therefore at the point of failure

$$\rho_{tf} = \frac{p_f}{\bar{\sigma}_y t\left(1+\dfrac{\bar{m}A_{sz}}{bt}\right)} = 1 - \frac{\pi^2 w_{0f}(w_{0f}+2y_0)B_z(1-\bar{\nu}^2)}{8tL^2\bar{\sigma}_y\left(1+\dfrac{\bar{m}A_{sz}}{bt}\right)} \tag{6.3.97}$$

where p_f has been found from eqns (6.3.95) and (6.3.93). This expression is the general formula for the effectiveness of an orthotropic plate at the point of failure. For the special case of an isotropic flange plate $j_1 = j_2$, $\mu_1 = \mu_2$, $B_z = B_x$, $\bar{\nu} = \nu$, $b = L$, and $A_{sz} = 0$. When these values and eqn (6.3.20) are substituted into eqn (6.3.97) the following expression is obtained.

$$(\rho_{tf})_{\text{isotropic plate}} = 1 - \frac{\pi^2(m^2-1)}{8}\left(\frac{E}{\bar{\sigma}_y}\right)\left(\frac{y_0}{t}\right)^2\left(\frac{t}{b}\right)^2 \tag{6.3.98}$$

This expression agrees with that for K_{bs} (eqn (6.3.23)) obtained by Horne and Narayanan (1975) except that they use the true yield stress and not a fictitious value. However, in the case of a flange plate with longitudinal stiffeners the effectiveness is further reduced by another factor which accounts for local buckling of the plate panels and stiffeners (Fig. 6.3.36). This factor in the global effectiveness ρ_{gf} of the flange is given the symbol ρ'_f. The relationship

$$\rho_{gf} = \rho_{tf}\rho'_f \tag{6.3.99}$$

expresses the concept that at failure the global effectiveness is the product of the effectiveness due to the smeared flange and that caused by local buckling.

If the stiffeners are stocky it can be shown from eqn (6.3.97) that $\rho_{tf} \approx 1$ and the stress distribution in the flange at the failure load is that shown in the lower diagram of Fig. 6.3.36. For this case Maquoi and Massonnet (1971) recommend the use of an empirical rule due to Faulkner for the ratio of effective width b'_e at failure to actual width b' of the plate panels. For present purposes this rule is written in the following notation.

$$\frac{b'_e}{b'} = \frac{2t}{b'}\sqrt{\left(\frac{E}{\rho_{tf}\sigma_y}\right) - \frac{t^2 E}{b'^2\rho_{tf}\sigma_y}} \tag{6.3.100}$$

where ρ_{tf} is found from eqn (6.3.98) with $b = b'$. Hence, for stocky stiffeners

$$\rho_{gf} = \rho'_f = \frac{\bar{m}b'_e\sigma_{zf}t + \bar{m}A_{sz}\sigma_{zf}}{(\bar{m}b't + \bar{m}A_{sz})\sigma_{zf}} = \frac{\dfrac{b'_e}{b'} + \dfrac{A_{sz}}{b't}}{1 + \dfrac{A_{sz}}{b't}} \qquad (6.3.101)$$

where σ_{zf} is the axial stress at failure.

When the stiffeners are weaker the stress distribution is that shown in the upper diagram of Fig. 6.3.36. For this case Maquoi and Massonnet again recommend the use of the Faulkner formula but it is first necessary to find the distribution of stress in the smeared flange panel at failure. This is done by solving eqn (6.3.95) for w_{0f}, establishing Φ from eqns (6.3.93) and (6.3.89) and the distribution of σ_{zf} from eqn (5.2.4). The effective width b'_e for each plate panel is found from eqn (6.3.98) by using the average stress σ_{zf} along a line midway between each pair of stiffeners. This enables the effectiveness factor ρ'_f to be calculated and the global effectiveness ρ_g is then found from eqn (6.3.99). The average stress at failure in the flange panel is

$$(\sigma_z)_{av} = \rho_g\sigma_y \qquad (6.3.102)$$

and hence the bending moment of the box girder can be evaluated.

Maquoi and Massonnet have checked the above method against test results published by Dubas (1972) and three results published by Massonnet and Maquoi (1973). The b'/t ratio of these tests were in the range 48 to 71 and their analysis suggests that the method predicts failure with a small margin of safety.

There is generally a shortage of test results to ascertain the reliability of their method at present. Eight box girders were tested by Dowling et al. (1973) and their results highlight the fact that collapse may also arise from buckling of the web plates or by a complex interaction of buckling in the flange, webs, and other components. Roderick and Ings (1977) report on a box girder tested as a cantilever and compare the experimental collapse load with the stiffened-plate theories of Horne and Narayanan (1975), Dwight and Little's (1976) and Little's (1976a) methods described in Section 6.3.1, and Murray's theory (1975). It appears that the first method gave best agreement with test results but firm conclusions cannot be drawn from a single test.

Models of box-girder bridges similar to those of Dubas (1971) were tested by Steinhardt (1973, 1975) and his co-worker Valtinat (1975). They developed an energy method for determining the collapse load of a box girder also based upon Faulkner's expression for effective width. Diagrams have been developed to assist designers to evaluate the stresses at the edge of the plate or at the free edges of the stiffeners so that the failure criteria, that is, yield at these points causing failure, can be checked.

Six box girders, measuring 8.6 m span by 1.5 m high by 0.1 m wide and using stiffened plates, were tested to failure by Mikami, Dogaki and Yonezawa (1980). They compared the experimental failure loads of two girders only with those predicted by twenty-two published theories. It is shown that the theories described in this book, i.e. those due to Horne and Narayanan, Dwight and Little, Maquoi and Massonnet, and Allen, predict the failure load fairly accurately and generally on the safe side.

In closing this section it is worth emphasising again that as for box columns the number of experiments carried out on box girders is relatively small and further test results are needed to increase confidence in the theories described.

7

THIN-WALLED STRUCTURES IN CODES OF PRACTICE

7.1. INTRODUCTION

The purpose of this chapter is to provide a link between some of the theories presented in the earlier chapters and some codes of practice which deal with thin-walled structures. It is not possible here to review in detail all or even one of the codes relevant to thin-walled structures. It is only intended to discuss the general way that the theories have been used by some codes. However, before doing this the general philosophy and purpose of codes are briefly discussed. A feature of all codes is that they are not permanent but rather they are like organic things undergoing changes during their transitory life. Smith (1980) describes the life of a code as follows.

> 'The life of a code may, without too much parody, be roughly divided into three parts, respectively those of misunderstanding, acceptance, and preservationist. In the first stage, endless arguments take place in the design office as to the exact meaning and purpose of this clause or that. In the second stage the arguers reach a common understanding, and a body of case law, or lore, is built up. In the third stage, older engineers have forgotten the ambiguities, and younger ones have never been aware of them. All want to cling to the good old book.'

Codes are often written or interpreted as if they are both a legal document, which must be followed in detail by the designer, and a document which embodies all that is considered to be 'good' practice and theory within its scope. The better codes clearly separate these two aspects and define some parts as mandatory and others as non-mandatory. For example, the mandatory parts of the Merrison rules (1973) for box girders are those dealing with the various kinds of loading, with certain aspects of the way the analysis is to be carried out, with the materials used, and with the standard of workmanship to be achieved. The other aspects dealing with the design rules and theory and general design information are not mandatory but the rules state that they (the rules)

> 'are not necessarily comprehensive and that they must be used only under the direction of designers competent and experienced in design of steel-plated structures. They must not be used outside the limits given in this document.
>
> Alternative methods of strength assessment may be used, provided that they

are soundly based on theoretical or experimental evidence demonstrating that they provide lower-bound estimates of strength or unserviceability as appropriate (approximating to the mean, minus two standard deviations). Theoretical methods must be verified by comprehensive comparison, either with test results or with the methods contained in Parts II and III (of the Rules), or other methods well tried against the results of properly conducted tests. Empirical methods must be based on the results of a sufficient number of suitable and comprehensive tests, due account being taken of scale, actual material properties and imperfections, and of variability of results.'

The Foreword to BS 5400 similarly states that some parts of it

'are in specific terms. It is intended that the application and interpretation of BS 5400 in design is entrusted to appropriately qualified and experienced chartered engineers, and that construction is carried out under the direction of appropriately qualified supervisors.'

In Germany it was the practice for many years to use the codes rigidly as both legal documents and as documents of good practice but there has been a reaction against tying the hands of designers and builders so tightly. Scheer, Nölke, and Gentz (1979) writing in the Introduction to the 1979 DASt-Richtlinie 012 state that,

'The expert commission responsible for the building code decided in 1979 to dispense with an Introduction to the DASt-Richtline 012 which controlled building. This step is to be welcomed if it leads to a liberalization of building permit procedures. Assuming this is so, it is possible to relax the requirement that the parts of the code dealing with exact technical matters must be considered as rules in the art of building.' [Author's translation.]

The purpose of a code is therefore to guide the designer, showing him what is current good practice but allowing him to use more advanced theories and methods where these can be justified by comparison with the results of properly conducted tests or with other well-established methods.

Design rules are usually a much simplified version of the more exact theories upon which they are based and two aspects are always uppermost in the minds of drafting committees. The first of these is that the rules must be safe and err on the conservative side. This leads to structures which may be slightly heavier than they need to be. The second is that the rules should be easy to use. This means that many shortcuts may be taken, empirical rules may be employed, and the links with the theory behind the rules in the code may become tenuous and difficult to follow. It is an unfortunate feature of most codes that they omit even a list of references to source material, much less a commentary upon how the rules were developed from it.

The links with the background theory fall into roughly three classes. In the first the theory is followed exactly. Examples of this class are few but they include simple tension, compression, and bending members in which buckling problems are deliberately excluded. In the second the theory is simplified by the introduction of empirical rules in a fairly minor way. The Perry–Robertson formula for column failure which is presented in Section 6.3, is an example of this. The empiricism arises because it is assumed that failure occurs when the yield stress is reached at one point in the column and because the amount of imperfection assumed is related empirically to the slenderness (L/r) of the column. Both of these rules are conservative approximations to the truth but the remainder of the theory is exact to a first order of accuracy. In the third class the theory is either lost in antiquity, the theory has only been used to develop simple rules by some experimental or parametric studies, or the theory has not been developed. Some of the empirical rules for calculating residual stresses (see Chapter 6) fall into this class. Some of the restrictions on the dimensions of various structural elements are based upon past experience rather than rigorous analysis. This does not mean that the methods incorporated into papers and codes should not be used. They are, in fact, the best that the current state-of-the-art can offer the designer. They should be used with caution and their range of validity should not be exceeded. This last class of link usually occurs in those aspects of design which are not a prime safety consideration or which have a large number of variables influencing structural response to loads.

In the codes considered in this chapter the reader will find all three categories of link between theory and design rule.

Ever since codes were first drafted, the question of safety has held a prime place. The safety of a steel structure used to be allowed for in codes of practice by restricting the level of maximum stress to a value less than the yield stress. Thus

$$\sigma_{\max} = \sigma_y/f \qquad (7.1.1)$$

where f was called the factor of safety. This approach led to a considerable variation from one structure to the next in the margins of safety they have against actual collapse. It seemed desirable that all structures should have the same load factor λ, where the load factor is defined as the following ratio

$$\lambda = \text{collapse load/design load} \qquad (7.1.2)$$

The advent of plastic design in the late 1940s with its emphasis on the determination of collapse load allowed this approach to design to flourish. However, it was recognized in the 1960s that it is more logical to introduce two fundamental changes.

The first of these was to require designers to check their designs for two limiting conditions, or so-called *limit states*. These are the *serviceability limit state* and the *collapse limit state*. The serviceability limit state occurs when the worst combination of working loads in the design life of the structure is acting. Under these conditions there should be no excessive deflection, permanent deformation, failure of finishes, and other unsightly effects. This means that the designer must have a good understanding of linear elastic analysis and, for thin-walled structures, of non-linear elastic buckling theory. A check on the collapse limit state is made by comparing the estimated collapse load or loads with the worst combination of design loads which have first been multiplied by suitable load factors. Thus the designer is also required to have a good understanding of the plastic theory of steel structures.

The second fundamental change was introduced because of the variations which inevitably occur, firstly, in the material, e.g. variations in strength, size of members, and time-dependent properties; secondly, in the loading, e.g. variability in traffic, wind and temperature loads, accidental overloads, and design inaccuracies; and thirdly, in other parameters such as the assumed life of the structure, accuracy of construction, and changes to the form of the structure during its life. These matters have been dealt with by using probability theory. By considering the variability of all of these parameters it is possible to arrive at partial safety factors, some of which are applied to the calculated strength (or resistance) of the structure and others to the calculated design loads. Figure 7.1.1 shows for example the partial factor format adopted in Part 1 of the British Standard BS 5400 Bridge Code (1978). It is based upon the principles of limit state design presented in ISO 2394. The designer must ensure that the factored resistance R^* of the structure exceeds the factored loads S^* where the partial factors γ_{m1}, γ_{m2}, etc. are given in that code. The details of the derivations of these partial factors from probability theory are explained in a paper by Flint, Smith, Baker, and Manners (1980). For the section (i.e. Part 3) of BS 5400 which deals with steel bridges, γ_{f3}, γ_{m1}, γ_{m2} are combined into a single factor γ_m, for the sake of simplicity. Hence the particular form of the design equation adopted is simplified to

$$R^* = \frac{1}{\gamma_m} \text{fn} \cdot \{\sigma_y, \text{ other parameters}\} > S^* = \text{effects of } \gamma_{fL} Q_K \qquad (7.1.3)$$

γ_m has been derived for different kinds of structures and structural elements, an average value being about 1.3, and γ_{fL} depends upon the type of loading (dead load, live load, wind load, etc.) and the limit state (serviceability or collapse) being considered. γ_{fL} tends to be about 1.0 for the serviceability limit state and varies between 1.05 (for dead loads) and 1.75 (for superimposed dead loads) for the collapse limit state.

The next two sections discuss problems of common interest to all codes.

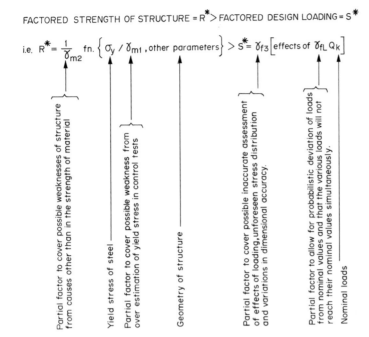

FACTORED STRENGTH OF STRUCTURE = $R^* >$ FACTORED DESIGN LOADING = S^*

i.e. $R^* = \dfrac{1}{\gamma_{m2}}$ fn. $\left\{ \sigma_y / \gamma_{m1} , \text{other parameters} \right\} > S^* = \gamma_{f3} \left[\text{effects of } \gamma_{fL} Q_k \right]$

Partial factor to cover possible weaknesses of structure from causes other than in the strength of material

Yield stress of steel

Partial factor to cover possible weakness from over estimation of yield stress in control tests

Geometry of structure

Partial factor to cover possible inaccurate assessment of effects of loading, unforeseen stress distribution and variations in dimensional accuracy.

Partial factor to allow for probabilistic deviation of loads from nominal values and that the various loads will not reach their nominal values simultaneously.

Nominal loads

Fig. 7.1.1. Partial factor format for the design of bridges given in BS 5400 Part 1 (1978).

7.1.1. Adjustment of graphs for variations in yield stress

In presenting formulae and graphs for the evaluation of failure loads it will often be seen that parameters such as $b/t\sqrt{(\sigma_y/355)}$ appear. The following is a brief explanation and justification of the factor $\sqrt{(\sigma_y/355)}$ which obviously adjusts b/t when the yield stress differs from a standard value of 355 MPa. Reference to Fig. 6.3.1 shows that the failure stress of a plate panel is a function of b/t. The value of b/t at which the yield stress is equal to the critical stress is $(b/t)_y$, given by the following equation (see eqn (5.3.24))

$$\sigma_y = \frac{k\pi^2 E}{12(1-\nu)^2} \left(\frac{t}{b} \right)_y^2$$

i.e. $\left(\dfrac{b}{t} \right)_y \sqrt{\sigma_y} = \pi \sqrt{\left[\dfrac{kE}{12(1-\nu^2)} \right]}.$ \hfill (7.1.4)

Since the right-hand side is a constant it is seen that failure curves like those in Fig. 6.3.1 can be drawn for the standard yield stress of 355 MPa and the abscissa becomes $b/t\sqrt{(\sigma_y/355)}$. A similar technique is used for the overall buckling of panels which are assumed to behave like pin-ended

Euler columns. Thus

$$\sigma_y = \frac{\pi^2 E}{(L/r)_y^2}$$

i.e. $\left(\dfrac{L}{r}\right)_y \sqrt{\sigma_y} = \pi \sqrt{E}.$ (7.1.5)

7.1.2. Elastic buckling of plate panels with complex in-plane stresses

The critical stress of a perfectly flat thin steel plate can be found by using one of the methods described in Chapter 5. Even though the pattern of in-plane stresses is quite complex the finite difference method can be used. However, designers require charts or formulae which give them the critical stress directly. This section reviews the results of studies of this problem for simply supported rectangular plates.

The most general combination of stresses considered here is that shown in Fig. 7.1.2 where it is seen that the stress distributions are at most linear and that bending stresses in the x-direction are not considered. In the following analysis it is assumed that these stresses increase in the same ratio as the plate is loaded. Thus suppose the applied stresses are σ_x, σ_z, σ_b and τ then the problem is to find the critical value of a factor λ by which each of these stresses is multiplied so that the plate is just on the point of buckling. At this point the following equations can be written

$$\sigma'_{x\,cr} = \lambda_{cr}\sigma_x; \qquad \sigma'_{z\,cr} = \lambda_{cr}\sigma_z; \qquad \sigma'_{b\,cr} = \lambda_{cr}\sigma_b; \qquad \tau'_{cr} = \lambda_{cr}\tau \qquad (7.1.6)$$

where $\sigma'_{x\,cr}$, etc. are the critical values of σ_x, etc. when they are acting together.

There are many existing solutions of classical buckling problems which are the values of the critical values of σ_x, σ_z, etc. when they act alone.

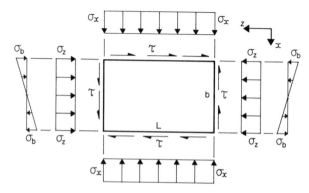

Fig. 7.1.2. Combination of in-plane stresses acting on a rectangular plate.

Here those values are given the symbols $\sigma_{x\,cr}$, $\sigma_{z\,cr}$, $\sigma_{b\,cr}$, and τ_{cr} and they are related to $\sigma'_{x\,cr}$, $\sigma'_{z\,cr}$, etc. by the ratios R_x, R_z, R_b, and R_s, thus

$$R_x = \frac{\sigma'_{x\,cr}}{\sigma_{x\,cr}}\,; \qquad R_z = \frac{\sigma'_{z\,cr}}{\sigma_{z\,cr}}\,; \qquad R_b = \frac{\sigma'_{b\,cr}}{\sigma_{b\,cr}}\,; \qquad R_s = \frac{\tau'_{cr}}{\tau_{cr}}\,. \tag{7.1.7}$$

A relationship between R_x, R_z, R_b, and R_s will establish a surface in four-dimensional space which intersects the coordinates axes at $R_x = 1$, $R_z = 1$, $R_b = 1$, and $R_s = 1$. This surface defines the combinations of σ_x, σ_z, etc. which will just cause buckling of the plate. Because these stresses are increased in constant ratios to one another as the loading parameter λ is increased the stress combination is represented by points on a straight line through the origin. When the point is reached at which this line intersects the buckling surface the plate will buckle. Unfortunately a four-dimensional diagram cannot be drawn, but as an example Fig. 7.1.3 suggests the form of the buckling surface (when $\sigma_x = 0$) and the meaning of λ_{cr}, $\sigma'_{z\,cr}$, etc. Point A represents a point when the plate is in its pre-buckled state. When the stresses at A are factored by λ_{cr} the point B on the buckling surface is reached and buckling occurs.

The form of the buckling surface for the four stress components has not yet been established but it has been for two stress components at a time. In other words the curves such as CD, DE, and CE have been developed. The shape of these curves is a function of the aspect ratio b/L. Figure 7.1.4 shows a typical comparison between a set of theoretical buckling

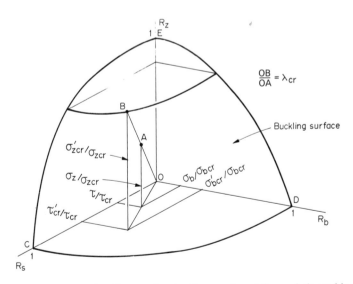

Fig. 7.1.3. Typical interaction diagram showing the meaning of the symbols used in the text (Section 7.2).

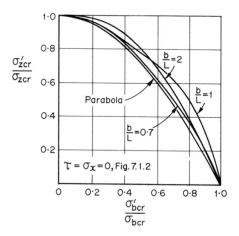

Fig. 7.1.4. Comparison of theoretical interaction curves for buckling of a rectangular plate and a parabola. (Stüssi, Kollbrunner, and Wanzenried 1953.)

curves derived by Stüssi, Kollbrunner, and Wanzenried (1953) and a parabola which obviously can be used as a good approximation to the curve FD in Fig. 7.1.3. Table 7.1.1 presents a list of formulae which give this and similar relationships. This table covers all possible cases of in-plane stresses taken two at a time but it may be necessary to rotate the

Table 7.1.1. Interaction formulae for buckling of a rectangular plate (refer Figs. 7.1.2 and 7.1.3).

Case	Approx. formula	Source	Eqn no.
$\tau = \sigma_x = 0$	$\dfrac{\sigma'_{z\,cr}}{\sigma_{z\,cr}} + \left(\dfrac{\sigma'_{b\,cr}}{\sigma_{b\,cr}}\right)^2 = 1$ (Curve ED)	Stüssi, Kollbrunner, and Wanzenried (1953)	(7.1.8)
$\sigma_b = \sigma_x = 0$	$\dfrac{\sigma'_{z\,cr}}{\sigma_{z\,cr}} + \left(\dfrac{\tau'_{cr}}{\tau_{cr}}\right)^2 = 1$ (Curve EC)	Chwalla (1936)	(7.1.9)
$\sigma_z = \sigma_x = 0$	$\left(\dfrac{\sigma'_{b\,cr}}{\sigma_{b\,cr}}\right)^2 + \left(\dfrac{\tau'_{cr}}{\tau_{cr}}\right)^2 = 1$ (Curve CD)	Timoshenko (1935) and Schleicher (1955)	(7.1.10)
$\sigma_b = \tau = 0$	$\dfrac{\sigma'_{z\,cr}}{\sigma_{z\,cr}} + \dfrac{\sigma'_{x\,cr}}{\sigma_{x\,cr}} = 1$	Schaefer (1934) and Kollbrunner and Meister (1958)	(7.1.11)

coordinate axes through 90° to obtain the two cases which at first sight seem to be missing.

The general analysis of plates with more than two stresses has not yet been carried out. Following a suggestion by Stüssi (1955) one may be able to combine the equations. For example, the first three cases can possibly be combined to give the empirical equation

$$\frac{\sigma'_{z\,cr}}{\sigma_{z\,cr}} + \left(\frac{\sigma'_{b\,cr}}{\sigma_{b\,cr}}\right)^2 + \left(\frac{\tau'_{cr}}{\tau_{cr}}\right)^2 = 1 \tag{7.1.12}$$

for the case when only σ_x is always zero, but this has not yet been fully proved. Some of the results of the analyses by Harding, Hobbs, and Neal referred to in Chapter 6 have been used as a check (Horne, 1980) and it has been shown that eqn (7.1.12) gives satisfactory estimates of buckling stresses.

An attempt to develop a general formula arises from the work of Schaefer (1934) who proved that the buckling surface when seen from outside is always convex. This means that it is always conservative to use the Dunkerly formula

$$\frac{\sigma'_{x\,cr}}{\sigma_{x\,cr}} + \frac{\sigma'_{z\,cr}}{\sigma_{z\,cr}} + \frac{\sigma'_{b\,cr}}{\sigma_{b\,cr}} + \frac{\tau'_{cr}}{\tau_{cr}} = 1 \tag{7.1.13}$$

This is the equation of a plane surface and it often gives very conservative estimates of the buckling stresses. Equation (7.1.11) is a special case of eqn (7.1.13) and it is known as Dunkerly's line. Kollbrunner and Meister (1958) show that in the middle-range Dunkerly's line underestimates the buckling stresses by a substantial margin and suggest other formulae. However, Dunkerly's line has been adopted by some codes, notably the German Code DIN 4114, because of its simplicity and because it is known that it always gives safe results. It will be seen later that some codes treat this problem in other ways. Dunkerly's formula can be written in a general form for a plate of any shape with N applied stresses.

$$\sum_{n=1}^{N} \frac{\sigma'_{n\,cr}}{\sigma_{n\,cr}} = 1 \tag{7.1.14}$$

where $\sigma'_{n\,cr}$ and $\sigma_{n\,cr}$ are the critical values of the nth stress component when the stresses are applied simultaneously and when the nth component is applied alone, respectively. In this form, Dunkerly's formula is very powerful because it can be used by designers to obtain a conservative estimate of the buckling stresses of plates with quite complicated geometry.

7.2. A BRIEF REVIEW OF SOME DESIGN RULES IN THE MERRISON RULES (1973)

The Merrison rules (1973) for the design of box girder bridges were drafted in Britain as a result of the need for a comprehensive box-girder code which hitherto had not existed. They were described as interim rules and have now been superseded by the code BS 5400 which covers not only box girder bridges but all other kinds of bridge. Although the use of the Merrison rules is in decline it is instructive to discuss some aspects of them because they represented a great step forward when they were first published in 1973. They introduced many advanced concepts of plate buckling theory, the effects of initial stresses and plate imperfections, tension field theory, fatigue, and so on into codes and they paved the way for the more ready acceptance of BS 5400. The Merrison rules (1973) have been criticized for being so complicated. Among other things there are thirty eight pages of notation which must be justified on the grounds that the problems they tackle are themselves complicated and beyond the scope of normal structural engineering.

The Merrison rules (1973) consist of four parts as follows.

Part I   Loading and general design requirements.
Part II   Design rules.
Part III   Basis for the design rules in Part II and for the design of special structures not within the scope of Part II.
Part IV   Materials and workmanship.

This section is concerned mainly with some aspects of Parts II and III, which are the non-mandatory sections of the Rules.

In the design of a box girder one of the first steps is to carry out a so-called global analysis whose purpose is to evaluate the bending moments, torques, stress distributions, and so on in the girder at working loads. Because of certain special effects, such as shear lag, in box girders, not all of the cross-section may be effective in resisting the applied forces. Furthermore the plating of a box girder which is effective carries not only bending and shear stresses but warping, distortional, and other stresses. Section 7.2.1 describes the way these effects are allowed for in the Merrison rules (1973). In later parts of a design it is necessary also to check each plate panel for its maximum stress and stiffness, and Section 7.2.2 shows how this is carried out. In Section 7.2.3 the method adopted by the Merrison rules (1973) for analysing panels which carry complex in-plane stress patterns (see Section 7.1.2) is described. The transverse stiffeners in a stiffened panel carrying predominantly longitudinal stresses control the buckling length and therefore the buckling load. Their stiffness must satisfy certain requirements in order to do this and this problem is treated in Section 7.2.4.

7.2.1. Structural action of box girders

The structural action of a box girder when it is loaded has been comprehensively described by Horne (1977) who was a member of the drafting committee for the Merrison rules (1973). Many of the formulae quoted in the rules have also been justified by Horne (1975). This section is a brief summary only of the general behaviour of box girders when they carry their more usual loading patterns. In order to describe the various components of structural action considered by the Merrison rules (1973) the single box girder illustrated in Fig. 7.2.1(a) is considered. The serviceability check for yielding is carried out on such a girder by calculating eight elastic stress components and comparing their combined effect against a von Mises–Hencky yield criterion (eqn (6.3.44)). The load applied to the web in Fig. 7.2.1(a) is first replaced by the three equivalent sets of loads shown in Fig. 7.2.1(b) and the eight components of stress arise as follows.

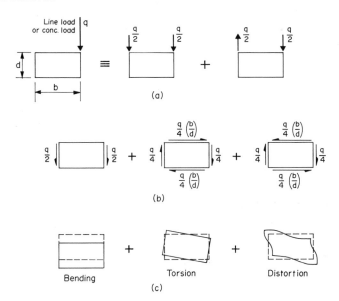

Fig. 7.2.1. An eccentric load acting on a box girder results in bending, torsion, and distortion stresses.

(a) *Direct stresses due to axial forces and bending moments including the influence of shear lag*

These direct bending stresses are found by using engineers' theory for bending (Section 1.4.1) and the direct axial stresses by calculating the axial force and the area of cross-section. In each of these calculations the properties of the effective cross-section only should be used because of the influence of shear lag (Section 1.4.1) in redistributing the axial stresses in the flanges towards their join with the webs. The calculation of the effective width of flange is difficult if elastic theory is used because it involves the use of Airy stress functions, finite elements, or similar methods. The Merrison rules (1973) overcome this difficulty by presenting graphs of an effective width factor Ψ against the distance along the girder. Figure 7.2.2 shows a typical set of graphs which show that Ψ is a function of the type of loading, the distance along the beam, the aspect ratio of the flange and the ratio of the cross-sectional areas of the stiffeners and flange plate. The graphs enable the inner flange or the outer flange to be treated separately. In the former case A_w is the area of both webs and in the latter case it is the area of one web only. The value of Ψ so obtained is multiplied by the actual area A_f of the flange to obtain the effective area of flange which is then used to derive the section properties

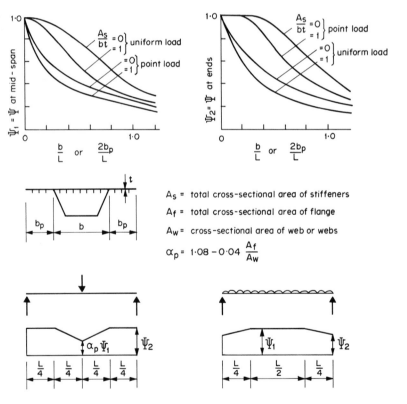

Fig. 7.2.2. Effective width factor Ψ, which allows for the effects of shear lag in a simply supported box girder (Merrison rules 1973), can be evaluated from these diagrams.

(second moment of area, section modulus, etc.) of the effective cross-section. These calculations are required for both the serviceability and fatigue checks. The curves of the effective width factor Ψ are based upon the finite element studies of Moffatt and Dowling (1972, 1975) on typical box girders.

(b) *Shear stresses other than those due to torsion*

These stresses may arise from the vertical and horizontal shear forces in the girder as a result of the applied traffic, wind, and other loads. The method described in Section 1.4.3 may be used.

(c) *Torsion shear stresses*

When the plates of a box girder are stocky or well-stiffened and the girder is long and slender it is reasonable to evaluate the torsional shear stresses using the Saint Venant theory (Section 1.4.2) only. In these calculations the effects of open and closed stiffeners should be ignored as they contribute little to the torsional stiffness of the box girder (see eqn (1.4.22) and Exercise 1.6). This theory is accurate if warping of the cross-section of the girder is freely allowed at every point in the span. The Merrison rules (1973) assume that in the first instance this is the case and quotes the formulae of Saint Venant torsion for calculating shear stresses. When warping is constrained torsional warping stresses (see (d) below) must be calculated.

(d) *Torsional warping stresses*

Direct and shear stresses which arise in a box girder from the restraint of warping (Fig. 7.2.3) can be found by the methods described in Chapters 2 and 3 of this book. However, for design purposes their evaluation is too long and tedious and the Merrison rules (1973) circumvent this problem by using the results of Gent and Shebini (1972). They studied twenty

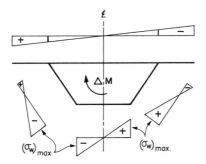

Fig. 7.2.3. Distribution of longitudinal stresses in a box girder due to the restraint of warping.

existing box girders and showed that the maximum longitudinal warping stress can be approximated by the following empirical formula

$$(\sigma_w)_{max} = \frac{d(\Delta M)}{J_p} \tag{7.2.1}$$

where d = depth of box between flanges

   J_p = Saint Venant torsion constant of the box

   ΔM = change of torque at a support or under a concentrated load.

In practice it is usually found that these stresses are not important. These stresses die out quite rapidly along the girder and the formula for this attenuation is given. The rules do not require the evaluation of the warping shear stresses.

(e) Distortional warping stresses

The distortional component of the loading (Fig. 7.2.1(b) and (c)) causes the cross-section of the girder to distort in the manner shown. The sideways movements of the upper and lower flanges mean that they must be bending about a vertical axis and associated with this there must be longitudinal stresses. They may be particularly important when the deck loads are not applied at a diaphragm. The method for evaluating these stresses in the Merrison rules (1973) is based upon the work of Wright, Abdel-Samed, and Robinson (1968) who developed an analogy between a box girder and a beam on an elastic foundation. The accuracy of the method was checked by comparison with Lim and Moffatt's (1971) finite element method and with a displacement method developed by Dalton and Richmond (1968).

(f) Stresses due to the distortion of the box cross-section

The distortion of the cross-section of the box (Fig. 7.2.1(b) and (c)) also results in transverse bending stresses whose maximum values occur at the junction of the vertical stiffeners in the webs and the transverse stiffeners in the flanges. The empirical formulae used to evaluate these stresses are also derived from the beam-on-elastic-foundation analogy (see (e) above) and from the work of Billington, Ghavami, and Dowling (1972) who carried out extensive parametric studies. The methods described in this paragraph and the previous one for evaluating the stresses due to the distortion of the cross-section of a box girder are approximate. Accurate methods of treating this problem exist but they are beyond the scope of this book. The works of Billington, Ghavami, and Dowling (1972) and Wright, Abdel-Samed, and Robinson (1968) have already been mentioned but readers will also find treatment of this important problem by Dabrowski (1965, 1968), Dalton and Richmond (1968), Richmond

(1969), Hees (1972), Janssen and Veldpaus (1972, 1973), Kristek (1970), and Roik, Carl, and Lindner (1972). Several of these investigators found that the over-use of diaphragms may lead to considerable increases in the shear stresses.

(g) *Transverse stresses on twin boxes interconnected by the upper deck*

When the cross-section consists of twin boxes interconnected by a deck (Fig. 2.4.5) transverse stresses can develop in the deck plate under the action of vertical loading.

(h) *Stresses due to creep and shrinkage*

These stresses arise in composite girders, i.e. girders whose webs and bottom flange are steel and whose upper flange is concrete. They occur because of the time-dependent properties of the concrete.

After the foregoing stresses are calculated for each panel it must be checked for yielding at working loads. This is achieved by calculating an equivalent stress, σ_e, thus,

$$\sigma_e = [(\sigma_x + \sigma_{bx})^2 + (\sigma_z + \sigma_{bz})^2 - (\sigma_x + \sigma_{bx})(\sigma_z + \sigma_{bz}) + 3\tau_{xz}^2]^{\frac{1}{2}} \qquad (7.2.2)$$

where σ_x, σ_z are the in-plane direct stresses in the transverse and longitudinal directions, respectively,

σ_{bx}, σ_{bz} are the bending stresses in the same directions,

τ is the shear stress.

Yielding is assumed to occur when $\sigma_e = \sigma_y$ where σ_y is the guaranteed minimum tensile yield stress.

Residual stresses due to welding may be treated in one of two ways by designers. Firstly, they may be calculated from simple empirical formulae which relate the tendon force in the weld to its area of cross-section. The compressive stress in the plate which balances this tendon force is simply added to σ_x and σ_z in eqn (7.2.2). Secondly, they can be added as an effective initial dishing to the assumed and acceptable initial dishing of each plate panel and empirical rules are given for this purpose. The total effective initial dishing is used in the analysis of a plate panel for stress distribution and stiffness as described in the next section.

7.2.2. Analysis of an imperfect plate panel for stress distribution and stiffness.

The secant effective width factor K_{bs} of a plate panel of width b subjected to an average longitudinal stress σ_{av} was discussed in Section 6.3.1. Equation (6.3.16) defines K_{bs} and it is seen that it is equal to both the ratio of effective width b_e to b and to the ratio of σ_{av} to the stress σ_e at the edge of the plate panel. Alternative expressions for K_{bs} are given in eqns (6.3.22) and (6.3.23) and in the Merrison rules (1973) they have been plotted as graphs of K_{bs} against σ_{av}/σ_{cr} for a family of y_0/t values. However, the maximum stress in an imperfect plate panel may occur at the edge (Fig. 6.3.16(a)) as a membrane stress or it may occur at the point of maximum lateral deflection where the bending stresses can be quite large. The Merrison rules (1973) contain graphs which enable the maximum stress in a plate panel to be evaluated and the serviceability requirement that the yield stress must not be exceeded at working loads is thereby readily checked.

The stiffness of an imperfect plate panel which is loaded longitudinally is given by the slope of the tangent to the graph of axial load against shortening. Figure 6.3.16(b) shows that at a typical point S on the graph of average axial stress against axial compressive strain the effective Young's modulus is the slope of the graph at S. The definition of the tangent stiffness factor K_{bt} is thus

$$K_{bt} = \frac{1}{E}\left(\frac{d\sigma}{d\varepsilon}\right)_S \qquad (7.2.3)$$

An expression for K_{bt} may be derived from eqn (5.3.56) and in the Merrison rules (1973) it has been graphed as a function of σ_{av}/σ_{cr} for a family of y_0/t values. K_{bt} is required for the buckling analysis of a deck panel.

7.2.3. Buckling analysis of plate panels which carry complex patterns of in-plane stresses

The determination of the critical stress in a plate panel which carries a complex pattern of stresses (Fig. 7.1.2) is considered in Section 7.1.2. The Merrison rules (1973) require that buckling of plate panels shall not occur at working loads and suggest three ways of treating this problem.

The first method is the same as that presented in Fig. 7.1.2 except that the interaction relationships between various combinations of applied in-plane stresses taken three at a time are presented as graphs. Figure 7.2.4 shows two typical sets of curves. It is seen that they allow for the fact that the critical stress generally increases as the aspect ratio L/b increases from 1 to ∞. This effect was considered to be small for most loading cases when the formulae in Table 7.1.1 were derived. For example, Fig. 7.1.4 shows that a parabola is a good approximation for a combination of axial and bending stresses. In Fig. 7.2.4(a) this case is represented by the contour heights along the vertical axis (i.e. R_b axis) and the two diagrams agree. It is true to say that all of the formulae in

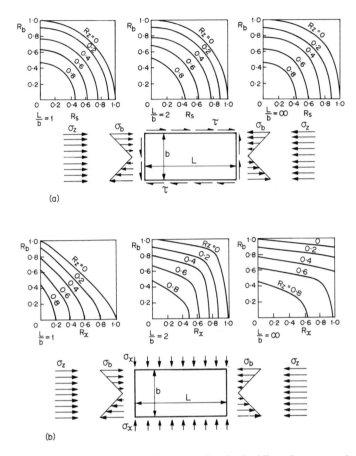

Fig. 7.2.4. Two typical sets of interaction curves for the buckling of a rectangular plate presented in the Merrison rules (1973).

Table 7.1.1 agree with the interaction buckling diagrams of the Merrison rules (1973) but the latter curves do show the strong influence of L/b in some cases, notably for those cases in which longitudinal bending and transverse compressive stresses are combined with one other stress component. Such a case is the stress combination shown in Fig. 7.2.4(b). The following example shows how Fig. 7.2.4 may be used to evaluate critical stresses.

Example 7.2.1

A simply supported plate measuring $300 \times 150 \times 3$ mm carries the stress distribution shown in Fig. 7.2.4(a) with $\sigma_z : \sigma_b : \tau = 1:2:3$. Calculate the critical values of σ_z, σ_b, and τ by using the appropriate chart in Fig.

7.2.4(a). $E = 206\,000$ MPa, $\nu = 0.3$

(a) Critical values of σ_z, σ_b, and τ when they act alone

From Fig. 5.1.4 with $L/b = 2$ and $\psi = 1$

$$\sigma_{z\,cr} = \frac{4\pi^2 \times 206\,000}{12(1-0.3^2)} \times \left(\frac{3}{150}\right)^2 = 298 \text{ MPa}.$$

From Fig. 5.1.4 with $L/b = 2$ and $\psi = -1$

$$\sigma_{b\,cr} = \frac{24\pi^2 \times 206\,000}{12(1-0.3^2)} \times \left(\frac{3}{150}\right)^2 = 1787 \text{ MPa}.$$

From eqns (6.3.39) and (6.3.40) with $d = 140$, $b = 300$, $t_w = 3$

$$\tau_{cr} = \left[5.35 + 4\left(\frac{150}{300}\right)^2\right] \frac{\pi^2 \times 206\,000}{12(1-0.3^2)} \left(\frac{3}{150}\right)^2 = 473 \text{ MPa}.$$

(b) Critical values of σ_z, σ_b, and τ when they act together

Taking a pre-buckling stress distribution of $\sigma_z = 10$ MPa, $\sigma_b = 20$ MPa, and $\tau = 30$ MPa it is required to find by trial the value of the common factor λ_{cr} so that the plate is on the point of buckling. Equations (7.1.6) and (7.1.7) are used for this purpose.

First   trial: $\lambda = 10$, $R_z = \dfrac{10 \times 10}{298} = 0.34$, $R_b = \dfrac{10 \times 20}{1787} = 0.11$, $R_s = \dfrac{10 \times 30}{473} = 0.63$.

These values are inconsistent with the middle chart of Fig. 7.2.4(a) because the stress combination chosen does not lie on the buckling surface. By trial it is seen that an increase in λ is required.

Second trial: $\lambda = 12$, $R_z = 0.41$, $R_b = 0.13$, $R_s = 0.76$

which give a consistent set of results. Hence the combination of stresses at the point of buckling is $\sigma'_{z\,cr} = 12 \times 10 = 120$ MPa, $\sigma'_{b\,cr} = 12 \times 20 = 240$ MPa, and $\tau'_{cr} = 12 \times 30 = 360$ MPa.

When the number of stress components exceeds three it is suggested that another method, which can deal with up to five stress components, be used. This method uses an algebraic interaction relationship from which it is possible to obtain for λ_{cr} by solving a quadratic equation.

When the stress distribution within a plate panel varies in a complex manner a third method is recommended. In this method, the distribution of in-plane stresses is first calculated, probably by using simple Airy stress functions. The panel is then divided into convenient rectangular elements to which the rules assign coefficients. The actual stresses in each element are multiplied by these coefficients and the results are summed for the whole panel. These sums are then called the *effective stress components*. As there are three of these effective stress components it is seen that this method enables any stress distribution to be resolved into an equivalent set of three stresses which can then be treated by the first method described above.

Each of the above methods requires the evaluation of the critical stress of the given plate panel when it carries only one stress component at a time. For rectangular, triangular, and even trapezoidal and other plates these critical stresses are often available in existing literature on plate buckling. When such information is not available, the Merrison rules (1973) recommend that for the purposes of buckling analysis the plate panel should be replaced by an equivalent rectangular plate which has the size and shape of the smallest circumscribed rectangle around the actual plate.

7.2.4. Stiffness of cross-frames and transverse stiffeners in box girders

Figure 7.2.5 shows a typical arrangement of the stiffening in a box girder. The diaphragms are for all practical purposes rigid within their own plane but they are flexible in the longitudinal direction. Their purpose is to preserve the shape of the box, and if they are located at supports, to transmit the applied loads into the bearings. In some bridges cross-frames are used instead of diaphragms and they have some flexibility in their own plane. The transverse stiffeners are usually rolled sections or fabrications from thin plating and their purpose is to transmit the loads on the upper deck of the box into the webs and to control overall buckling of the compression flange (Fig. 7.2.6). When the transverse stiffeners are very stiff in their own plane the half wavelength of buckling L_{cr} is equal to the

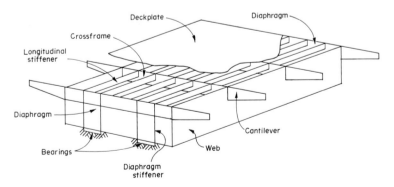

Fig. 7.2.5. Typical arrangement of stiffening of a box girder.

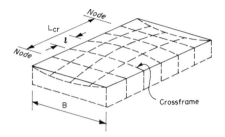

Fig. 7.2.6. Global buckling of the upper deck of a box girder is controlled by transverse stiffening when the diaphragms are a long way apart.

spacing l of the transverse stiffeners. If they are very light in their own plane the upper flange will probably buckle with a half wavelength equal to the spacing L_D of the diaphragms. If flexible cross-frames are used instead of diaphragms L_{cr} could exceed L_D. The purpose of this section is to present a method due to Horne (1975) which enables the required stiffness of the transverse stiffeners to be established. This method is now incorporated into both the Merrison rules (1973) and Part 3 of BS 5400 (1981).

The governing equations for an imperfect plate with stiffening in both orthogonal directions of z longitudinally and x transversely and with both in-plane loads, that is, N_x, N_z, and N_{xz}, and out-of-plane loading $Y(x, z)$ was presented as eqn (5.4.10). To obtain its critical stress it is only necessary to consider a perfect plate with no loading in the y-direction (i.e. $Y = 0$). Hence there is only one governing equation, that is,

$$D_x \frac{\partial^4 w}{\partial x^4} + 2H \frac{\partial^4 x}{\partial x^2 \partial z^2} + D_z \frac{\partial^4 w}{\partial z^4} - \sigma_z t_{eff} \frac{\partial^2 w}{\partial z^2} = 0 \tag{7.2.4}$$

where

$$t_{eff} = \frac{\text{total area of cross-section of flange}}{\text{width of flange } B}. \tag{7.2.5}$$

One way of dealing with the buckling problem described above is to use the following classical solution. This solution ignores the presence of the diaphragms and treats the upper flange as an infinitely long ortho-tropic plate which is simply supported along its sides. For this case (Timoshenko and Gere 1961) the critical stress and critical length are

$$\sigma_{cr} = -\frac{2\pi^2}{Bt_{eff}} [\sqrt{D_x D_z} + H] \tag{7.2.6}$$

$$L_{cr} = B(D_z/D_x)^{\frac{1}{4}}. \tag{7.2.7}$$

However, these formulae are too conservative as they do not take into account the constraint on the buckling length by the diaphragm and many

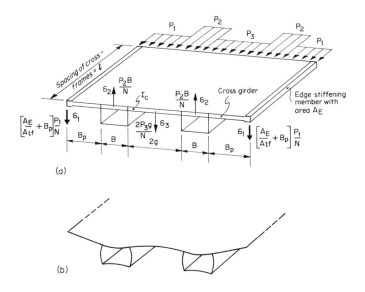

Fig. 7.2.7. (a) Stability analysis of twin box girder for global buckling mode (b).

other factors, and a more thorough analysis is required. The following analysis of a twin box girder (Fig. 7.2.7(a)) results in the formula in the Merrison rules (1973) and in BS 5400 (Part 3) for the required second moment of area of the transverse stiffeners across the upper deck of such a girder. A similar analysis is required for a single box girder. The notation used in the rules has been adopted for the following analysis.

The girder is divided into the longitudinal strips whose edges coincide with the locations of the webs. The in-plane loads in the strips are $P_1 B_p$, $P_2 B$, etc. It is assumed that when the panel is on the point of buckling in the mode shown in Fig. 7.2.7(b) the transverse stiffeners must resist a certain proportion of these loads in the vertical direction. A convenient way of applying these vertical loads is to let them act at the centre of each panel as shown and to assume that they have the magnitudes $P_1 B_p/N$, $P_2 B/N$, etc., where N is a convenient scaling factor. This combination of loads produces elastic deflections δ_1, δ_2, etc. at the corresponding points so that $\delta_1 N$, $\delta_2 N$, etc. are independent of N. Flexibility coefficients α_1, α_2, and α_3 are now defined as follows.

$$\alpha_1 = \frac{\delta_1 EI_{zz}N}{P_2 B_p^4}; \qquad \alpha_2 = \frac{\delta_2 EI_{zz}N}{P_2 B^4}; \qquad \alpha_3 = \frac{\delta_3 EI_{zz}N}{P_2(2g)^4} \qquad (7.2.8)$$

where I_{zz} is the effective second moment of area of the transverse stiffener and its associated width of plating (Fig. 7.2.5). The strain energy of the transverse stiffener is thus

$$U_i = \left(\frac{A_e}{A_{tf}} + B_p\right)\frac{P_1\delta_1}{N} + \frac{BP_2\delta_2}{N} + \frac{gP_3\delta_3}{N} \tag{7.2.9}$$

where A_{tf} is the average area of deck per unit width of deck. The work done U_e by the external forces P_1B_p/N, P_2B/N, etc. is found by integrating the axial shortening over the whole width of each panel of the girder and summing. The critical value of the axial force per unit width P_2 is found by equating U_i and U_e, whence

$$P_{2cr} = \frac{2EI_{zz}}{\pi^2\eta B^4 l}\left[\phi L^2 + \frac{\pi^4 I_{xx}B^4 l}{2L^2 I_{zz}}\psi\right] \tag{7.2.10}$$

where

I_{xx} = the second moment of area of the flange per unit width about its centroidal axis

$$\phi = 2\left(1 + \frac{A_e}{B_pA_{tf}}\right)\frac{P_1}{P_2}\alpha_1\lambda^5 + 2\alpha_2 + 32\mu^5\alpha_3\frac{P_3}{P_2} \tag{7.2.11}$$

$$\psi = \frac{4}{3}\left(1 + \frac{3I_{xxe}}{B_pI_{xx}}\right)\alpha_1^2\lambda^9 + 2\alpha_2^2 + 512\mu^9\alpha_3^2 \tag{7.2.12}$$

$$\eta = \frac{4}{3}\left(1 + \frac{3A_e}{B_pA_{tf}}\right)\alpha_1^2\frac{P_1}{P_2}\lambda^9 + 2\alpha_2^2 + 512\mu^9\alpha_3^2\frac{P_3}{P_2} \tag{7.2.13}$$

$$\lambda = \frac{B_p}{B} \tag{7.2.14}$$

$$\mu = \frac{g}{B} \tag{7.2.15}$$

I_{xxe} = second moment of area of edge member,

A_e = area of cross-section of edge member.

To obtain the minimum value of P_{2cr} the right-hand side of eqn (7.2.10) is differentiated with respect to L, and equated to zero whence

$$L_{cr} = \pi B\left(\frac{I_{xx}l\psi}{2I_{zz}\phi}\right)^{\frac{1}{4}}. \tag{7.2.16}$$

This equation is valid provided $2l < L_{cr} < L_D$ where L_D is the spacing of the rigid diaphragms. For this case the minimum critical load is

$$P_{2cr} = \frac{2.82E}{B^2\eta}\left[\frac{I_{xx}I_{zz}\phi\psi}{l}\right]^{\frac{1}{2}} \tag{7.2.17}$$

When a calculation using eqn (7.2.16) shows that $L_{cr} > L_D$, the critical

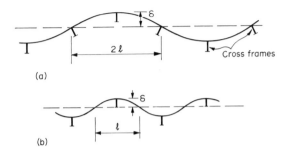

Fig. 7.2.8. Two buckling modes which must be checked when the critical buckling length $L_{cr} \leqslant 2l$.

load is found from eqn (7.2.10) with $L = L_D$. However, when the calculations shows that $L_{cr} < 2l$ it is necessary to investigate the two buckling modes shown in Fig. 7.2.8(a) and (b). Analyses of these modes are carried out in a similar manner to that leading to eqn (7.2.10) and result in the following equations for the critical loads.

For the mode shown in Fig. 7.2.8(a)

$$P_{2cr} = \frac{4EI_{zz}}{\pi^2 \eta B^4 l} \left[2\phi l^2 + \frac{\pi^4 I_{xx} B^4 \psi}{16 l^2 I_{zz}} \right] \tag{7.2.18}$$

and for the mode shown in Fig. 7.2.8(b)

$$P_{2cr} = \frac{4EI_{zz}}{\pi^2 \eta B^4 l} \left[\phi l^2 + \frac{\pi^4 I_{xx} B^4 \psi}{4 l^2 I_{zz}} \right]. \tag{7.2.19}$$

7.3. A BRIEF REVIEW OF SOME DESIGN RULES IN BS 5400 (1981)

BS 5400 is the British code for the design of steel, concrete, and composite bridges. It consists of ten parts as follows:

Part 1   General statement.
Part 2   Specification for loads.
Part 3   Code of practice for design of steel bridges.
Part 4   Code of practice for design of concrete bridges.
Part 5   Code of practice for design of composite bridges.
Part 6   Specification for materials and workmanship, steel.
Part 7   Specification for materials and workmanship, concrete reinforcement, and prestressing tendons.
Part 8   Recommendations for materials and workmanship, concrete, reinforcement, and prestressing tendons.

Part 9   Code of practice for bearings.
Part 10   Code of practice for fatigue.

In this section some of the rules in Parts 3 and 6 are discussed.

7.3.1. Limitations on geometry of cross-sections

In BS 5400 one approach to the problem of stability of thin-walled compression and bending members is that of restricting cross-section geometry so as to avoid certain types of local buckling. Some of the restrictions placed on open stiffeners are so severe that it is difficult to see how they can be used economically. For example the ratio of the span of a bulb flat stiffener between supporting members to the depth of the stiffener is restricted to $3(355/\sigma_y)^{\frac{1}{2}}$. As another example, curves of allowable depth of plain rectangular stiffeners (Fig. 7.3.1) show that when b/t exceeds 30 the depth of the stiffener cannot exceed $10t_s$. If deep stiffeners

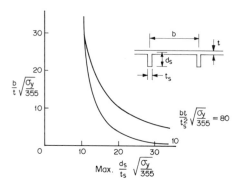

Fig. 7.3.1. Typical curves which limit the geometry of stiffened plates. (BS 5400: 1981.)

are desired by the designer he must reduce their spacing to quite small values or increase their thickness. This will encourage him to use other forms of stiffener such as closed stiffeners for which the restrictions are not so exacting.

7.3.2. Design of compression flanges of bending members

Another approach to the control of buckling in BS 5400 is to use the effective width of plating in the design of thin-walled compression and bending members. For example, graphs of effective width ratio b_e/b for the evaluation of the failure loads of compression members without longitudinal stiffeners [Fig. 7.3.2(a)] are presented as a function of $(b/t)\sqrt{(\sigma_y/355)}$. The failure load is then determined as the product of the

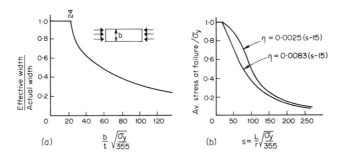

Fig. 7.3.2. Typical curves presented in BS 5400 for the design of compression members for the collapse limit state.

effective area of cross-section and the so-called 'ultimate stress'. The ultimate stress is found from a family of Perry–Robertson curves (Fig. 7.3.2(b)) which show the functional relationship between the ultimate stress and the slenderness ratio $(L/r)\sqrt{(\sigma_y/355)}$ for various values of the imperfection parameter η. The design capacity or resistance R^* in eqn (7.1.3) is then given by the formula

$$R^* = (\text{effective area} \times \text{ultimate stress})/\gamma_m \qquad (7.3.1)$$

A similar method is used for the design of flanges with longitudinal stiffeners but the parameters governing the form of the curves are different. When the compression flange and/or web of a plate girder or box girder has longitudinal stiffeners it is allowable to transfer up to 60% of the longitudinal stresses in the web to the compression flange. Thus the code recognises the significance of load shedding (Section 6.3.2). The longitudinal stresses may arise from axial loads as well as from a share of the bending moment which is carried by the web. If the designer chooses to shed some of the longitudinal stresses to the compression flange he is obliged to make it correspondingly stronger.

7.3.3. Design of webs

The code distinguishes between webs without longitudinal stiffeners and those with them. In the former case the code rules for the collapse limit state are based upon the tension field theory of Rockey, Evans and Porter (1978) which is described in Section 6.3.2. For these webs the code also requires a serviceability check which ensures that the web does not buckle at working loads. In the case of webs with longitudinal stiffeners the code specifies that every panel shall be checked for the collapse limit state but, if the code method is used, a serviceability check is not required.

Beams without longitudinal stiffeners

The shear capacity V_S of the web of a girder is, in the absence of any bending moment, given by the point S in Figure 6.3.27 and by eqns (6.3.63) and (6.3.65) where τ_{cr} is given by eqns (6.3.39) and (6.3.40). It will be recalled that the theory was developed for the failure mode (e) shown in Fig. 6.3.25 and that eqn (6.3.65) was derived for the special case shown in Fig. 6.3.25(c). In this case τ_{cr} is less than τ_{yw} so elastic buckles appear before yielding starts and therefore τ_m is taken as τ_{cr}. For mode (d) when τ_{cr} exceeds τ_{yw} yielding occurs first as indicated in Fig. 6.3.25(d) and in this case τ_m is taken as τ_{yw}. Hence in the theory of Rockey, Evans, and Porter V_S is given by eqn (6.3.65) with τ_{cr} replaced by τ_m where τ_m has the form AED shown in Fig. 7.3.3. In this graph the elastic arm ED, which represents the buckling mode (c) in Fig. 6.3.25, is derived from eqn (6.3.39), thus

$$\frac{\tau_m}{\tau_{yw}} = \frac{\tau_{cr}}{\tau_{yw}} = \frac{k\pi^2 E}{12(1-\nu^2)}\left(\frac{\sqrt{3}}{\sigma_{yw}}\right)\left(\frac{t_w}{d}\right)^2 = 904k\left(\frac{t_w}{d}\right)^2\left(\frac{355}{\sigma_{yw}}\right) \qquad (7.3.2)$$

where k is given by eqn (6.3.40) and

$$\beta = \frac{d}{t_w}\sqrt{\left(\frac{\sigma_{yw}}{355}\right)}. \qquad (7.3.3)$$

The plastic arm AE represents the failure mode (d) in Fig. 6.3.25.

In the code this theory is modified by the introduction of a straight transition curve BC in Fig. 7.3.3 to allow for the strong influence of imperfections in the region where yielding and elastic buckling interact.

For convenience the code presents the solution of eqn (6.3.65) (with τ_{cr} replaced by τ_m shown in Fig. 7.3.3) as a series of graphs of V_S/V_{yw} against the variable $d/t_w\sqrt{(\sigma_{yw}/355)}$ for different values of the aspect ratio b/d and the parameter $M_{pf}\gamma_m/(d^2 t_w\sigma_{yw})$ which compares the plastic moments of the flange and web. Figure 7.3.4 shows three such typical curves. The code defines how much of the weaker of the two flanges and of the web is

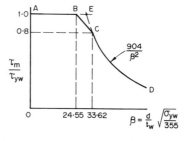

Fig. 7.3.3. Graph showing the value of τ_m used for the design of webs in BS 5400 (1981).

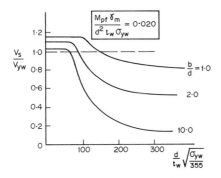

Fig. 7.3.4. Typical curves which are solutions of eqn (6.3.65) and allow V_s (Fig. 7.3.5) to be evaluated.

to be taken as effective in calculating M_{pf}. The safety factor γ_m is another modification to the original theory which has ben introduced by the code.

When the loading is a combination of shear and bending a slightly modified form of the interaction diagram in Fig. 6.3.27 is used by the code. Figure 7.3.5 shows that used by BS 5400 and it is seen that although the parabolae are replaced by straight lines the forms of the two diagrams are very similar.

In the case of web panels designed by this method it is necessary to carry out a check on the serviceability limit state (Clause 4.6) which ensures that the web does not buckle at working loads.

In order to do this the interaction buckling curve between the applied shear stress τ and the average axial stress σ_1 in the web panel is assumed to be the empirical interaction formula (eqn (7.1.9))

$$\left(\frac{\tau}{\tau_{cr}}\right)^2 + \frac{\sigma_1}{\sigma_{cr}} \leqslant 1 \tag{7.3.4}$$

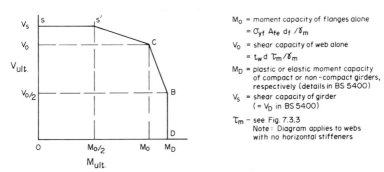

Fig. 7.3.5. Interaction diagram for failure of web of a plate girder used in BS 5400 (compare Fig. 6.3.27).

From eqn (6.3.39) and

$$\sigma_{cr}\left[=\frac{\pi^2 E}{3(1-\nu^2)}\left(\frac{t_w}{d}\right)^2\right]$$

for a uniformly loaded square panel, eqn (7.3.4) can be changed into the form given in the code, thus

$$\tau \leqslant \tau_{cr}\sqrt{\left(1-\frac{\sigma_1}{\sigma_{cr}}\right)} \tag{7.3.5}$$

$$V \leqslant \frac{0.9EK_\phi t_w^3 d_w}{\gamma_m d^2}\sqrt{\left[1-\frac{\gamma_m\sigma_1}{3.62E}\left(\frac{d}{t_w}\right)^2\right]} \tag{7.3.6}$$

where the code replaces k in eqn (6.3.40) by the approximation (see Fig. 6.3.25)

$$K_\phi = 5+5\left(\frac{d}{b}\right)^2. \tag{7.3.7}$$

Design of the webs of beams with longitudinal stiffeners

The code requires (Clause 6.3.2) that all of the individual panels in a web with longitudinal stiffeners shall be checked so as to avoid both yielding and buckling. While this is admitted to be a conservative approach it is argued by Horne (1980) that the tension-field theory of such girders is not yet sufficiently developed. Yielding is avoided in the following way.

It is assumed that an element in a plate which is subjected to a combination of direct and shear stresses will start to yield when the von Mises–Hencky criterion (eqn (6.3.44)) is satisfied. Figure 7.1.2 shows the general stress condition considered by BS 5400. The case when $\sigma_x = 0$ is considered here first.

One possible criterion of failure is when the point of first yield is reached at the boundary. From the von Mises–Hencky criterion this occurs when

$$\left[\frac{\sigma_z+\sigma_b}{\sigma_{yw}}\right]^2+\left[\frac{\tau}{\tau_{yw}}\right]^2 = 1. \tag{7.3.8}$$

Another possibility is to allow a fully plastic distribution of stress (Fig. 7.3.6). Because of the presence of the shear stress τ in the web the apparent yield stress σ_y' will have a lower value that when τ is absent. From the von Mises–Hencky criterion

$$\sigma_{yw}' = \sigma_{yw}\sqrt{\left[1-\left(\frac{\tau}{\tau_{yw}}\right)^2\right]}. \tag{7.3.9}$$

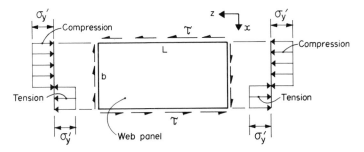

Fig. 7.3.6. Combined stress distribution for fully plastic failure. (Horne 1980.)

This effective yield stress is substituted into eqn (1.2.14)

$$\frac{M}{\sigma'_{yw}\dfrac{b^2 t_w}{4}} + \left(\frac{\sigma_z}{\sigma'_{yw}}\right)^2 = 1. \tag{7.3.10}$$

When a fictitious elastic bending stress σ_b defined by the equation $M = \sigma_b(d^2 t_w/6)$ is introduced the following criterion of failure by fully plastic stresses is obtained (Horne 1980).

$$\left(\frac{\sigma_z}{\sigma_{yw}}\right)^2 + \frac{2}{3}\left(\frac{\sigma_b}{\sigma_{yw}}\right)\sqrt{\left[1 - \left(\frac{\tau}{\tau_{yw}}\right)^2\right]} + \left(\frac{\tau}{\tau_{yw}}\right)^2 = 1. \tag{7.3.11}$$

The first criterion is too conservative and the second is unsafe because elasto-plastic buckling would probably occur before this criterion could be satisfied. Therefore the code takes the failure criterion as

$$\left[\frac{\gamma_m\sigma_z + 0.77\gamma_m\sigma_b}{\sigma_{yw}}\right]^2 + \left[\frac{\gamma_m\tau}{\tau_{yw}}\right]^2 = 1. \tag{7.3.12}$$

A slightly different form of this equation is used when $\sigma_x \neq 0$. To deal with panel buckling under a combination of stresses σ_b, σ_z, and τ (Fig. 7.1.2) Stüssi's equation (eqn (7.1.12)) has been incorporated into BS 5400 in a slightly modified form.

7.3.4. Design of cross-frames and other stiffening of compression flanges

The purpose of this section is to show that the formulae quoted in BS 5400 for the design of cross-frames are essentially the same as those quoted in the Merrison rules (Section 7.2.4).

For the design of transverse stiffening BS 5400 allows part of the upper flange plate to be considered as effective with the cross-frames. The formula given in the code for the second moment of area of a cross-frame is explained here by again referring to a typical panel of width B (Fig. 7.2.7(a)) but to a girder without edge stiffening members. The stiffness requirement for this effective cross beam is that its second moment of area I_{zz} should satisfy the following inequality.

$$I_{zz} \geq \frac{9P_2B^4l}{16YE^2I_{xx}} \tag{7.3.13}$$

where Y is generally a factor which is a function of the ratios B_p/B and (I_{zz} of the cantilevers)/I_{zz}. Charts are available for the easy evaluation of Y. For an internal panel such as that of width B shown in Fig. 7.2.2(a), $Y = 24$. For other panels the appropriate parameters must be used.

Equation (7.3.13) is derived from eqn (7.2.17) as follows.

To prevent buckling

$$I_{zz} \geq \frac{9P_2^2B^4l}{16E^2I_{xx}} \left[\frac{\eta^2}{2.82^2\phi\psi} \times \frac{16}{9} \right] \tag{7.3.14}$$

and it is seen that the term in square brackets is the inverse of Y.

7.4. A BRIEF REVIEW OF SOME DESIGN RULES IN DIN 4114 (1952) AND DASt-RICHTLINIE 012 (1978)

The German standard DIN 4114 (1952) is used for the stability analysis of steel structures. The types of instability covered are column buckling, overturning and plate buckling. Although this section is mainly concerned with plate buckling it is necessary to consider column buckling and its interaction with plate buckling when stiffened plates are loaded axially. The rules in DIN 4114 (1952) have been explained and elaborated by a working party PLATTEN of the Deutsche Ausschuss für Stahlbau (DASt). They have produced a document DASt-Richtlinie 012 (1978) which gives more detailed direction to designers of thin-walled structures than does DIN 4114 (1952). In 1979 a detailed explanation of DASt-Richtlinie 012 (1978) was written by Scheer, Nölke, and Gentz (1979) and their document should be studied by readers wishing to obtain further information.

In this section the notation used in the earlier parts of this book is adhered to but Table 7.4.1 has been included to enable readers to refer to the above three German publications.

7.4.1. General remarks on the analysis of a plated compression member

The analysis of a plate by the German method is divided into two parts: firstly, the available safety factor (γ_{avail}) against failure is calculated and,

Table 7.4.1. Main notation used in this book and in German literature.

German	This book	Explanation with notation of this book
Stresses		
σ_F	σ_y	Yield stress
σ_V	σ_e	Equivalent stress $= \sqrt{(\sigma_x^2 + \sigma_z^2 - \sigma_x\sigma_z + 3\tau^2)}$
σ_e		$= \dfrac{\pi^2 E}{12(1-\nu^2)}\left(\dfrac{t}{b}\right)^2$
σ_{VK}	σ_m	Average stress at failure of plate
		For $\left(\dfrac{b}{t}\right)\Big/\left(\dfrac{b}{t}\right)_y > 1.291$
		$\sigma_{VK} = \sigma_m = \sigma_{cr} = \dfrac{k\pi^2 E}{12(1-\nu^2)}\left(\dfrac{t}{b}\right)^2$
σ_{Ki}	σ_E	Euler stress $= \dfrac{\pi^2 E}{(L/r)^2}$
σ_{xKi}, etc.	$\sigma_{x\ cr}$, etc.	Critical value of σ_x, etc. when it acts alone
σ_{xKi}^*, etc.	$\sigma'_{x\ cr}$, etc.	Critical value of σ_x, etc. when it acts with other stresses (see Section 7.1.2)
σ_{VKi}	$\sigma_{e\,cr}$	Critical value of equivalent stress $= \sqrt{[(\sigma'_{x\ cr})^2 + (\sigma'_{z\ cr})^2 - \sigma'_{x\ cr}\sigma'_{z\ cr} + 3(\tau'_{cr})^2]}$
σ_{BK}	σ_{BK}	The stress at which a compression member fails due to interation of plate and column buckling
σ_G	σ_a	Maximum allowable stress
$\bar{\sigma}_V, \bar{\sigma}_{VKi}$, etc.	$\sigma_e/\sigma_y, \sigma_{e\,cr}/\sigma_y$, etc.	The bar indicates the stress ratio $\sigma_e/\sigma_y, \sigma_{e\,cr}/\sigma_y$, etc.
Slenderness ratios		
λ_K	L/r	Slenderness ratios of a column
λ_V		Slenderness factor of a plate $= \sqrt{\left(\dfrac{\pi^2 E}{\sigma_{cr}}\right)} = \sqrt{\left[\dfrac{12(1-\nu^2)}{k}\right]}\dfrac{b}{t}$
$\bar{\lambda}_V$	$\left(\dfrac{b}{t}\right)\Big/\left(\dfrac{b}{t}\right)_y$	$(b/t)_y$ (see Fig. 6.3.1) is the b/t ratio at which $\sigma_{cr} = \sigma_y$. Also $\bar{\lambda}_V = \sqrt{(\sigma_y/\sigma_{cr})}$
$\bar{\lambda}_K$	$\left(\dfrac{L}{r}\right)\Big/\left(\dfrac{L}{r}\right)_y$	$(L/r)_y$ is slenderness ratio of a column at which $\sigma_E = \sigma_y$. Also $\bar{\lambda}_K = \sqrt{(\sigma_y/\sigma_E)}$
Safety factors		
vorh ω_B^*	γ_{avail}	Available safety factor against buckling failure
erf ν_B^*	γ_{reqd}	Safety factor required by DASt 012
Dimensions		
b	b	Width of isolated plate
b_{ik}	b	Width of plate element in a larger stiffened plate
a	L	Length of plate
b'_{ik}	b_e	Effective width of plate

secondly, the safety factor (γ_{reqd}) required by the codes is evaluated. The design is safe if the former exceeds the latter.

The flow diagram illustrated in Fig. 7.4.1 (Scheer, Nölke, and Gentz 1979) shows the steps to be followed in the calculation of γ_{avail}. This pattern will be followed here in an attempt to explain the rules used in the codes. The evaluation of γ_{reqd} follows a quite complicated procedure which has been described by Scheer, Nölke, and Gentz (1979). Its philosophy will be briefly outlined in Section 7.4.4. A central issue in the method is the criterion of failure and that is described first.

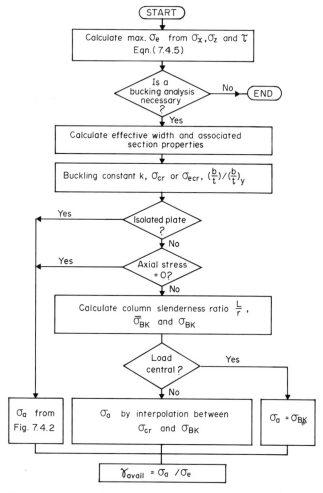

Fig. 7.4.1. Flow diagram for calculation of available factor of safety. (Scheer, Nölke, and Gentz 1979.)

7.4.2. Criterion of failure

The criterion of failure of a plate adopted by DASt 012 is shown in Fig. 7.4.2 where it is seen that there are plastic, elasto-plastic, and elastic regions. Whereas the British codes allow for some post-buckling strength for plates with high b/t ratios the German codes define failure as the start of elastic buckling in this region so the approach is more conservative. The abscissa can be expressed in two ways. Firstly, it is the ratio of (b/t) to $(b/t)_y$ (see Fig. 6.3.1), where $(b/t)_y$ is the value of b/t at which $\sigma_{cr} = \sigma_y$, i.e.

$$\left(\frac{b}{t}\right)_y = \pi \sqrt{\left[\frac{kE}{12(1-\nu^2)\sigma_y}\right]}. \tag{7.4.1}$$

For an arbitrary value of b/t the associated critical stress is σ_{cr}, and hence

$$\left(\frac{b}{t}\right) = \pi \sqrt{\left[\frac{kE}{12(1-\nu^2)\sigma_{cr}}\right]}. \tag{7.4.2}$$

Therefore, the second way of writing the abscissa is

$$\left(\frac{b}{t}\right) \Big/ \left(\frac{b}{t}\right)_y = \sqrt{\frac{\sigma_y}{\sigma_{cr}}} \tag{7.4.3}$$

The criterion of failure of a plated structure by column buckling used by DASt 012 is also shown in Fig. 7.4.2 when the abscissa can also be expressed in two ways, that is,

$$\left(\frac{L}{r}\right) \Big/ \left(\frac{L}{r}\right)_y = \sqrt{\frac{\sigma_y}{\sigma_E}} \tag{7.4.4}$$

When the structure fails by interaction between plate and column buckling the interaction diagram shown in Fig. 7.4.3 is used. The graphs along the axes are those in Fig. 7.4.2. The diagram itself is divided into

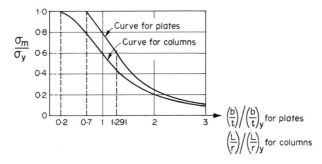

Fig. 7.4.2. Curves of failure stress for plates and columns. (DASt-012 1978.)

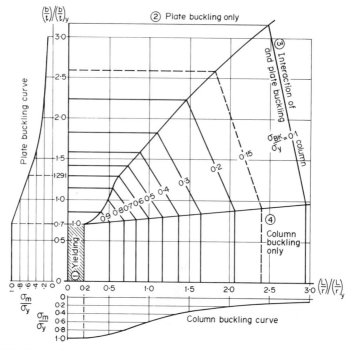

Fig. 7.4.3. Graph of failure stress for interaction between column and plate buckling. (DASt-012 1978.)

four regions, of which 1, 2, and 4 are, respectively, pure yielding without buckling, plate buckling, and column buckling. In Region 3 the interaction between plate and column buckling is allowed for. The diagram is used by first locating the point with coordinates $\left(\dfrac{L}{r}\right)\bigg/\left(\dfrac{L}{r}\right)_y, \left(\dfrac{b}{t}\right)\bigg/\left(\dfrac{b}{t}\right)_y$. If it lies in Regions 2 or 4 the curves along the vertical or horizontal axes, respectively, are used to establish the ratio of the failure stress to the yield stress. If the point lies in Region 3 this ratio is read from the diagonal lines. This type of interaction diagram is proposed for the forthcoming American specification for steel box girders (Wolchuk 1981) but it is based upon Little's (1976) investigations.

7.4.3. Procedure for buckling analysis

After calculating σ_x, σ_z, and τ at every point in the plated structure an equivalent axial stress σ_e is evaluated by the von Mises–Hencky formula (eqn (6.3.44))

$$\sigma_e = \sqrt{(\sigma_x^2 + \sigma_z^2 - \sigma_x\sigma_z + 3\tau^2)}. \qquad (7.4.5)$$

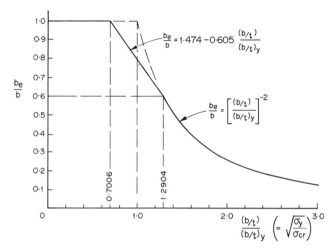

Fig. 7.4.4. The effective width ratio b_e/b of a plate panel is related empirically to the ratio $(b/t)/(b/t)_y$ in DASt-Richtlinie 012.

If σ_e, L/b, and b/t simultaneously satisfy certain conditions, which are sufficiently severe to exclude the possibility of buckling, then a buckling analysis is not required. If they do not satisfy these conditions the effective width of plating is next calculated from the empirical relationship shown in Fig. 7.4.4 between the effective width ratio b_e/b and $(b/t)/(b/t)_y$. It will be recalled (eqn (6.3.16)) that the former ratio is also that of $\sigma_{av}/\sigma_{edge}$. From these calculations the section properties of the effective cross section of a stiffened plate and hence its Euler stress σ_E may be found.

The critical elastic stress is found for simple combinations of stress by using a table in DASt 012 but for more complicated combinations of stress the designer is referred to the books by Klöppel and Scheer (1960) and Klöppel and Möller (1968) and to computer programs available in universities. From these critical values of $\sigma'_{x\,cr}$, $\sigma'_{z\,cr}$, σ'_{bcr}, and τ'_{cr} (see Section 7.1.2) the critical value of the equivalent stress is obtained from the equation

$$\sigma_{ecr} = \sqrt{[(\sigma'_{x\,cr})^2 + (\sigma'_{z\,cr} + \sigma'_{bcr})^2 - \sigma'_{x\,cr}(\sigma'_{z\,cr} + \sigma'_{bcr}) + 3(\tau'_{cr})^2]} \qquad (7.4.6)$$

For the cases when $\sigma_x = 0$ approximate values of σ_{ecr} can also be found by using a buckling surface (Fig. 7.1.3) with a form similar to that given in eqn (7.1.12). In DASt 012 the equations have been solved to give a closed form solution, that is,

$$\sigma_{ecr} = \frac{\sqrt{[(\sigma_z + \sigma_b)^2 + 3\tau^2]}}{\dfrac{1+\psi}{4}\dfrac{\sigma_z + \sigma_b}{(\sigma_z + \sigma_b)_{cr}} + \sqrt{\left\{\left[\dfrac{3-\psi}{4}\dfrac{\sigma_z + \sigma_b}{(\sigma_z + \sigma_b)_{cr}}\right]^2 + \left(\dfrac{\tau}{\tau_{cr}}\right)^2\right\}}} \qquad (7.4.7)$$

where

$$\psi = \frac{\sigma_z - \sigma_b}{\sigma_z + \sigma_b} \qquad (7.4.8)$$

and

$$(\sigma_z + \sigma_b)_{cr} = \frac{k\pi^2 E}{12(1 - \nu^2)} \left(\frac{t}{b}\right)^2. \qquad (7.4.9)$$

k depends upon the stress distribution, the condition of support along the boundaries, and the details of the stiffening. It can be estimated from Fig. 5.1.4 or the books referred to earlier or Murray and Thierauf (1981). For the case when $\sigma_b = \tau = 0$, Dunkerly's line (eqn (7.1.11)) is recommended.

For an isolated plate panel the maximum allowable stress σ_z can be evaluated by using the plate curve of Fig. 7.4.2. (This is also the curve along the vertical axis of Fig. 7.4.3). For a centrally loaded stiffened plate consisting of several plate panels and stiffeners it is necessary to use the interaction diagram (Fig. 7.4.3). With the two ratios $(b/t)/(b/t)_y$ and $(L/r)/(L/r)_y$ evaluated from eqns (7.4.3) and (7.4.4) the coordinates of a point in the diagram are established and this enables the allowable stress to be found, but the procedure depends upon whether the point lies in Regions 1–4. If the load is not centrally applied an interpolation procedure as indicated in Figure 7.4.1 is required. Finally, the available safety factor γ_{avail} $(=\sigma_a/\sigma_e)$ is evaluated.

7.4.4. Determination of required safety factor

In the second part of the process the required factor of safety γ_{reqd} is found by a somewhat elaborate procedure which is designed to yield different factors for different loading cases and for different stress combinations. For example, under the main loading case for pure shear stress, γ_{reqd} is 1.32 and for pure compression 1.50. The basic values are presented in tabular form as $\gamma_{reqd}(\sigma_z)$, $\gamma_{reqd}(\tau)$, etc. for the cases of simple stress components acting alone. The required safety factor for a combination of stresses is then evaluated from the empirical rule

$$\gamma_{reqd} = \left[\frac{(\sigma_z/\sigma_{z\,cr})^2 + (\sigma_x/\sigma_{x\,cr})^2 + (\tau/\tau_{cr})^2}{\left\{\frac{\sigma_z/\sigma_{z\,cr}}{\gamma_{reqd}(\sigma_z)}\right\}^2 + \left\{\frac{\sigma_x/\sigma_{x\,cr}}{\gamma_{reqd}(\sigma_x)}\right\}^2 + \left\{\frac{\tau/\tau_{cr}}{\gamma_{reqd}(\tau)}\right\}^2} \right]^{\frac{1}{2}} \qquad (7.4.10)$$

When bending in the z-direction also occurs, σ_z and $\sigma_{z\,cr}$ are replaced by $(\sigma_z + \sigma_b)$ and $(\sigma_z + \sigma_b)_{cr}$, respectively.

The description presented here of the analysis of a plated compression member by the German method has been simplified for the purposes of explanation. Readers who wish to study the application of the German codes to actual structures can refer to the examples presented by Scheer, Nölke, and Gentz (1979) and other text books.

7.5. A BRIEF REVIEW OF SOME DESIGN RULES IN AS 1538 (1974)

The Australian code AS 1538 (S.A.A. 1974) deals with the design of cold-formed steel structures. One example of such a structure is a strut or beam whose cross-section is shown in Fig. 4.5.9(a) or (b). The code is a typical example of such documents drafted for the purpose of helping designers with the special problems associated with this class of thin-walled steel structure. One reason for reviewing it here is because of its treatment of euler and torsional buckling of columns and of lateral buckling of beams. As is the case with the other codes reviewed in this book the general approach to local plate buckling, column buckling, and the possible effects arising from their interaction is in the first place to establish an effective width of the plate panels, and therefore, an effective cross-section to carry the axial loads as a column. The rules for determining effective widths are described in Section 7.5.1 and the application of these results to euler, torsional, and lateral buckling is explained in Section 7.5.2 where the questions of allowable stresses and load factors are also discussed. The theories underlying this code are all elastic because it does not allow the use of plastic design methods except where it can be demonstrated that the plastic hinges will be global and not local (see Fig. 6.1.1).

7.5.1. Effective width and section properties

In a similar manner to the Merrison rules (1973) this code distinguishes between the effective width b_e which is required for estimating strength (i.e. against the collapse limit state) and that required for stiffness and deflection calculations. Here only the strength calculations are discussed.

In eqns (6.3.22) and (6.3.23) alternative expressions are given for the secant effective width factor K_{bs} of a plate with initial imperfections. In these equations the magnitude of the initial imperfection is one of the parameters which determine the value of K_{bs}. Also from eqn (6.3.16) it is seen that K_{bs} is both the ratio of b_e to the actual width b of the plate and the ratio of the average stress σ_{av} to that at the edge, that is, σ_e. For the collapse condition, which is assumed to occur when σ_e first reaches the yield stress, AS 1538 (S.A.A 1974) quotes an empirical formula for K_{bs}

which is based upon extensive experimental results. This formula is written here in the following form to conform with the style of such equations found elsewhere in this book.

$$K_{bs} = \frac{b_e}{b} = \frac{\sigma_{av}}{\sigma_y} = \frac{A_1}{\frac{b}{t}\sqrt{\frac{\sigma_y}{355}}}\left[1 - \frac{A_2}{\frac{b}{t}\sqrt{\frac{\sigma_y}{355}}}\right] \tag{7.5.1}$$

where A_1 and A_2 are constants defined by the code as follows. The figures in brackets refer to closed rectangular cross-sections.

(i) For $\frac{b}{t}\sqrt{\frac{\sigma_y}{355}} < 23.88(25.63)$, $K_{bs} = 1$

(ii) For $23.88(25.63) < \frac{b}{t}\sqrt{\frac{\sigma_y}{355}} < 60\sqrt{\frac{\sigma_y}{355}}$, $A_1 = 35.24(35.24)$, $A_2 = 8.17(7.01)$

(iii) For $\frac{b}{t}\sqrt{\frac{\sigma_y}{355}} > 60\sqrt{\frac{\sigma_y}{355}}$, as for case (ii) above but reduce b_e by $(b - 60t)/10$ if one edge is not fully supported.

It is seen that these rules contain no factor specifically incorporated to account for the influence of initial imperfections. Their influence is, however, included because the formulae represent a lower bound around the experimental data used.

The code also gives rules for the effective area of stiffeners to be used in the calculation of the section properties of the effective cross-section.

7.5.2. Treatment of euler, torsional, and lateral buckling, and safety factors

As in other codes it is assumed that a stiffened compression panel behaves like an euler column with an effective cross-section. In AS 1538 (S.A.A. 1974) the rules for defining the effective cross-section are those set out in Section 7.5.1. Also the code recognised that some columns may fail by torsional buckling and the Perry–Robertson formula (see eqn (6.3.13)) for the collapse load of a strut has been modified accordingly. The modified formula is written here in the following form.

$$P_{fail} = \tfrac{1}{2}[1.25P'_y + P'_{cr}] - \tfrac{1}{2}[(1.25P'_y + P'_{cr})^2 - 4P'_yP'_{cr}]^{\frac{1}{2}} \tag{7.5.2}$$

where P'_y is the squash load of the effective cross-section ($=\sigma_y \times$ effective area of cross-section) and P'_{cr} is the least of

(i) the euler buckling load of the effective cross-section about the x-axis;

(ii) the euler buckling load of the effective cross-section about the y-axis;

(iii) the torsional buckling load P_ϕ of the column using the complete cross-section to calculate the section properties. For doubly-symmetric cross-sections P_ϕ is given by eqn (3.5.16) and for cross-sections with only one axis of symmetry the lower of the two roots of a quadratic equation (see Example 3.5.1) is used to obtain P_ϕ. For cross-sections with no symmetry P_ϕ is taken as the lowest root of eqn (3.5.19). The code also defines the effective length to be used when P_ϕ is calculated.

The code then applies a load factor λ (eqn (7.1.2)) of $1/0.6$ to evaluate the design load. This factor is the one generally adopted throughout the code both as a load factor and as a factor of safety on stress.

The lateral buckling of deep beams is controlled by using the theory set out in Section 3.6.2 but with some elaborations and with some simplifications. For deep beams with equal and opposite end-moments M_0 their critical value is given by the expression

$$(M_0)_{cr} = \frac{\pi (EGF_{xx}J_p)^{\frac{1}{2}}}{L} \left\{ \frac{\pi\delta}{2} + \left[\left(\frac{\pi\delta}{2}\right)^2 + \left(1 + \frac{\pi^2 EF_{\omega\omega}}{FJ_pL^2}\right) \right]^{\frac{1}{2}} \right\}$$ (7.5.3)

where

F_{xx} = second moment of area about the y-axis,
$F_{\omega\omega}$ = warping constant with pole at the shear centre,
F = area of cross-section

$$\delta = \pm \frac{\delta_x}{L} \sqrt{\frac{EF_{xx}}{GJ_p}}$$ (7.5.4)

with

$$\delta_x = \frac{1}{F_{yy}} \left[\int x^2 y \, dF + \int y^3 \, dF \right] - 2y_0.$$ (7.5.5)

The integrals are taken over the whole cross-section and y_0 is the coordinate of the shear centre (Fig. 3.5.2). For cross-sections with double symmetry $\delta = 0$; for those with the large flange in compression the positive value of δ is used.

For deep beams which carry transverse concentrated loads an empirical method is used to evaluate their critical values. When the member is required to carry simultaneously axial and bending about both principal axes the appropriate form of Dunkerly's formula is used (eqn (7.1.14)).

REFERENCES

CHAPTER 1

Bernoulli, J. (1789). *Essai theoretique sur les vibrations des plaques elastiques rectangulaires et libres.* Nova Acta. St. Petersburg.

Croll, J. G. A. and Walker, A. C. (1972). *Elements of structural stability.* Macmillan, London.

Hadji-Argyris, J. and Cox, H. L. (1944). Diffusion of load into stiffened panels of varying section. Br. Aeron. Res. Council Repts Mem. No. 1969.

Khoo, P. S. (1979). Plastic local buckling of thin-walled structures. Ph.D. thesis, Monash University.

Malcolm, D. J. and Redwood, R. G. (1970). Shear lag in stiffened box-girders. *J. Struct. Div. ASCE* **96**, ST7, 1403–15.

Matheson, J. A. L. (1971). *Hyperstatic structures*, (2nd edn), Vol. 1. Butterworth, London.

Moffatt, K. R. and Dowling, P. J. (1975). Shear lag in steel box-girder bridges. *Struct. Engineer* **53**, 439–48.

Reissner, E. (1946). Analysis of shear lag in box beams by the principle of minimum potential energy. *Q. appl Math.* **4**, 268–78.

Sewell, M. J. (1965). The static perturbation technique in buckling problems. *J. Mech. phys. Solids* **13**, 247–65.

Thompson, J. M. T. and Hunt, G. W. (1973). *A general theory of elastic stability.* J. Wiley and Sons.

Vlasov, V. Z. (1961). *Thin-walled elastic beams* (2nd edn). Israel Program for Scientific Translations, Jerusalem.

Wood, R. H. (1958). The stability of tall buildings. *Proc. ICE.* **11,** Paper No. 6280, 69–102.

CHAPTER 2

Benscoter, S. U. (1954). A theory of torsion bending for multi-cell beams. *J. appl. Mech.* **21,** Tr. ASME, (1), 25–34.

Kollbrunner, C. F. and Hajdin, N. (1965). *Wölbkrafttorsion dünnwandiger Stabe mit geschlossenem Profil.* Schweizer Stahlbau Vereinigung, Heft 32, Zurich.

Matheson, J. A. L. (1971). *Hyperstatic structures* (2nd edn), Vol. 1. Butterworth, London.

Roik, K., Carl, J., and Lindner, J. (1972). *Biegetorsionsprobleme gerade dünnwandiger Stäbe.* Ernst & Sohn.

Vlasov, V. Z. (1961). *Thin-walled elastic beams* (2nd edn). Israel Prog. for Scientific Translations, Jerusalem.

von Karman, T. and Christensen, N. B. (1944). Methods of analysis for torsion with variable twist. *J. Aero. Sci.* (II), **2**, 110–24.

CHAPTER 3

Khan, A. H. and Tottenham, H. (1977). The method of bimoment distribution for the analysis of continuous thin-walled structures subject to torsion. *Proc. ICE* (2), **63**, 843–63.

Murray, N. W. and Grundy, P. (1970). An experiment on the lateral buckling of beams. *Bull. mech. Eng. Ed.* **9**, 45–8.

Sokolnikoff, I. S. and E. S. (1941). *Higher mathematics for engineers and physicists.* McGraw-Hill.

Spiegel, M. R. (1965). *Theory and problems of Laplace transforms.* Schaum Outline Series (McGraw-Hill).

Timoshenko, S. P. (1907). *Bull. Polytech. Inst. Kiev* (in Russian). Published later as: Einige Stabilitätsprobleme der Elasticitätstheorie. *Zeitsch. für Math. und Physik* **58**, [1910], 337–85.

Timoshenko, S. P. and Gere, J. M. (1961). *Theory of elastic stability* (2nd edn). McGraw-Hill, Kogakusha.

Walker, A. C. (1975). *Design and analysis of cold-formed sections.* Intertext Books, London; and Heath and Reach.

Zbirohowski-Koscia, K. (1967). *Thin-walled beams—from theory to practice.* Crosby Lockwood and Son Ltd.

CHAPTER 4

American Institute of Steel Construction (1963). *Design manual for orthotropic steel plate deck bridges.* AISC Publication.

Cheung, Y. K. (1976). *Finite strip method in structural analysis.* Pergamon.

Föppl, A. (1907). *Vorlesungen über technische Mechanik.* **5**, 132.

Klöppel, E. K. and Möller, K. H. (1968). *Beulwerte ausgesteifter Rechteckplatten,* Vol. II. W. Ernst, Berlin.

Klöppel, E. K. and Scheer, J. (1960). *Beulwerte ausgesteifter Reckteckplatten,* Vol. I. W. Ernst, Berlin.

Loo, Y. C. (1976). Analysis of continuous highway box bridges with intermediate stiffening. Eighth Aust. Road Res. Board Conf., pp. 13–20.

Manko, Z. (1979). *Statische Analyse von Stahlfahrbahnplatten.* Der Stahlbau. Heft 6, 176–82.

Meyer, C. and Scordelis, A. C. (1970). Computer program for prismatic folded plates with plate and beam elements. SESM Report No. 70–3, Univ. of Calif., Berkeley.

Murray, N. W. and Thierauf, G. (1981). *Tables for the design and analysis of stiffened steel plates.* Vieweg.

Pelikan, W. and Esslinger, M. (1957). Die Stahlfahrbahn, Berechnung und Konstruktion. M.A.N. Forschungsheft No. 7.

Stanley, C. R. and Sved, G. (1975). Critical buckling stress of longitudinally stiffened plates. Fifth Australasian Congress on the Mechanics of Structures and Materials.

Timoshenko, S. P. and Woinowsky-Krieger, S. (1959). *Theory of plates and shells* (2nd edn), McGraw-Hill, Kogakusha.

Williams, F. W. and Wittrick, W. H. (1969). Computational procedures for a matrix analysis of the stability and vibration of thin flat-walled structures in compression. *Int. J. mech. Sci.* **11**, 979–88.

Wittrick, W. H. (1968). General sinusoidal stiffness matrices for buckling and vibration analysis of the stability and vibration of thin flat-walled structures in compression. *Int. J. mech. Sci.* **10**, 949–66.

CHAPTER 5

Aalami, B. and Chapman, J. C. (1969). Large deflection behaviour of rectangular orthotropic plates under transverse and in-plane loads. *Proc. ICE.* **42**, 347–82.
Aalami, B. and Williams, D. G. (1975). *Thin plate design for transverse loading.* Constrado Monographs, Crosby Lockwood Staples, London.
Abdel-Sayed, G. (1969). Effective width of thin plates in compression. *J. struct. Div. ASCE* (Oct.) 2183–202.
American Institute of Steel Construction (1963). *Design manual for orthotropic steel plate deck bridges.* AISC publication.
Basu, A. K. and Chapman, J. C. (1966). Large deflection behaviour of transversely loaded rectangular orthotropic plates. *Proc. ICE* **35**, 79–110.
Bernoulli, J. (1789). *Essai theoretique sur les vibrations des plaques elastiques rectangulaires et libres.* Nova Acta, Vol. 5. St. Petersburg.
Bilstein, W. (1974). *Anwendung der nichtlinear Beultheorie auf vorverformte, mit diskreten Langsteifen verstärkte Rechteckplatten unter Längsbelastung.* Publ. Inst. für Statik und Stahlbau, T. H. Darmstadt.
Boussinesq, J. V. (1879). Complements a une etude sur la theorie de l'equilibre et du movement des solides elastiques dont certaines dimensions sont tres-petites par rapport a l'autre. *J. Math.* (2), **16**, [1871], 125–274; (3), **5**, [1879], 329–44.
British Standard Institution (1979). Draft BS 5400: *Steel, concrete, and composite Bridges.* BSI Code.
Bryan, G. H. (1891). On the stability of a plane plate under thrusts in its own plane, with application to the buckling of the sides of ships. *Proc. London Math. Soc.* **22**, 54–67.
Bulson, P. S. (1970). *The stability of flat plates.* Chatto and Windus, London.
Coan, J. M. (1951). Large deflection theory for plates with small initial curvatures loaded in edge compression. *J. appl. Mech.* **18**, Tr. ASME, (73), 143–51.
Column Research Committee of Japan (1971). *Handbook of structural stability.* Corona Publ. Co. Ltd., Tokyo.
Dawson, R. G. (1971). Local buckling of thin-walled structural forms. Ph.D. Thesis, University of London.
Falconer, B. H. and Chapman, J. C. (1953). Compressive buckling of stiffened plates. *The Engineer* **195**: (June 5), 789–91; (June 12), 822–5.
Fok, C. D. (1980). Effects of initial imperfections on the elastic post-buckling behaviour of flat plates. Ph.D. Thesis, Monash University.
Gehring, F. (1860). *De aequationibus differentiatibus quibus aequilibrium et motus laminae crystallinae definiuntur.* Diss. Tech. Hoch, Berlin.
Green, J. R. and Southwell, R. V. (1946). Relaxation applied to engineering problems VIII A: Problems relating to large transverse displacements of thin elastic plates. *Phil. Trans. Roy. Soc. London* **A239**, 539–78.
Hoff, N. J., Boley, B. A., and Coan, J. M. (1948). Development of a technique for testing stiff panels in edgewise compression. *Proc. Soc. Exp. Stress Anal.* **5**, (2), 14–24.
Huber, M. T. (1914). Die Grundlagen einer rationeller Berechnung der kreuzweise bewehrten Eisenbetonplatten. *Zeitschr. Österreichischen Ing. und Arch. Verein* **66**, 557.

Kaiser, R. (1936). Rechnerische und experimentelle Ermittlung der Durchbiegungen und Spannungen von quadratischen Platten bei freier Auflagerung an den Rändern, gleichmässig vertielter Last und grossen Ausbiegungen. *Zeitschr. für angewandte Mathematik und Mechanik* **16,** 73–98.

Keays, R. H. and Williams, D. G. (1975). Curves for the elastic analysis of initially deformed plates subject to combined uni-axial compression and shear. *IEA civil Eng. Trans.* **C.E. 17,** (1), 18–21.

Khoo, P. S. (1979). Plastic local buckling of thin-walled structures. Ph.D. Thesis, Monash University.

Kirchhoff, G. R. (1877). Vorlesungen über mathematische Physik. *Mechanik* (2nd edn), p. 450.

Klöppel, E. K. and Möller, K. H. (1968). *Beulwerte ausgesteifter Rechteckplatten.* Vol. II. W. Ernst, Berlin.

Klöppel, E. K. and Scheer, J. (1960). *Beulwerte ausgesteifter Rechteckplatten.* Vol. I. W. Ernst, Berlin.

Klöppel, E. K. and Unger, B. (1969). Das Ausbeulen einer am freien Rand versteiften dreiseitig momentfrei gelagerten Platte unter Verwendung der nichtlinearen Beultheorie Teil I: Analytische Behandlung. *Der Stahlbau* **38,** (Heft 10), 289–99.

Koiter, W. T. (1945). The stability of elastic equilibrium. Thesis, Delft. (English translation NASA TT-F-10833, 1967.)

Kollbrunner, C. F. and Meister, M. (1958). *Ausbeulen Theorie und Berechnung von Blechen.* Springer, Berlin.

Levy, S. (1942a). Bending of rectangular plates with large deflections. NACA Rept. 737.

Levy, S. (1942b). Buckling of rectangular plates with built-in edges. *J. appl. Mech.* **9,** Tr. ASME, 171–4.

Levy, S. (1942c). Bending of rectangular plates with large deflections. NACA tech. Note No. 846.

Levy, S. (1942d). Square plates with clamped edges under normal pressure producing large deflections. NACA tech. Note No. 847.

Levy, S. and Greenman, S. (1942). Bending with large deflection of a clamped rectangular plate with length–width ratio of 1.5 under normal pressure. NACA tech. Note No. 853.

Marguerre, K. (1938). Zur Theorie der gekrümmmter Platte grosser Formänderung. Proc. Fifth int. Congress appl. Mech., p. 93.

Merrison rules (1973). *Inquiry into the basis of design and method of erection of steel box-girder bridges.* Her Majesty's Stationery Office, London.

Michelutti, W. (1976). Stiffened plates in combined loading. Ph.D. Thesis, Monash University.

Navier, L. (1823). *Bull. soc. Phil. Math.,* Paris.

Otter, J. R. H., Cassel, A. C. and Hobbs, R. E. (1966). Dynamic relaxation. *Proc. ICE* **35,** 633–56.

Pelikan, W. and Esslinger, M. (1957). Die Stahlfahrbahn, Berechnung und Konstruktion. M.A.N. Forschungsheft No. 7.

Piaggio, H. T. H. (1954). *Differential equations.* G. Bell and Son.

Rhodes, J. and Harvey, J. M. (1971). The post-buckling behaviour of thin flat plates in compression with unloaded edges elastically restrained against rotation. *J. mech. Engng. Sci.* **13,** (2), 82–91.

Rostovtsev, G. G. (1940). Calculation of a thin plate sheeting supported by rods. Tondy, Leningrad, Inst. Inzkenerov, Grazhdanskogo Vasdushnogo Flota, No. 20.

Rushton, K. R. (1969). Dynamic relaxation solutions of plate problems. *Int. J. Strain Anal.* **3**, 23–32.

Saint Venant (1883). Discussion in *Theorie de l'elasticite des corps solides*, Clebsch, p. 704.

Shaw, F. S. (1953). *An introduction to relaxation methods.* Dover.

Sokolnikoff, I. S. and E. S. (1941). *Higher mathematics for engineers and physicists* (2nd edn). McGraw-Hill.

Stein, M. (1959). Load and deformations of buckled rectangular plates. NASA tech. Rept. R-40.

Thompson, J. M. T. (1962). The elastic instability of a complete spherical shell. *Aero. Q.* **13**, (2), 189–201.

Thompson, J. M. T. (1964). The rotationally symmetric branching behaviour of a complete spherical shell. *Koninklijke Nederlandse Akademie van Wetenschapen, Proc.* (B), **67**, 295–311.

Thompson, J. M. T. and Hunt, G. W. (1973). *A general theory of elastic stability.* J. Wiley and Sons.

Thompson, J. M. T., Tulk, J. D., and Walker, A. C. (1974). An experimental study of imperfection-sensitivity in the interactive buckling of stiffened plates. *IUTEM symposium buckling of structures,* (ed. Bulliansky (1976)). Springer Verlag, Cambridge, Mass.

Timoshenko, S. P. (1907). *Bull. Polytech. Inst. Kiev.* (in Russian). Published later as: Einige Stabilitätsprobleme der Elasticitätstheorie. *Zeitsch. für Math. und Physik* [1910], 337–85.

Timoshenko, S. P. (1913). Sur la stabilite des systemes elastiques. *Ann. Ponts et Chaussees.*

Timoshenko, S. P. and Gere, J. M. (1961). *Theory of elastic stability.* McGraw-Hill, Kogakusha.

Timoshenko, S. P. and Woinowsky-Krieger, S. (1959). *Theory of plates and shells.* McGraw-Hill, Kogakusha.

Troitsky, M. S. (1976). *Stiffened plates—bending, stability, and vibrations.* Elsevier Scientific Publishing Co.

Tulk, J. D. and Walker, A. C. (1976). Model studies of the elastic buckling of a stiffened plate. *J. Strain Anal.* **11**, (3), 137–43.

von Karman, T. (1910). Festigkeitsprobleme im Maschinenbau, *Encyklopädie der mathematischen Wissenschaften.* Vol. 4. B. G. Teubner, Leipzig.

Walker, A. C. (1969). The post-buckling behaviour of simply supported square plates. *Aero. Q.* **XX**, 203–22.

Walker, A. C. (1980). Analysis of plates under complex loadings. ICE Conf. on new BS 5400 Code.

Walker, A. C. and Murray, N. W. (1975). A plate collapse mechanism for compressed plates. *Publs. Int. Assn. for Bridge and Struct. Eng.* **35–I**, 217–36.

Wang, C. T. (1948a). Non-linear large deflection boundary value problems of rectangular plates. NACA tech. Note No. 1425.

Wang, C. T. (1948b). Bending of rectangular plates with large deflection. NACA tech. Note No. 1462.

Way, S. (1938). Uniformly loaded clamped rectangular plates with large deflections. Proc. Fifth intl. Congr. of app. Mech., pp. 123–8.

Williams, D. G. (1971). Some examples of the elastic behaviour of initially deformed bridge panels. *Civ. Eng. and Publ. Wks. Rev.* 1107–12.

Williams, D. G. (1973). The elastic design of initially deformed plates. Aust. Inst. of Steel Constr. Conf. on Steel Developments, pp. 251–60.

Williams, D. G. and Aalami, B. (1979). *Thin plate design for in-plane loading.* Constrado Monographs, Granada Publ. Ltd.

Williams, D. G. and Walker, A. C. (1975). Explicit solutions for the design of initially deformed plates subject to compression. *Proc. ICE* (2), **59,** 763–87.

Williams, D. G. and Walker, A. C. (1977). Explicit solutions for plate buckling analysis. *Jl. Eng. Mech. Div., ASCE,* 103 (EM4), Proc. Paper 13122, pp. 549–68.

Williams, F. W. and Wittrick, W. H. (1969). Computational procedures for a matrix analysis of the stability and vibration of thin flat-walled structures in compression. *Int. J. mech. Sci.* **11,** 979–88.

Wittrick, W. H. (1968). General sinusoidal stiffness matrices for buckling and vibration analyses of thin flat-walled structures. *Int. J. mech. Sci.* **10,** 949–66.

Wittrick, W. H. and Williams, F. W. (1974). Buckling and vibration of anisotropic or isotropic plate assemblies under combined loadings. *Int. J. mech. Sci.* **16,** 209–39.

Yamaki, N. (1959). Post-buckling behaviour of rectangular plates with small initial curvatures loaded in compression edge. *J. appl. Mech.* **26,** 407–14.

CHAPTER 6

Allen, D. (1975). Discussion on 'Analysis and design of stiffened plates for collapse load' by N. W. Murray. *Struct. Engr.* **53,** 381–2.

Barbré, R., Grassl, H., Schmidt, H., and Kruppe, J. (1976). Traglastversuche an Ausschnitten gedrückter Gurte mehrerer Hohlkastenbrücken. Institut für Stahlbau Tech. Uni. Braunschweig.

Bares, R. and Massonnet, C. (1968). *Analysis of beam grids and orthotropic plates by the Guyon–Massonnet–Bares method* SNTL, Praque and Crosby–Lockwood, London.

Basler, K. (1961). Strength of plate girders in shear. *J. Struct. Div. ASCE* ST7, (2967), 151, Oct.

Box, T. (1883). *A practical treatise on the strength of materials.* Spon, London.

British Standards Institution (1977). *BS 2573 Permissible stresses in cranes and design rules.* B.S.I. Code.

British Standards Institution (1978–1981). *BS 5400 Steel, concrete and composite bridges.* [Appearing in parts.]

Chan, K. S., Law, C. L., and Smith, D. W. (1977). Stability of stiffened compression flanges under in-plane forces and wheel loads. Rept. Wolfson Bridge Res. Unit, Univ. of Dundee.

Chatterjee, S. and Dowling, P. J. (1977). The design of box-girder compression flanges. Steel plated structures symposium (eds Dowling, Harding, and Frieze) pp. 196–228. Crosby Lockwood.

Chern, C. and Ostapenko, A. (1969). Ultimate strength of plate girders under shear. Fritz Eng. Lab. Report, No. 328.7.

Cooper, P. B. (1965). Bending and shear strength of longitudinally stiffened plate girders. Fritz Eng. Lab. Report, No. 304.6, Lehigh Univ.

Cooper, P. B. (1971). The ultimate bending moment for plate girders. *Proc. Int. Assn. of Bridge and Struct. Engng.,* **30,** 113–48.

de George, D. (1979). Collapse behaviour of trough stiffened steel plates. Ph.D. Thesis, Monash University.

de George, D., Michelutti, W. M. and Murray, N. W. (1979). Studies of some steel plates stiffened with bulb-flats or with troughs. Thin-walled structures conference, pp. 86–99. Granada Publ.

Dowling, P. J. (1971). *The behaviour of orthotropic steel deck bridges.* Crosby Lockwood.

Dowling, P. J., Chatterjee, S., Frieze, P., and Moolani, F. (1973). The experimental and predicted collapse behaviour of rectangular stiffened steel box birders. Int. Conf. on steel box girder bridges, pp. 77–94. Inst. Civil Engrs.

Dubas, P. (1972). Tests about post-critical behaviour of stiffened box girders. Proc. of Colloquium of Design of Plate and Box Girders for Ultimate Strength, pp. 367–79. Int. Assn. Bridge and Struct. Eng.

Dwight, J. B. (1971). Collapse of steel compression panels. Conf. on developments in bridge design and construction. Crosby Lockwood.

Dwight, J. B. and Little, G. H. (1976). Stiffened steel compression flanges—a simpler approach. *Struct. Engineer* **54,** (12), 501–9.

Dwight, J. B. and Moxham, K. E. (1969). Welded steel plates in compression. *Struct. Engineer* **47,** (2), 49–66.

Evans, H. R., Porter, D. M., and Rockey, K. C. (1976). A parametric study of the collapse behaviour of plate girders. Univ. College Cardiff Rept.

Evans, H. R., Porter, D. M., and Rockey, K. C. (1978). The collapse behaviour of plate girders subjected to shear and bending. *Proc. Int. Assn. Bridge and Struct. Eng. P*–18/78, 1–20.

Graves-Smith, T. P. (1967). The ultimate strength of locally buckled columns of arbitrary length. Symposium on thin-walled steel structures (ed. K. C. Rockey and H. V. Hill). Crosby Lockwood.

Harding, J. E. and Hobbs, R. E. (1979). The ultimate load behaviour of box-girder web panels. *Struct. Engineer* **75B,** (3), 49–54.

Harding, J. E., Hobbs, R. E., and Neal, B. G. (1977a). The elasto-plastic analysis of imperfect square plates under in-plane loading. *Proc. I.C.E.* **63,** (2), Paper No. 7981, pp. 137–58.

Harding, J. E., Hobbs, R. E. and Neal, B. G. (1977b). Ultimate load behaviour of plates under combined direct and shear in-plane loading. Steel plated structures symposium (eds Dowling, Harding, and Frieze), pp. 369–403. Crosby Lockwood.

Herzog, M. (1976). Die Traglast einseitig lansversteifter Bleche mit Imperfektionen und Eigenspannungen unter Axialdruck nach Versuchen. *VDI-Z* **118,** (7) 321–6.

Hill, R. (1967). *The Mathematical Theory of Plasticity.* (Clarendon Press, Oxford.

Hillerborg, A. (1956). Theory of equilibrium for reinforced concrete slabs (in Swedish). *Betong* **41,** (4), 171–82.

Horne, M. R. (1971). *Plastic theory of structures* (1st edn). Thomas Nelson and Sons Ltd.

Horne, M. R. (1977). Structural action in steel box girders. CIRIA Guide 3. CIRIA Publication, London.

Horne, M. R. (1980). Basic concepts in the design of webs. Paper presented at conference on British Standard BS 5400 Bridge Code.

Horne, M. R., Montague, P., and Narayanan, R. (1976a). Ultimate capacity of axially loaded stiffened plates collapsing by outstand failure. Simon Eng. Lab. Report (January). Manchester Univ.

Horne, M. R., Montague, P., and Narayanan, R. (1976b). The influence of stiffener spacing and weld/gap ratio on the ultimate capacity of axially loaded stiffened plates. Simon Eng. Lab. Report (February). Manchester Univ.

Horne, M. R. and Narayanan, R. (1974). Ultimate load capacity of longitudinally stiffened panels with $b/t = 48$ and $L/r = 20$ and 40. Simon Eng. Lab. Report (January). Manchester Univ.

Horne, M. R. and Narayanan, R. (1975). An approximate method for the design of stiffened steel compression panels. Simon Eng. Lab. Report (February). Manchester Univ.

Horne, M. R. and Narayanan, R. (1976a). Ultimate capacity of longitudinally stiffened plates used in box girders. Proc. Inst. civil Engrs. (2), **61**, 253–80.

Horne, M. R. and Narayanan, R. (1976b). Strength of axially loaded stiffened plates. Publs. Int. Assn. for Bridge and Struct. Eng. **36–I**, 125–57.

Kamtekar, A. G. (1975). Theoretical determination of welding residual stresses. Cambridge Univ. Eng. Dept. Report CUED/C–Struct/TR.45.

Khoo, P. S. (1979). Plastic local buckling of thin-walled structures. Ph.D. Thesis, Monash University.

Klöppel, K., Schmied, R., and Schubert, J. (1969). Die Traglast mittig und aussermittig gedrückter dünnwandiger Stützen mit kastenförmigem Querschnitt im überkritischen Bereich unter verwendung der nichtlinearen Beultheorie. Teil II. Experimentelle Untersuchungen, Vergleich der experimentellen und theoretischen Ergebnisse. Der Stahlbau **38**, 9–19.

Korol, R. M. and Sherbourne, A. N. (1972). Strength predictions of plates in uniaxial compression. J. Struct. Div. Proc. ASCE **98**, (ST 9), Paper No. 9239, pp. 2223–34.

Little, G. H. (1976a). Stiffened steel compression panels—theoretical failure analysis. Struct. Engineer. **54**, (12), 489–500.

Little, G. H. (1976b). Local and overall buckling of square box columns. Cambridge Univ. Dept. of Eng. Report, CUED/C–Struct, TR. 56.

Little, G. H. (1979). The strength of square steel box columns—design curves and their theoretical basis. Struct. Engineer **57A**, (2), 49–61.

Little, G. H. (1980). The collapse of rectangular steel plates under uniaxial compression. Struct. Engineer **58B**, (3), 45–61.

Maquoi, R. and Massonnet, C. (1971). Theorie non-lineaire de la resistance postcritique des grandes poutres en caisson raidies. Publs. Int. Assn. Bridge and Struct. Eng. **31**, (2) 91–140.

Marguerre, K. (1939). Zur Theorie der gekrümmten Platte grosser Formänderung. Jahrbuch der deutschen Luftfahrt-forschung.

Massonnet, C. (1975). Neue Erkenntnisse und Theorien aus europaischen Forschungsarbeiten. Deutscher Ausschuss für Stahlbau, **3**, 15–26.

Massonnet, C. and Maquoi, R. (1973). New theory and tests on the ultimate strength of stiffened box girders. In Steel box-girder bridges, pp. 131–43. Clowes and Sons, London.

Merrison rules (1973). Inquiry into the basis of design and method of erection of steel box-girder bridges. Her Majesty's Stationery Office, London.

Michelutti, W. (1976). Stiffened plates in combined loading. Ph.D. Thesis, Monash University.

Mikami, I., Dogaki, M., and Yonezawa, H. (1980). Ultimate load tests on multi-stiffened steel box girders. Technology reports of Kansai University, Osaka, Japan, (2) (March), 157–69.

Moxham, K. E. (1971a). Theoretical prediction of the strength of welded steel plates in compression. Cambridge Univ. Dept. of Eng. Report, CUED/C–Struct/TR.2.

Moxham, K. E. (1971b). Buckling tests on individual welded steel plates in compression. Cambridge Univ. Dept. of Eng. Report, CUED/C–Struct/TR.3.

Murray, N. W. (1973a). Das aufnehmbare Moment in einem zur Richtung der Normalkraft schräg liegenden plastischen Gelenk. *Die Bautechnik*, **2/1973**, 57–8.

Murray, N. W. (1973b). Buckling of stiffened panels loaded axially and in bending. *Struct. Engineer*. **51**, 285–301.

Murray, N. W. (1973c). Das Stabilitätsverhalten von axial belasteten in der Langsrichtung ausgesteiften Platten im plastichen Bereich. *Der Stahlbau* **12**, 372–9.

Murray, N. W. (1975). Analysis and design of stiffened plates for collapse load. *Struct. Engineer*. **53**, 153–8.

Murray, N. W. and Khoo, P. S. (1981). Some basic plastic mechanisms in thin-walled steel structures. *Int. J. mech. Sci.* **23**, (12), 703–13.

Murray, N. W. and Thierauf, G. (1981). *Tables for the design and analysis of stiffened steel plates*. Vieweg.

Ostapenko, A. and Chern, C. (1971). Ultimate strength of longitudinally stiffened plate girders under combined loads. Proc. Int. Assn. Bridge and Struct. Eng. Colloquium. London.

Ractliffe, A. T. (1966). The strength of plates in compression. Ph.D. Thesis, Cambridge Univ.

Rawlings, B. and Shapland, P. (1975). The behaviour of thin-walled box sections under gross deformation. *Struct. Engineer* **53**, (4), 181–6.

Robertson, A. (1928). The strength of tubular struts. *Proc. Roy. Soc. Lond.* **A121**, (A788), 558–85.

Rockey, K. C. (1968). Factors influencing ultimate behaviour of plate girders. Proc. conference on steel bridges, pp. 31–8.

Rockey, K. C., Evans, H. R., and Porter, D. M. (1973). Ultimate load capacity of stiffened webs subjected to shear and bending. In *Steel box-girder bridges*, pp. 45–61. Clowes and Sons, London.

Rockey, K. C., Evans, H. R., and Porter, D. M. (1978). A design method for predicting the collapse behaviour of plate girders. *Proc. ICE* (2), **65**, Paper No. 8086, pp. 85–112.

Rockey, K. C. and Skaloud, M. (1968). Influence of flange stiffness upon the load-carrying capacity of webs in shear. Final Report, 8th Congress Intl. Assn. of Bridge and Struct. Eng. New York.

Rockey, K. C. and Skaloud, M. (1972). The ultimate load behaviour of plate girders loaded in shear. *Struct. Engineer* **50**, (1), 29–47.

Roderick, J. W. and Ings, N. L. (1977). The behaviour of small-scale box-girders of stiffened plate construction. *Aust. Welding Res.*, Dec. 77, 15–29.

Schmidt, H. (1975). Zum Tragverhalten axial gedrückter, geschweisster, Längsversteifter Blechfelder. Vier Vorträge zum Plattenbeulproblem Tech. Uni. Hannover, Heft 9.

Schubert, J. (1975). Das Verhalten gedrückter dünnwandiger Stützen mit kastenformigen Querschnitt im überkritischen Bereich. Vier Vorträge zum Plattenbeulproblem Tech. Uni. Hannover, Heft 9.

Sherbourne, A. N. and Korol, R. M. (1971). Ultimate strength of plates in uniaxial compression. *ASCE Nat. Struct. Eng. Mtg*. Baltimore. Preprint No. 1386. April.

Sherbourne, A. N. and Korol, R. M. (1972). Postbuckling of axially compressed plates. *J. Struct. Div. Proc. ASCE* **98**, (ST 10), 2223–34.

Smith, C. S. (1975). Compressive strength of welded steel ship grillages. Royal Inst. of Naval Arch., Spring Meeting, Paper No. 9, pp. 1–23.

Steinhardt, O. (1973). Recent revisions to German standard DIN 4114. In *Steel box-girder bridges*, pp. 203–8. Clowes and Sons, London.

Steinhardt, O. (1975). Berechnungsmodelle für ausgesteifte kastenträger. *Deutscher Ausschus für Stalbau*, **3**, 27–35.

Valtinat, G. (1975). Theorie und Berechnung beulgerfährdeter Vollwand-und Kastenträger. Vier Vorträge zum Plattenbeulproblem Tech. Uni. Hannover, Heft 9.

van Iterson, F. K. T. (1947). *Plasticity in engineering*. Blackie and son Ltd.

White, J. D. (1977a). Longitudinal shrinkage of a single-pass weld. Cambridge Univ. Dept. of Eng. Report CUED/C–Struct/TR.57.

White, J. D. (1977b). Longitudinal stresses in welded T-sections. Cambridge Univ. Dept. of Eng. Report CUED/C–Struct/TR.60.

White, J. D. (1977c). Longitudinal stresses in a member containing non-interacting welds. Cambridge Univ. Dept. of Eng. Report CUED/C–Struct/TR.58.

White, J. D. (1977d). Longitudinal shrinkage of multi-pass welds. Cambridge Univ. Dept. of Eng. Report CUED/C–Struct/TR.59.

Wood, R. H. and Jones, L. L. (1967). *Yield line analysis of slabs*. Thames and Hudson, London.

CHAPTER 7

Billington, D. J., Ghavami, K., and Dowling, P. J. (1972). Steel box girders—parametric study of cross-sectional distortion due to eccentric loading. Eng. Struct. Lab. Report, CE SLIC Rept. No. BG 16. Imperial College, London.

British Standards Institution. BS 5400: *Steel, Concrete, and Composite Bridges*. Appearing in parts from 1978–1981.

Chwalla, E. (1936). *Beitrag zur Stabilitätstheorie des Stegbleches vollwandiger Träger, Der Stahlbau*, **9**, 161–6.

DASt-Richtlinie 012 (1978). *Beulsicherheitsnachweise für Platten*. Publication of Working party PLATTEN of Deutsche Ausschuss für Stahlbau.

DIN 4114 (1952). Die Berechnungsgrundlagen für Stabilitatsfalle im Stahlbau DIN 4114.

Dabrowski, R. (1965). Näherungsberechnung der gekrümmten Kastenträger mit verformbaren Querschnitt. *Seventh Int. Ass. of Bridge and Struct. Engng. Congress Preliminary Publications*. Zurich. pp. 299–306.

Dabrowski, R. (1968). *Gekrümmte dünnwandige Träger*. Springer. [A Cement and Concrete Association translation (Transl. No. 144) of this book is available.]

Dalton, D. C. and Richmond, B. (1968). Twisting of thin-walled box girders of trapezoidal cross-section. *Proc. I.C.E.* **39**, 61–73.

Flint, A. R., Smith, B. W., Baker, M. J., and Manners, W. (1980). The derivation of safety factors for design of highway bridges. Paper presented at conference on British Standard BS 5400 Bridge Code.

Gent, A. R. and Shebini, V. K. (1972). *Parametric study: Report on torsional warping*. Report to DOE (U.K.). Imperial College, London.

Hees, G. (1972). Querschnittsverformung des einzelligen Kastenträgers mit vier Wanden in einer zur Wölbkrafttorsion analogen Darstellung. *Die Bautechnik*, Teil I, Heft II (1971), 370–7; Teil II, Heft I, 21–8.

Horne, M. R. (1975). *The behaviour of box girder bridges and the Merrison Design Rules*. Post-graduate course notes, University of Sydney School of Civil Eng.

Horne, M. R. (1977). Structural action in steel box girders. CIRIA Guide 3. CIRIA Publication, London.

Horne, M. R. (1980). Basic concepts in the design of webs. Paper presented at conference on British Standard BS 5400 Bridge Code.

International Standards Organisation. ISO 2394: *General principles for the verification of the safety of structures.*

Janssen, J. D. and Veldpaus, F. E. (1972). Über die Stärke und Steifigkeit von Kastenträgern mit Rechteckquerschnitt. *Proc. Int. Assn. Bridge and Struct. Eng.* **31–II,** 85–106.

Janssen, J. D. and Veldpaus, F. E. (1973). Der Einfluss von Querschotten auf das Verhalten von Kastenträgern mit Rechtechquerschnitt. *Proc. Int. Assn. Bridge Struct. Eng.* **33–I,** 65–88.

Klöppel, E. K. and Möller, K. H. (1968). *Beulwere augesteifter Rechteckplatten.* Vol. II. W. Ernst, Berlin.

Klöppel, E. K. and Scheer, J. (1960). *Beulwerte ausgesteifter Rechteckplatten.* Vol. I. W. Ernst, Berlin.

Kollbrunner, C. F. and Meister, M. (1958). *Ausbeulen Theorie und Berechnung von Blechen.* Springer, Berlin.

Kristek, V. (1970). Tapered box girders of deformable cross section. *J. Struct. Div., ASCE* **96,** (ST8), 1761–93.

Lim, P. T. K. and Moffatt, K. R. (1971). A general purpose finite element program. Proc. PTRC Bridge Program Review Symposium, London.

Little, G. H. (1976). Stiffened steel compression panels—theoretical failure analysis. *Struct. Engineer.* **54,** (12), 489–500.

Merrison rules (1973). *Inquiry into the basis of design and method of erection of steel box-girder bridges.* Her Majesty's Stationery Office, London.

Moffatt, K. R. and Dowling, P. J. (1972). Steel box girders, parametric study on the shear lag phenomenon in steel box girder bridges. Engineering Struct. Labs., Civil Engng. Dept., Imperial College, London. CESLIC Report BG 17. Sept.

Moffatt, K. R. and Dowling, P. J. (1975). Shear lag in steel box-girder bridges. *Struct. Engineer* **53,** (10), 439–48.

Richmond, B. (1969). Trapezoidal boxes with continuous diaphragms. *Proc. I.C.E.* **43,** 641–50.

Rockey, K. C., Evans, H. R., and Porter, D. M. (1978). A design method for predicting the collapse behaviour of plate girders. *Proc. I.C.E.,* (2), **65,** (Paper No. 8086), 85–112.

Roik, K., Carl, J., and Lindner, J. (1972). *Biegetrosionsprobleme gerader dünnwandiger Stäbe.* W. Erust & Sohn.

Schaefer, H. (1934). Beitrag zur Berechnung des kleinsten Eigenwertes eindimensionaler Eigenwertprobleme. Diss. Hannover 1934 and *Zeitsch. fur angew. Math. Mech.* Heft **14,** 367.

Scheer, J., Nölke, H., and Gentz, E. (1979). Beulsicherheitsnachweise für Platten, DASt-Richtlinie 012 Grundlagen, Erläuterungen, Beispile Stahlbau, p. 147. Verlags GmbH Köln.

Schleicher, F. (1955). *Taschenbuch für Bauingenieure,* Bd. I, p. 1029. Springer, Berlin, Göttingen and Heidelberg.

Smith, D. W. (1980). Guide to the use of the code. Paper presented at conference on British Standard BS 5400 Bridge Code.

Standards Association of Australia (1974). AS 1538–1974. S.A.A. Code: Cold formed steel structures.

Stüssi, F. (1955). *Tragwerke aus Aluminium*. Springer, Berlin, Göttingen, and Heidelberg.

Stüssi, F., Kollbrunner, C. F., and Wanzenried, H. (1953). Ausbeulen rechteckiger Platten unter Druck, Biegung und Druck mit Biegung. Mitt. Nr. 26 aus Inst. Baustatik E.T.H. Zurich. Leeman, Zurich.

Timoshenko, S. P. (1935). Stability of webs of plate girders. *Engineering* **238,** 207.

Timoshenko, S. P. and Gere, J. (1961). *Theory of elastic stability* (2nd edn). McGraw-Hill, Kogakusha.

Wolchuk, R. (1981). Design rules for steel box-girder bridges. *Proc. Int. Assn. of Bridge and Struct. Eng.* **P–41/81,** 49–60.

Wright, R. N., Abdel-Samed, S. R., and Robinson, A. R. (1968). Beam on an elastic foundation analogy for analysis of box girders. *J. Struct. Div.*, *ASCE* **94,** (ST 7), 1719–43.

NAME INDEX

SUBJECT INDEX